Aromatic and Medicinal Plants of Drylands and Deserts

The description and analysis of the Mexican and other countries' desertic plants from the point of view of their use in traditional medicine and their potential use in integrative medicine is the overall theme of this book. *Aromatic and Medicinal Plants of Drylands and Deserts: Ecology, Ethnobiology and Potential Uses* describes the historic use of drylands plants, botanical and geological classification as well as the endemic plants used in traditional medicine, going through the most relevant aspects of biomedicine and integrative medicine. The chemical and bioactive compounds from desertic medicinal and aromatic plants and the analytic techniques to determine chemical and bioactive compounds are reviewed. Ethnobiology is detailed in the present book as well as the importance of the integrative medicine for the ancient and actual cultures. The book represents an effort to keep the ethnobiological knowledge of communities for the use of traditional desertic plants with the actual analytical techniques to unveil the chemical molecules responsible for the biological or biomedical applications.

Features:

- Describes the endemic plants used in traditional medicine
- Includes the chemical and bioactive compounds from desertic medicinal plants
- Addresses the analytic techniques to determine chemical and bioactive compounds
- Represents an effort to keep the ethnobiological knowledge of communities

To execute this book, there are collaborations by authors from different institutions in northern Mexico, which is where the arid and semi-arid ecosystems of the country are found. Although the subject of medicinal plants has been treated from different angles, this book offers a holistic and comprehensive vision of these important organisms of the Mexican desert, thus resulting in an updated work for specialized readers and for those who are beginning in this exciting theme.

EXPLORING MEDICINAL PLANTS

Series Editor

Azamal Husen

Wolaita Sodo University, Ethiopia

Medicinal plants render a rich source of bioactive compounds used in drug formulation and development; they play a key role in traditional or indigenous health systems. As the demand for herbal medicines increases worldwide, supply is declining as most of the harvest is derived from naturally growing vegetation. Considering global interests and covering several important aspects associated with medicinal plants, the Exploring Medicinal Plants series comprises volumes valuable to academia, practitioners, and researchers interested in medicinal plants. Topics provide information on a range of subjects including diversity, conservation, propagation, cultivation, physiology, molecular biology, growth response under extreme environment, handling, storage, bioactive compounds, secondary metabolites, extraction, therapeutics, mode of action, and healthcare practices.

Led by Azamal Husen, PhD, this series is directed to a broad range of researchers and professionals consisting of topical books exploring information related to medicinal plants. It includes edited volumes, references, and textbooks available for individual print and electronic purchases.

Traditional Herbal Therapy for the Human Immune System, *Azamal Husen*

Environmental Pollution and Medicinal Plants, *Azamal Husen*

Herbs, Shrubs and Trees of Potential Medicinal Benefits, *Azamal Husen*

Phytopharmaceuticals and Biotechnology of Herbal Plants, *Sachidanand Singh, Rahul Datta, Parul Johri, and Mala Trivedi*

Omics Studies of Medicinal Plants, *Ahmad Altaf*

Exploring Poisonous Plants: Medicinal Values, Toxicity Responses, and Therapeutic Uses, *Azamal Husen*

Plants as Medicine and Aromatics: Conservation, Ecology, and Pharmacognosy, *Mohd Kafeel Ahmad Ansari, Bengu Turkyilmaz Unal, Munir Ozturk, and Gary Owens*

Sustainable Uses of Medicinal Plants, *Learnmore Kambizi and Callistus Bvenura*

Medicinal Plant Responses to Stressful Conditions, *Arafat Abdel Hamed Abdel Latef*

Aromatic and Medicinal Plants of Drylands and Deserts: Ecology, Ethnobiology and Potential Uses *David Ramiro Aguillón-Gutiérrez, Cristian Torres-León, and Jorge Alejandro Aguirre-Joya*

For more information about this series, please visit: https://www.routledge.com/Exploring-Medicinal-Plants/book-series/CRCEMP

Aromatic and Medicinal Plants of Drylands and Deserts

Ecology, Ethnobiology and Potential Uses

Edited by
David Ramiro Aguillón-Gutiérrez, Cristian Torres-León, and
Jorge Alejandro Aguirre-Joya

CRC Press
Taylor & Francis Group
Boca Raton London New York

CRC Press is an imprint of the
Taylor & Francis Group, an **informa** business

First edition published 2023

by CRC Press
6000 Broken Sound Parkway NW, Suite 300, Boca Raton, FL 33487-2742

and by CRC Press
4 Park Square, Milton Park, Abingdon, Oxon, OX14 4RN

CRC Press is an imprint of Taylor & Francis Group, LLC

ISBN: 9781032169729 (hbk)
ISBN: 9781032169804 (pbk)
ISBN: 9781003251255 (ebk)

DOI: 10.1201/9781003251255

Typeset in Kepler Std
by Deanta Global Publishing Services, Chennai, India

Contents

Preface

Historically, humans have used plants for medicinal purposes, sometimes serving as the only alternative to treat diseases, formerly because the pharmacological industry did not exist, but even today, in communities without access to commercial medicines, and in some cases due to lack of trust in them or lack of financial resources to purchase them. Even so, most of the drugs that are manufactured today have the extract of a plant as their active compound.

The knowledge that different cultures have developed on the use of medicinal plants is part of an invaluable cultural wealth that, both in Eastern and Western societies, form links between different peoples and the natural resources that surround them, but also, establish ethnic or traditional health systems that have allowed these societies to face their main diseases and illnesses.

Plants in arid and semi-arid environments have, over time, developed adaptation mechanisms for the adverse conditions of these ecosystems. In addition to generating spines in some species, they also produce substances that help them keep predators away or carry out important biochemical and physiological processes for the survival of the plant. These substances are those that have biomedical and pharmacological potential once they are isolated, identified and medicinal properties verified at the laboratory level.

It is of the utmost importance that plants that have medicinal use are conserved in their natural environments. Unfortunately many of them are in danger of extinction because of very different factors. It is necessary not only to understand the functioning of the active compounds that these plants generate in nature but also to propose sustainable use systems that allow future generations to obtain a benefit from these organisms. Knowledge generated at the molecular, cellular, physiological, biochemical and genetic levels have a conservation impact on the population, community and ecosystem level.

In this book, the case of the medicinal plants of the deserts in Mexico and other countries is treated in a particular way, from its biological and ecological generalities to its ethnobotanical history, passing definitely through the biochemical analysis of its most relevant compounds and its biomedical applications. In the book, there are collaborations by authors from different institutions in northern Mexico, which is where the arid and semi-arid ecosystems of the country are found. Although the subject of medicinal plants has been treated from different angles, this book offers a holistic and comprehensive vision of these important organisms of the Mexican desert, thus resulting in an updated work for specialized readers and for those who are beginning in this exciting theme.

Editors

David Ramiro Aguillón-Gutiérrez, PhD, graduated from the Faculty of Veterinary Medicine and Animal Sciences, Autonomous University of Nuevo Leon, Mexico, in 2003. He earned his Specialist degree in Occupational Health from the Autonomous University of Noreste, Mexico, MSc degree in Biological Sciences (Specialist in Embryology) from Lomonosov Moscow State University, Russia, in 2009, and his PhD in Biological Sciences (Specialist in Embryology and Zoology) also from Lomonosov Moscow State University, Russia, in 2012. From 2013 to 2014 he was a Postdoctoral Researcher in Biodiversity and Conservation at the Autonomous University of the State of Hidalgo, Mexico. Currently, he is a full-time Research Professor at the Autonomous University of Coahuila, Mexico, and a member of the National System of Researches, Mexico. He has published 25 peer-reviewed papers, five book chapters and three books (two as author and one as editor).

Cristian Torres-León, PhD, is Professor at the Research Center and Ethnobiological Garden, Autonomous University of Coahuila, Mexico. He completed a degree in Agroindustrial Engineering (2013) from the National University of Colombia before earning his Master's (2016) and PhD (2019) in Food Science and Technology, both from the Autonomous University of Coahuila, Mexico. In addition, he completed a PhD (2019) in Biotechnology at the Federal University of Pernambuco, Brazil. He worked as a Young Researcher at the Colombian Ministry of Science between 2013 and 2014 and has undertaken a research stay at the University of Minho in Portugal (2016). He has evaluated projects for government organizations, and his activities and accomplishments include conferences, seminars, workshops, and the publication of 26 scientific papers in indexed journals and 16 book chapters. He is now a full Professor at the Autonomous University of Coahuila, Mexico, and mainly works in the Food Science and Technology area with an emphasis on biodegradable packaging, phenolic compounds, use of agro-industrial waste, fermentation technologies, ethnopharmacology and food security.

Jorge Alejandro Aguirre-Joya, PhD, is full-time Professor at the Autonomous University of Coahuila, Mexico. He is a chemical pharmacologist with a degree from the Autonomous University of Coahuila (2011). He earned a Master's degree in Food Science and Technology from the Autonomous University of Coahuila, Mexico (2014) and earned his PhD in Food Science and Technology in 2018 from the same university. Currently he is Dean of the Research Center and Ethnobiological Garden from the Semidesert of Coahuila. He has published 12 peer-reviewed papers, 15 book chapters and one book as an editor. Also, he has participated in and coordinated 10 technology transfer projects. Dr. Aguirre-Joya is member of the National System of Reserches (Mexico) and serves as reviewer for several notable journals.

Contributors

Aguilar-González, C. N
Autonomus University of Coahuila
Coahuila, México

Aguillón-Gutiérrez, D. R
Autonomous University of Coahuila,
Coahuila, México

Aguirre-Joya, J. A
Autonomous University of Coahuila
Coahuila, México

Anguiano-Cabello, J. C
La Salle Saltillo University
Coahuila, México

Arredondo-Valdés, R
Autonomus University of Coahuila
Coahuila, México

Ascacio-Valdés, J. A
Autonomus University of Coahuila
Coahuila, México

Barroso-Ake, A. C
Autonomus University of Coahuila
Coahuila, México

Beltrán-Martínez, M. E
Universidad de Sonora
Sonora, México

Cándido-del Toro, C. A
Autonomus University of Coahuila
Coahuila, México

Cano-Villegas, O
Universidad Juárez del Estado de Durango
Durango, México

Charles-Rodríguez, AV
Universidad Autónoma Agraria Antonio Narro
Coah Coahuila, México

Encina-Domínguez, J. A
Universidad Autónoma Agraria Antonio Narro
Coahuila, México

Flores-López, M. L
Universidad Interserrana del Estado de Puebla Ahuacatlán
Puebla, México

García-Campoy, A. H
Instituto Potosino de Investigación Científica y Tecnológica
San Luis Potosí, México

García-Dávila, J
LIPMAN Family Farms
Sonora, México.

González-Machado, A. C
Autonomous University of Coahuila
Coahuila, México

González-Zamora, A
Universidad Juárez del Estado de Durango
Durango, México

Govea-Salas, M
Autonomus University of Coahuila
Coahuila, México

Guía-García, J. L
Autonomus University of Coahuila
Coahuila, México

Hench-Cabrera, A. F
Universidad Interserrana del Estado de Puebla Ahuacatlán
Puebla, México

Hernández-Castillo, F. D
Autonomous Agrarian University Antonio Narro
Coahuila, México

Hernández-Herrera, J. A
Universidad Autónoma Agraria Antonio Narro
Coahuila, México.

Iliná, A
Autonomus University of Coahuila
Coahuila, México

Ireta-Paredes, A. R
Universidad Interserrana del Estado de Puebla Ahuacatlán
Puebla, México

Lara-Reimers, E. A
Universidad Autónoma Agraria Antonio Narro
Coahuila, México

Laredo-Alcalá, E. I
Autonomus University of Coahuila
Coahuila, México

Lima-Soares, I
Universidade Federal do Ceará
Fortaleza, Brazil

López-García, R
Universidad Autónoma Chapingo
Durango, México

López-Romero, J. C
Universidad de Sonora
Sonora, México

Martínez-Sifuentes, A. R
Instituto Nacional de Investigaciones Forestales, Agrícolas y Pecuarias
Durango, México

Medeiros-Bandeira, M. A
Universidade Federal do Ceará
Foraleza, Brazil

Muñiz-Ramírez, A
Instituto Potosino de Investigación Científica y Tecnológica
San Luis Potosí, México
Consejo Nacional de Ciencia y Tecnología
Ciudad de México, México

Muro-Pérez, G
Universidad Juárez del Estado de Durango
Durango, México

Nascimento-Magalhães, K
Universidade Federal do Ceará
Fortaleza, Brazil.

Navarrete-Molina, C
Universidad Tecnologica de Rodeo
Durango, México

Ochoa-Fuentes, Y. M
Autonomous Agrarian University Antonio Narro
Coahuila, México

Olaitan-Balogun, S
Universidade Federal da Grande Dourados
Mato Grosso do Sul, Brazil

Olivas-Calderón, E
Universidad Juárez del Estado de Durango
Durango, México

Peña-Ramos, F.M
Universidad Autónoma Agraria Antonio Narro
Coahuila, México

Pérez-Miranda, S
Consejo Nacional de Ciencia y Tecnología
Ciudad de México, México

Pérez-Morales, R
Universidad Juárez del Estado de Durango
Durango, México

Ramírez-Hernández, A
Instituto Potosino de Investigación Científica y Tecnológica
San Luis Potosí, México.
Consejo Nacional de Ciencia y Tecnología
Ciudad de México, México.

Ramírez-Moreno, A.
Autonomous University of Coahuila
Coahuila, México

Ramírez-Rodríguez, Y
Instituto Potosino de Investigación Científica y Tecnológica
San Luis Potosí, México.

Ramos-González, R
Autonomus University of Coahuila
Coahuila, México

Reyes-de la Luz, M
Universidad Interserrana del Estado de Puebla Ahuacatlán
Puebla, México

Robledo-Olivo, A
Universidad Autónoma Agraria Antonio Narro
Coahuila, México

Rodríguez-Martínez, K. L
Universidad Estatal de Sonora
Sonora, México.

Sánchez-Salas, J
Universidad Juárez del Estado de Durango
Durango, México

Segura-Ceniceros, E. P
Autonomus University of Coahuila
Coahuila, México

Solís-Quiroz, O. S
Autonomous University of Coahuila
Coahuila, Mexico

Suárez-García, A
Universidad de Sonora
Sonora, México

Téllez-López, M. A
Universidad Juárez del Estado de Durango
Durango, México

Torres-León, C
Autonomous University of Coahuila
Coahuila, México

Torres-Moreno, H
Universidad de Sonora
Sonora, México

Trujillo, J
Instituto Potosino de Investigación Científica y Tecnológica
San Luis Potosí, México.
Consejo Nacional de Ciencia y Tecnología
Ciudad de México, México.

Tucuch-Peréz, M. A
Greencorp Biorganiks de México
Coahuila, México.

Valenzuela-Núñez, L. M
Universidad Juárez del Estado de Durango
Durango, México

Chapter 1

Introduction to Plant Taxonomy: Vascular and Non-vascular Plants with Medicinal Use

Gisela Muro-Pérez, Jaime Sánchez-Salas, Omag Cano-Villegas, Raúl López-García, and Luis Manuel Valenzuela-Nuñez

CONTENTS

1.1 THE ARID AND SEMI-ARID AREAS OF MEXICO

The American continent is known for the great breadth of its desert surface. In North America, it is estimated that one-third of the surface can be considered arid or semi-arid. Understanding arid and semi-arid as the total rainfall expressed in millimeters (mm) per year, and considering some socioeconomic indicators, the National Commission of Arid Zones (CONAZA, 1994) considers that a little more than 40% of Mexico are arid and semi-arid zones in which approximately 18% of the national population lives. There is an erroneous idea that these areas lack both plant and fauna diversity. In the Mexican territory, the Sonoran Desert is distinguished, covering the lower parts of the states of Sonora and Baja California Norte and Baja California Sur, extending into part of the United States. The other is the Chihuahuan Desert Region (RDCH), which is located in the states of SLP, Zacatecas, Coahuila, Nuevo León, Chihuahua, Tamaulipas, Durango and extends into part of the United States. The Tamaulipeca semi-arid zone encompasses the states of Tamaulipas, part of Nuevo León and Coahuila. Other semi-arid zones are the Hidalguense, which encompasses the states of Querétaro and Hidalgo and the Poblano-Oaxaqueña, encompassing the states of Puebla and Oaxaca (Rzedowski, 1978). Many of the communities settled in the arid and semi-arid zones have vast knowledge of the natural resources that surround them, mainly of the flora of each place, including what plants grow in the area, what they can be used for, where they can be found and how the plants can be consumed. However, there is also a loss of this knowledge due to the lack of transculturation. In this sense, these communities have gradually changed part of their economy, and many of the needs are satisfied from outside their environment; however, they still depend, to a large extent, on spontaneous plant resources for their

DOI: 10.1201/9781003251255-1

economy, either directly or indirectly, through the exploitation of wild fauna or species under domestication.

It is common that in many of the areas where agricultural activities were carried out, after removal and cultivation (be it beans, corn, etc.), the growth of species such as the "gobernadora" (*Larrea tridentata*) is normal, establishing a plant community where the dominance of this species is almost absolute. Other species that appear immediately are the "Hojasen" (*Flourensia cernua*), the "Sangre de Grado" (*Jatropha dioica*) and others.

There is a vast collection of different plants with different uses, which drastically affects the natural populations. As we well know, "lechuguilla" (*Agave lechuguilla*) in northern Mexico is exploited for the extraction of "ixtle" that is used for the manufacture of sacks, brooms, rope, etc.; from "candelilla" (*Euphorbia antisyphilitica*), wax is extracted for use in the food industry to give shine to fruits and in the pharmaceutical industry for the production of capsules, etc.; from "guayule" (*Parthenium argentaum*), a latex is extracted that is known as "natural rubber" and is used to make car tires, gloves, condoms, tubes, etc. Guayule resins are used to make wood preservatives, pesticides and plasticizers, and the latex extraction residues are used as fuel in mixtures to produce paper.

1.2 PLANT TAXONOMY

Also known as taxonomic botany, some authors restrict the field of plant taxonomy to the study of species classification; however, we will approach taxonomic botany as the study that deals with the diversity of plants and their identification, nomenclature and classification (Jones & Luchsinger, 1986). According to Jones and Luchsinger, taxonomic botany is based on five objectives: 1) inventorying the flora of each region, 2) providing a method for the identification of plant species, 3) producing a universal classification system, 4) demonstrating the evolutionary implications of plant diversity and 5) providing a scientific name for each group of plants through nomenclature.

"Flora" is understood as the set of plants that grow in a certain area. *"Identification"* is the recognition of certain characteristics of a plant and its application. Recognition occurs when a floristic specimen is similar to a previously known plant. If the characteristics of the specimen coincide with those of the studied specimen, we are talking about the same species, and if it differs in some of the characteristics, we are talking about a new one. This is where the term *"classification"* comes from, which is nothing more than the orderly arrangement of plants in groups that share common characteristics (Chiang, 1989). Therefore, a hierarchical system of ranks or categories results, which we commonly know as families, genera, species, subspecies, varieties, etc.

1.3 PLANT RESOURCES IN ARID AND SEMI-ARID ZONES AND LEVELS OF USE

The arid and semi-arid zones of Mexico occupy a little more than 40% of the country's total. Plant diversity and the genetic richness of plants is reflected in the large number of endemisms represented in families such as Cactaceae, Asteraceae and Fabaceae. Many of the substances produced by some of the desert plants are resins, latex, waxes and rubbers, which are concentrated in some part of the plant such as fruits, seeds, stems, roots, etc. The inhabitants of these areas take advantage of them in different ways:

a) Collection and direct use (edible, ornamental, medicinal, fodder).
b) Collection and incipient transformation of use (producers of fibers, other plant species are soaked, carved or dyed for handicrafts).
c) Collection and more processed treatment (such as that required by the "cerote" process of candelilla wax, making basketry, hat making or *jarcería*).
d) Industrialization of the product, as in the case of "jojoba" for the extraction of oil for shampoo.

1.3.1 Medicinal Uses of Vascular Plants

According to Villaseñor (2004), there are 23,424 species of vascular plants registered in Mexico that are grouped into 2,804 genera included in 304 families. In addition, 618 introduced and naturalized plant species are recognized (Villaseñor & Espinosa-García, 2004). The exploitation of plant species in the arid and semi-arid zones of northern Mexico, such as oregano, mesquite, lechuguilla, sotol, candelilla, "queen of the night" (also known as "deer egg"), "chamiza" (or "cow rib"), lagrima de San Pedro (or "thunderstorm"), guayule, granjeno, among others, is carried out intensively, which places the country in first place of use of plants.

Lippia graveolens, commonly known as oregano, belongs to the Verbenaceae family; it is a shrub with aromatic leaves, and when the leaves are dried, it is used as a condiment. There are many wild plant species considered as oregano (Robledo, 1990). Almeida (1991) mentions that the fundamental importance of the species known as oregano lies mainly in their organoleptic properties derived from their attributes as flavorings or food seasonings. Other species that are also important in Mexico, from an economic point of view, are *Lippia berlandierii* and *Lippia palmeri*. Martínez (1997) reports that the 16 most commonly used species in Mexico belong to different families such as Verbenaceae, Labiadas, Compuestas and Legumiosas (Maldonado, 1991).

Prosopis sp., known as mesquite, has been a valuable resource since ancient times for the inhabitants of arid zones, as multiple benefits were found in the species because all parts of the plant can be used. The species has been considered a common cultural denominator for the nomadic hunter-gatherer peoples who inhabited northern Mexico and the southern United States (CONAZA, 1994). The most common species in northern Mexico are *Prosopis laevigata*, *P. glandulosa*, *P. juliflora*, among others. The irrational and excessive exploitation to which it has been subjected has led to the accelerated degradation of mesquite communities, which is reflected not only in the loss of the resource but also in the deterioration of the soil. Therefore, these phenomena have led to the alteration of the ecological balance of the thickets known as mosqueles, which, in turn, has greatly affected the rural communities of those sites (CONABIO, 2000).

Agave lechuguilla is an agave or maguey from which the "ixtle" (fiber obtained from the leaves of the lechuguilla) is extracted to make brushes for industrial use, cords, furniture padding and car seats, rugs and carpets, ropes, hats, fabrics, among others (Rössel et al., 2003). When mixed with resins, it is used in the manufacture of doors, ceilings, walls, sheets, shelves and furniture (Mayorga et al., 2004). In addition, due to its detergent properties, it is used in the manufacture of liquid soaps or shampoo (Zapien, 1981). According to Sheldom (1980) in Baca (2000), archaeological finds are documented in different regions of northern Mexico (mainly in Coahuila). Lechuguilla fiber (ixtle) is exported to Europe, Asia and North and South America. The plant is bushy in shape, composed of a crown of 20 to 30 thick and fibrous leaves arranged in a whorled shape, giving a rosette appearance (Martínez, 2013; De la Garza, 1985; De la Cruz & Medina, 1988). It is a semelparous species, as it presents a single sexual reproductive event (it flowers only once) during its life cycle (Begon et al., 1986). It is distributed in xeric scrub in the Chihuahuan Desert at altitudes below 1,000 meters above sea level and up to 24,000 meters above sea level (Rzedowski, 1978).

Dasylirion leiophyllum, known as "sotol", is a plant that lives in the Chihuahuan Desert and blooms in spring or early summer (Benson, 1981) and bears fruit in the months of August–September (Cano, 2006). It is a species used in an artisanal alcoholic beverage known as "aguardiente" (Melgoza & Sierra, 2003). The distribution area of *Dasylirion* is reported in the physiographic zone of the Mexican Altiplano, in the range of 1,000–2,000 meters above sea level. Its inflorescence is used in construction as fencing to demarcate property lines, on roofs for shade and as fodder for livestock (Cano, 2006). It is also possible to extract inulin sugar, which is of great value to the pharmaceutical industry (Ibave et al., 2001). It is a common species in rosetophilous desert scrub, in grassland areas and in submontane scrub. Henrickson and Johnston (1977) report nine species in Mexico. According to Olivas and Rivera (1984), sotol plants are selected by their weight (10 to 12 kilograms for extraction), and the foliage is first removed with a "Bowie knife" and then the pineapple is removed from the ground.

Euphorbia antisyphilitica, commonly known as "candelilla", was exploited around 1905, when Connek and Landress investigated the composition, whitening and properties of candelilla wax. It is a plant with a wide range of distribution in the region of the Chihuahuan Desert. The importance of this species lies in the waxy covering of the stems, which is made up of hydrocarbons, esters, lactones and resins. It is mainly used in the cosmetic, food and textile industries as a coating for chocolates, sweets and chewing gum as well as for the manufacture of candles, leather products, paints, polishes, matches, spark plugs, tires, etc. (CONABIO, 2000).

Peniocereus gregii known as "queen of the night" or "deer egg" (due to the shape of the root tuber), has between 18 and 20 known species (Guzmán et al., 2003; Gómez-Hinostrosa & Hernández, 2005). It is distributed in northern Mexico and part of Texas in the United States at elevations from 1,200 meters above sea level (Sánchez-Salas et al, 2009; Hernández & Gomez, 2005). It is common to find the species with associations of *Prosopis laevigata* (Perroni, 2007). The illegal collection of this species is taking it to brink of disappearance and has been classified as needing special protection according to NOM-059-SEMARNAT-2010 (SEMARNAT, 2010). It is highly sought after by foreign collectors for its nocturnal and beautiful flowering, is also collected by locals because they attribute medicinal properties to the root of this plant with benefits for kidney problems.

Atriplex canescens, commonly known as "cow rib" or "chamizo", belongs to the Chenopodiaceae family. It is distributed in northern Mexico and the southeastern United States. It is a branched bush of 1 to 1.5 meters (m) with ash green leaves. In some indigenous populations of southwestern Texas and New Mexico, they chew the leaves with a pinch of salt, followed by a drink of water to relieve stomach pains. The leaves are also cooked and used as an emetic (disgorge) (Grajales, 2015). The plant is also used as fodder for sheep and goats (Kearney et al. 1960).

Tecoma stans also known as "tronadora" or San Pedro's tear, belongs to the Bignoniaceae family. It is considered a shrub, although it also has the growth form of a tree. It has compound leaves elongated with serrated margins and funnel-shaped flowers arranged in clusters. It is distributed from central Mexico to the southern United States. The leaves are used for the treatment of syphilis and diabetes. Made into an infusion, the leaves also help to calm cough and asthma and respiratory problems as well as stabilize temperature. It is considered a natural analgesic and anti-inflammatory (González, 1998; Adame & Adame, 2000; Naranjo et al., 2003).

Parthenium argentaum, known as guayule, belongs to the Asteraceae family, this family includes about 16 species that are distributed in Mexico (Rollins, 1950). The leaves are grayish with yellow flowers, it has been used as a source of natural rubber since the nineteenth century. By extrusion of the plant material, the fibrous material is separated from the latex, which is of high quality for the manufacture of medical products such as surgical gloves, catheters and condoms. It is known as a plant that produces natural rubber and is considered hypoallergenic. The presence of latex in the cells of the guayule plant has been demonstrated (Wood, 2002; Coffelt et al., 2009; Rodríguez, 2011).

Celtis palida, better known as the granjeno, belongs to the Ulmaceae family. It is a deciduous shrub that grows up to 3 m with small, rough leaves and small, white flowers. Its fruit is fleshy, ovoid, orange in color and edible. The leaves are used as poultices on pimples or inflammations. In northeastern Mexico, the ground leaves are applied to alleviate headaches. It is found at altitudes of 1,000 to 2,300 m above sea level. It is distributed in Mexico from Baja California, Baja California Sur, Chihuahua, Coahuila, Durango, Guanajuato, Hidalgo, Jalisco, Michoacán, Nuevo León, Oaxaca, Querétaro, San Luis Potosí, Sonora, Sinaloa, Tamaulipas, Veracruz and Zacatecas (Merla, 1990).

1.3.2 Medicinal Uses of Non-vascular Plants

Given the general lack of commercial value, their small size and their inconspicuous role in ecosystems, many bryophytes seem to have no use. However, there is evidence of Stone Age men living in present-day Germany collecting the moss *Neckera crispa* (Grosse-Brauckmann, 1979). A few other bits of evidence suggest a variety of uses by cultures around the world (Glime & Saxena, 1991). Barnett (1987) cited evidence indicating that wounds covered with a dressing of *Sphagnum sp.* moss recovered much faster than wounds that were not covered by the moss.

There is evidence of some moss species, such as *Polytrichum commune*, that, in ancient times, was mixed with an oil with calyptra extract to beautify the hair (Crum, 1973; Smith, 2007). The use of bryophytes as medicinal plants has been common in China and India as well as among Native Americans since time immemorial. Numerous compounds, including oligosaccharides, polysaccharides, polyols, amino acids, fatty acids, aliphatic compounds, prenylquinones and phenolic and aromatic compounds occur in bryophytes, but few links between medical effects have been established (Pant & Tewari, 1989, 1990).

In North American and Indian cultures, the use of bryophytes, such as *Bryum*, *Mnium*, *Philonotis* and *Polytrichum*, is common. Bryophytes are used as indicators, have medicinal uses, control pollution and fix nitrogen in the substrate among others. The liverwort *Marchantia polymorpha* is known to treat liver diseases and, in countries such as China, jaundice from hepatitis and is used as a topical ointment to reduce inflammation (Rachna & Vashishtha, 2015).

REFERENCES

Adame, J. & Adame, H. 2000. *Plantas curativas del Noreste Mexicano*. 1a. edición. (Castillo, editor). México, pp. 1–386.

Baca, M. S. 2000. Determinación del potencial productivo de la lechuguilla (*Agave lechuguilla* Torr.) en el Municipio de San Juan de Guadalupe, Dgo. Tesis Profesional Unidad Regional Universitaria de Zonas Áridas Universidad Autónoma Chapingo. Bermejillo, Durango. México. 49 p.

Chiang, F. 1989. La taxonomía vegetal en México: Problemas y perspectivas. In CIENCIA UNAM, Revista de Difusión. México. 07 p.

Coffelt, T. A., Nakayama, F. S., Ray, D. T., Cornish, K., McMahan, C. M. & Williams, C. F. 2009. Plant population, planting date, and germplasm effects on guayule latex, rubber, and resin yields. *Industrial Crops and Products* 29: 255–260.

CONABIO. 2000. Disponible en: http://www.conabio.com.mx.

Glime, J. M. & Saxena, D. K. 1991. *Uses of Bryophytes*. New Delhi: Today and Tomorrow's Printers and Publishers, p. 100.

Gómez Hinostrosa, C. & Hernández, H. M. 2005. A new combination in *Peniocereus* (Cactaceae). *Anales del Instituto de Biología, Universidad Nacional Autónoma de México, Serie Botánica*. 76: 129–135.

González, F. M. 1998. *Plantas Medicinales del Noreste de México*, 1ª. Edición. México: IMSS, pp. 1–128.

Grajales Tam, K. M. 2015. *Plantas de la Reserva de la Biosfera de Mapimí*.

Guzmán, A., Arías, S. & Dávila, P. 2003. *Catálogo de cactáceas mexicanas*. México, Distrito Federal, México: Universidad Nacional Autónoma de México, Comisión Nacional Para el Conocimiento y Uso de la Biodiversidad.

Hernández, H. M. & Gómez Hinostrosa, C. 2005. Cactus diversity and endemism in the Chihuahuan Desert region. In *Biodiversity and Conservation in Northern Mexico* (J. L. Cartron, R. Felger, & G. Ceballos, editors). New York: Oxford University Press, pp. 264–275.

Jones, S. B. & Luchsinger, A. E. 1986. *Plant Systematics*. 2ª. edition. Nueva York: McGraw-Hill.

Kearney, T. & Peebles, R. H. et al. 1960. *Arizona Flora*. Berkeley, Los Angeles, London: University of California Press.

Merla, R. G. 1990. *Geografía Regional de Nuevo León*. Universidad Autónoma de Nuevo León. Centro de Información de Historia Regional. Serie Bibliográfica de Nuevo León.

Naranjo, J., Corral, A., Rivero, G., Fernández, M. & Pérez, P. 2003. Efecto hipoglicemiante del extracto fluido de *Tecoma stans* Linn en roedores. *Rev. Cubana Milit.* 32(1): 13–17.

Pant, G. & Tewari, S. D. 1989. Various human uses of bryophytes in the Kumaun region of northwest Himalaya. *Bryologist*. 92: 120–122.

Pant, G. & Tewari, S. D. 1990. Bryophytes and mankind. *Ethnobotany*. 2: 97–103.

Perroni, V. Y. 2007. Islas de fertilidad en un ecosistema semiárido: Nutrimentos en el suelo y su relación con la diversidad vegetal. Tesis de doctorado, Instituto de Ecología, Universidad Nacional Autónoma de México, México, Distrito Federal, México.

Rachna, P. & Vashishtha, B. D. 2015. Bryophytes-boon to mankind a Review. *Agri Review*, 36(1): 77–79.

Rodríguez, F., Posadas, L., Vilchez, J., Ivorra, A. & Lahora, A. 2011. El guayule, *Parthenium argentatum* A. Gray (Asteraceae), asilvestrado en Almería (España). Nota breve. *Anales de Biología*. 33: 103–105.

Rollins, R. C. 1950. The guayule rubber plant and its relatives. *Contributions of the Gray Herbarium of Harvard University*. 172: 1–73.

Sánchez Salas, J., Flores Rivas J. D., Muro Pérez G. & Martínez Adriano C. 2009. El reinado desconocido de *Peniocereus greggii*. *Boletín de la Sociedad Latinoamericana y del Caribe de Cactáceas y otras Suculentas*. 6: 21–24.

Secretaria de Medio Ambiente y Recursos Naturales. 2010. Oficial Mexicana NOM-059-SEMARNAT-2010, protección ambiental especies nativas de México de flora y fauna silvestres categorías de riesgo y especificaciones para su inclusión, exclusión o cambio-lista de especies en riesgo. Segunda sección. Secretaría de Medio Ambiente y Recursos Naturales, Diario Oficial, 30 December 2010, México, Distrito Federal, México.

Sheldom, S. 1980. *Ethnobotany of Agave lechuguilla and Yucca carnerosana in Mexico's Zona Ixtlera In Economic Botany. 34 (4)*. Bronx, NY: New York Botanic Garden, pp. 376–390.

Smith Merrill, G. L. 2007. *Polytrichum en Flora of North America Editorial Committee, eds. 1993, ed., Flora of North America 27*. New York & Oxford: Oxford University.

Villaseñor, J. L. 2004. Los géneros de las plantas vasculares de la flora de México. *Bol. Soc. Bot. Méx.* 75: 105–135.

Villaseñor, J. L. & Espinosa-García, F.. 2004. The alien flowering plants of México. *Divers. Distrib.* 10: 113–123.

Wood, 2002. *Desert Shrub May Help Preserve Word*. ARS-USDA.

Chapter 2

Mexican Desertic Medicinal Plants

Biology, Ecology, and Distribution

José Antonio Hernández-Herrera, Luis Manuel Valenzuela-Núñez, Juan Antonio Encina-Domínguez, Aldo Rafael Martínez-Sifuentes, Eduardo Alberto Lara-Reimers, and Cayetano Navarrete-Molina

CONTENTS

DOI: 10.1201/9781003251255-2

2.1 INTRODUCTION

Medicinal plants are of great importance in human history. When the first humans learned to use the resources of their environment and took advantage of the biodiversity to satisfy their primary needs, such as food and shelter, plants were considered as an ecosystem service (Caballero-Serrano et al., 2019). Upon finding the healing properties of plants, humans have used the resources of the environment to relieve and cure diseases or pain. From ancient times to the present, there has been constant interest in learning about and finding properties of medicinal plants (Petrovska, 2012).

As plant use increased, people began to create different drinks (fermented drinks) and local dishes (gastronomy) and made use of them in their rituals and religious beliefs, but the most important was the use of plants as medicine (Tamang, 2010). The latter use developed into traditional herbal medicine (Pascual-Casamayor et al., 2014) and beliefs in each culture around the world, thus laying the foundations for the pharmaceutical industry. Many medicinal plants have various uses from which we can benefit through food or drink, baths, inhalations, ointments, gargles, massages and/or plasters (Magaña et al., 2010).

It is well known that communities have the greatest knowledge about their resources and their various uses, as this knowledge has been passed from generation to generation by the people of different cultures around the world (Magaña et al., 2010). It is important to understand that the knowledge of plants is developed to solve local problems within each community or culture, and this knowledge evolves as new plants and uses are incorporated and the use of others is discontinued because of side effects, toxicity, or inefficiency in solving the problems.

The semi-arid ecosystems of northern Mexico have their own importance and specific characteristics in terms of endemic and medicinal plants, covering more than 70 million hectare (ha) (Villavicencio-Gutiérrez et al., 2021). The arid and semi-arid regions of Mexico contribute to the richness of the Mexican flora with about 6,000 species of vascular plants, of which about 60% are endemic (Rzedowski, 1978, 1991, 2005).

Mexico and the United States form the most important xeric scrub region in the world: the Sonoran Desert and Chihuahuan Desert. The first region includes Sonora and Baja California, and the second region includes Tamaulipas, Nuevo León, Coahuila, Chihuahua, Durango, Zacatecas, and San Luis Potosí. Another arid area, called Tehuacán Valley matorral, is located in the center of the states of Puebla and Oaxaca (Dinerstein et al., 2017)

Despite Mexico's vast biocultural biodiversity, there are numerous regions, such as the north of the country, where the knowledge and use of medicinal plants are not recorded and have not yet been studied in-depth (Lara-Reimers et al., 2018). In addition, traditional knowledge of our plant resources that, for years, has supported the health of all those people who came to populate the semi-desert is fading because of patterns related to migration, poverty, loss of cultural identity, access to medicine, new health strategies applied by the Mexican government, industrialization and the lack of interest of younger generations.

The aim of this chapter is to document, describe, analyze and preserve traditional knowledge of the uses and applications of some of the most important medicinal plants used in the drylands of Mexico. The desert has many important plants that are used not only economically but also medicinally, industrially, gastronomically, religiously and culturally. These ecosystems are characterized by their high vulnerability to desertification and by the presence of extremely marginalized rural populations who exploit non-timber forest resources of various species for the production of raw materials.

2.2 CURRENTLY ACCEPTED SCIENTIFIC NAME *ACACIA SP.*

2.2.1 Biology

Acacia is a pantropical genus (second largest in the Fabaceae family) represented by 1,250 species, of which more than 950 are exclusively Australian, 120 African, and the rest American. In Mexico, we can find 85 species (commonly named *huizaches*) of which 46 are endemic, with the majority found in the arid and semi-arid regions of the country, which gives a clear idea of the ecological tolerance of the genus (Rico, 1984). *A. farnesiana* is present throughout the territory, encompassing a wide variety of climates and ecosystems (altitudes of 0–2,600 meters (m),

Figure 2.1 *Acacia* plant in the flowering period in the town of Buenavista, Saltillo, Coahuila.

temperatures of 5 °C –30 degrees Celsius (°C), and rainfall of 100–900 millimeters (mm) per year on average), also presenting great morphological variability. *A. schaffneri* is located mainly in the northern and eastern region of the country, commonly found together with *A. farnesiana*, so it is frequently confused with the latter (Rico, 1980) (Figure 2.1).

2.2.2 Active Principles or Bioactive Compounds

Trees are also known for their ethnopharmacological properties, and a number of scientific studies have been carried out previously. For example, the bronchodilator and anti-inflammatory effect of glycosidal fraction of *Acacia* were reported by Letizia et al., (2000). The root contains diterpenes and flavonoids (Mors, 2000).

Acacia bark contains a methanolic extract that exhibited antidiarrheal activity against castor oil and magnesium sulphate induced diarrhea along with antimicrobial activity against common pathogens responsible for diarrhea *in vitro*. Relatively simple alkaloids have been found in most species of the *Acacia* genus. From one of the most studied species, *Acacia berlandieri*, the presence of N-methyl-β-phenylethylamine (19) and nicotine (20) has been reported, and 2-methyl-1,2,3 was also isolated from *Acacia simplicifolia* 4-tetrahydro-β-carboline (Demole et al., 1969; Kjaer et al., 1961).

Further analysis indicated the presence of small amounts of more than 33 additional amines and alkaloids, including amphetamine, methylamphetamine, mescaline, mimosine methyl esters, nornicotine and isoquinoline alkaloids (Joshi et al., 1979).

2.2.3 Ethnobotanical Uses or Traditional Uses

Acacia is cultivated in several countries for its floral nectar which is used as an active ingredient in the production of cassie perfume (Lapornik, 2005). The branches and trunks of *Acacia* trees are widely used as fuelwood and for construction materials, particularly fence posts (Purata et al., 1999). Furthermore, the bark of some species has commercially useful tanning agents, and the root is reputedly an effective snake antivenom (Morse et al., 2000).

Acacia species are associated with many myths and religious beliefs among ancient peoples. They were conceived as plants highly valued for their healing properties and were considered extremely important and effective in driving away bad luck and evil spirits. *A. schaffneri* trees are used to treat gastric ulcers with infusions of the bark. It is also used to cure skin infections by directly applying the fresh fruit without the shell or by directly washing the affected part with the decoction of the dried fruit (Anderson, 1984).

2.2.4 Ecology and Distribution

Acacia trees are sparsely distributed in a number of xeric habitats such as deciduous forests, thorn scrub and savanna from Mexico to Colombia (McVaugh, 1987; Rzedowski, 1978; Siegler & Ebinger, 1988). Cattle are the principal factor for the formation of *Acacia* stands. The animals eat the seeds and spread them via their dung around the edges of existing patches of *Acacia*. There exists a very close association between *Acacia* and cattle, and this may have an evolutionary basis. The grazing and browsing of large mammals may favor the growth of this spiny tree over the rapid growth of other pioneer species (Brown, 1960). In addition, grazing reduces fire frequency which may favor the spread of *Acacia* in a fire- adapted grassland system. The indehiscent pods of *Acacia* trees are similar to those *Acacia* trees in Africa, which are dispersed by large mammals in natural savanna (Coe & Coe, 1987; Gwynne, 1969). Not long ago, the existing *Acacia* savannas were probably dominated by oaks until recent increases in wood cutting and the arrival of cattle, factors that ultimately favor the shift toward an acacia-dominated savanna. *Acacia* may also have been dominant during earlier periods with higher grazing and browsing pressures by the megaherbivores (Figure 2.2).

2.3 CURRENTLY ACCEPTED SCIENTIFIC NAME *AGAVE SALMIANA* OTTO EX. SALM-DICK

2.3.1 Biology

The *Agave salmiana*, or green maguey, belongs to the family Agavaceae. The plant has a large rosette of thick, fleshy leaves, each ending generally in a sharp point (Escamilla-Treviño, 2012). The plants have dark green leaves that distinctly valleculate toward the apex, the inflorescences are pyramid-shaped, and the bracts are rounded and appear fleshy (Smith & Figueiredo, 2012).

The *Agave salmiana* plants in northern Guanajuato present morphological variations such as different rosette diameters, leaf length or penca (Castañeda-Nava et al., 2019), indicating that there are different subspecies and varieties of this maguey (Chávez-Güitrón et al., 2019).

Agave salmiana is a specie with crassulacean acid metabolism (CAM) in which the leaves in a rosette arrangement use carbon dioxide (CO_2) fixation to allow maximal absorption of photosynthetically active radiation and other vital biochemical reactions during periods of drought (Stewart, 2015).

Asexual propagation by young rhizome suckers is the most common natural way of propagating *Agave*; however, sexual reproduction increases genetic variability (Díaz et al., 2011) (Figure 2.3).

2.3.2 Active Principles or Bioactive Compounds

The *Agave salmiana* sap mainly contains saponins and gentrogenin pentaglycoside, which has an apoptotic effect on colon cancer cells (Santos-Zea et al., 2016).

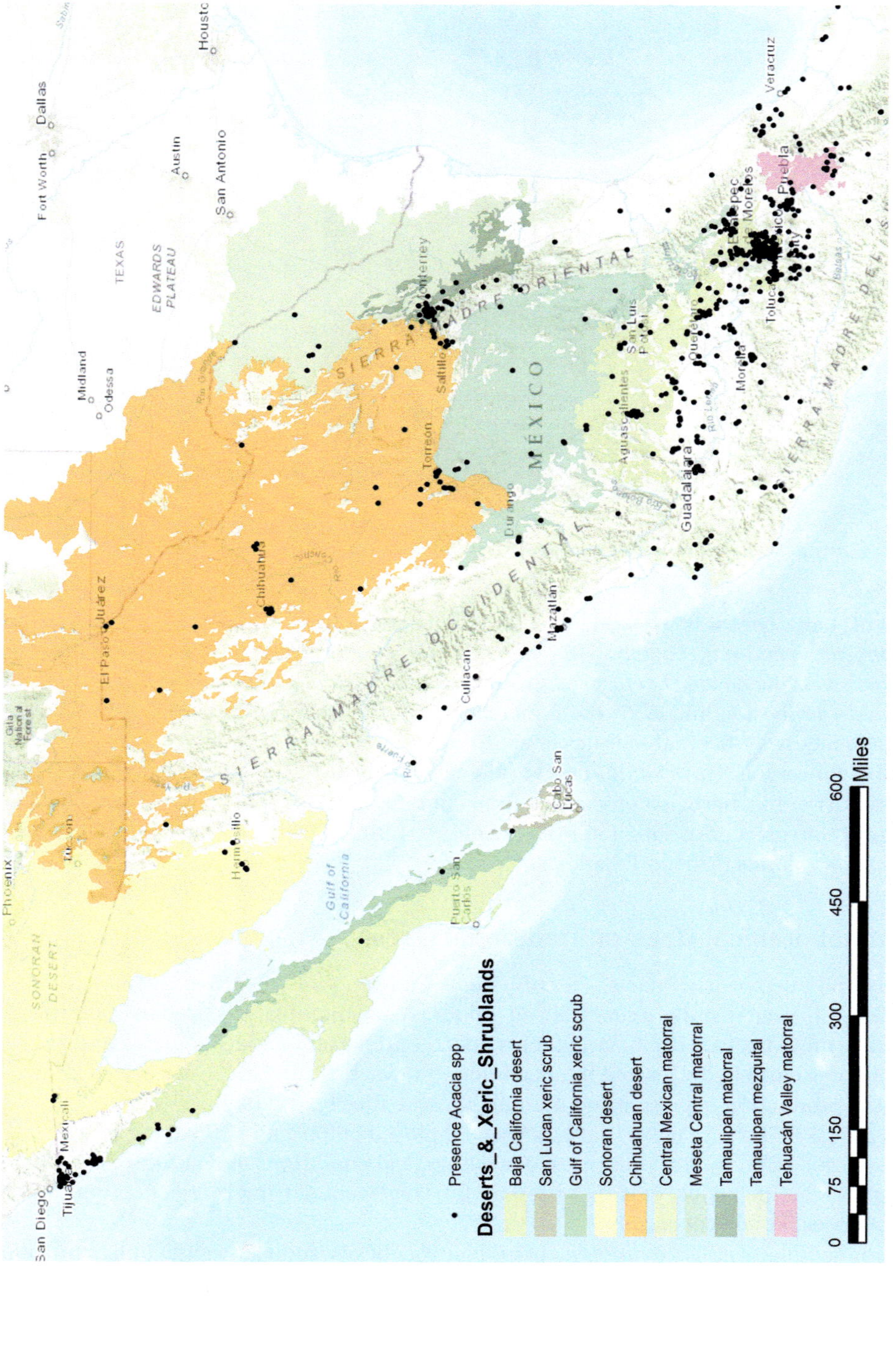

Figure 2.2 Map of the presence of *Acacia* with a wide distribution in Mexico.

Figure 2.3 *Agave salmiana* on the campus UAAAN in Buenavista, Saltillo Mexico.

The sap of *Agave* contains prebiotic and probiotic activity with the presence of *Leuconostoc sp.*, *Leuconostoc gelidum*, *Lactococcus lactis*, *Enterococcus casseliflavus*, *Pediococcus sp.*, *Trichococcus sp.*, *Leuconostoc*, *Lactococcus*, *Kazachstania zonata*, *Kluyveromyces marxianus* and *Saccharomyces cerevisiae* and with the concentration of fermentable sugars such as sucrose, fructose and glucose (Villarreal Morales et al., 2019).

The edible flowers of *Agave salmiana* contain carbohydrates (71.58 ± 0.92), protein (11.58 ± 0.70), macronutrients, microelements and a high content of carotenoids and ascorbic acid, and their consumption can contribute potential benefits to human health because they also contain antioxidant compounds (Pinedo-Espinoza et al., 2020).

2.3.3 Ethnobotanical Uses or Traditional Uses

Historically and currently, all parts – leaves, stems and sap – of the *Agave salmiana* are used. Non-distilled, it is used to make aguamiel and pulque. When distilled, it is used to make the alcoholic beverage mescal and syrup or sweeteners. Its potential uses include medicine and as bioenergy from soluble carbohydrates and lignocellulose (Stewart, 2015). This *Agave* is well known for its use for the production of pulque and mescal (Álvarez-Ríos et al., 2020).

The maguey is a Mexican plant that has been employed empirically for cancer treatment. Its popular use is for inflammatory diseases, dermatological conditions and skin tumors (Alonso-Castro et al., 2011). The penca, or leaf, is used for the treatment of tumors. The maguey has anticancer properties (Popoca et al., 1998).

The sap or aguamiel of *Agave salmiana* helps control obesity, reduces weight and fat mass and lowers serum glucose, insulin and LDL-cholesterol levels because the saponins and bacterial species modify the intestinal microbiota (Leal-Díaz et al., 2016).

The green maguey is used for pulque production, a typical Mexican fermented drink made from its sap or aguamiel (Muñiz-Márquez et al., 2015). In Saltillo, Coahuila and the communities of Nuevo Leon, Mexico, Agave syrup or aguamiel is used to make bread and produce a beverage (Estrada et al., 2007).

2.3.4 Ecology and Distribution

The green maguey is widely distributed in the highlands, the center and the south of Mexico and the southern United States in an arid and semi-arid ecosystem (Lara-Ávila & Alpuche-Solís, 2016). It is distributed in the highlands from an altitude of 1,800–2,400 m, in a dry steppes (BS) climate, with an average annual rainfall of 325.8 to 502.9 mm and average annual temperatures from 16 °C to 18.7 °C (Tello-Balderas & García-Moya, 2017).

The natural population of *Agave salmiana* in Zacatecas has densities of 653 to 3,064 magueys per ha in the municipalities of Pinos, Villa Hidalgo, Noria de Angeles, Loreto, and Villa García (Martínez Salvador et al., 2005) (Figure 2.4).

2.4 CURRENTLY ACCEPTED SCIENTIFIC NAME *CUCURBITA FOETIDISSIMA*

2.4.1 Biology

Cucurbita foetidissima, known as buffalo gourd, is a semi-xerophytic plant native to semi-arid lands of the southwestern United States and northern Mexico. It grows primarily as a weed in disturbed soils and in low areas (Dittmer & Talley, 1964). It is also known as Missouri gourd, chili coyote, calabacilla loca and fetid gourd. Its distribution is from the northern Great Plains to central Mexico in elevations of 1,000–7,000 feet (ft) (Scheerens et al., 1978).

The plant develops large and wide storage roots containing substantial quantities of starch (Berry et al., 1976). It is a plant with good yields that increase over the years and with good agricultural management, reaching up to 270 kg/ha in five years from planting (Bemis et al., 1978). It produces a dense groundcover vine growth and an abundant crop of fruit (Bemis et al., 1979). The fruits are 5–7 centimeters (cm) in diameter and contain 200–300 seeds. Each plant produces more than 270 fruits. The seeds contain 30%–40% edible oil and from 30% to 35% protein depending to seasonal growth fluctuations and competition (Bemis et al., 1978).

It has lanceolate leaves 12 to 25 cm long and 10 to 13 cm at the base with a high photosynthesis rate (Gómez-González et al., 2019). Its flowers are found at the nodes of the vines, and the most abundant flowers are male. Its asexual mode of reproduction is probably the major reason for its dense colony formation in areas of summer rainfall. Every node of the runner has the potential to develop an adventitious root if the moisture content of the soil is sufficient and coincides with an appropriate physiological stage of development (Beims et al., 1978) (Figure 2.5).

2.4.2 Active Principles or Bioactive Compounds

The fatty acid distribution of the seeds' oil content is as follows in percentages: palmitic 7.8%, stearic 3.6%, oleic 27.1% and linoleic 61.5%. About 30% of the seeds are protein with an amino acid content similar to other oilseeds (Scheerens et al., 1978). Nevertheless, both, amino acid supplementation and blending with complementary protein sources would be desirable for use in cattle feeding. The seed coats show a digestibility of 23%, which is similar to that of sunflower and cottonseed (Dreher et al., 1980). Further, mineral nutrients in the whole seeds compare favorably with those of dry cow rations (Lancaster et al., 1983).

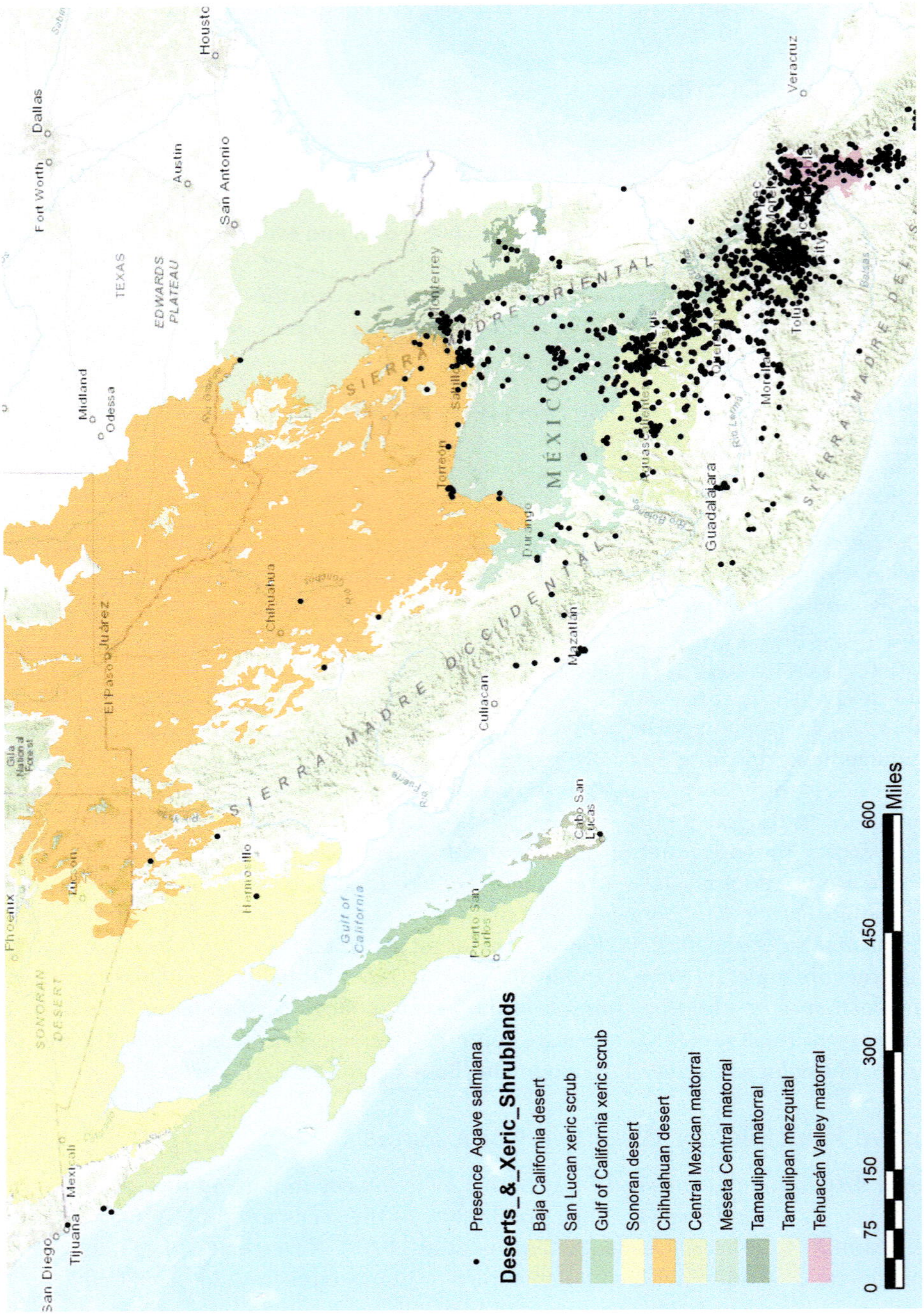

Figure 2.4 Map of the presence of *Agave salmiana* with a wide distribution in Mexico.

Figure 2.5 *Cucurbita foetidissima* fruit in the winter of 2022.

The seed meal contains about 45% protein with a comparable amount of fiber to of that of deoiled meal used in animal feeds (Bemis et al., 1978). Furthermore, seed flour contains 70% protein, and protein concentration of 80%–85% protein might be produced for the food industry by standard protein-fiber separation techniques.

The starch content varies from 50% to 65% on a dry-weight basis (Bemis et al., 1978; Berry et al., 1975, Nelson et al., 1983).

2.4.3 Ethnobotanical Uses or Traditional Uses

Plants are found primarily on disturbed land (roadsides and old agricultural fields), and it has been observed that they may have become dependent upon their long association with human populations (Whitaker & Bemis, 1975).

The whole plant has been used as soap and food (Niethammer, 1974). The fruit and the root have been used not only as hand and laundry soap as well as shampoo but also as stain remover (Hogan & Bemis, 1983). The seeds, when roasted, boiled, or cooked, are used to make a form of mush (Niethammer, 1974; Cutler & Whitaker, 1961). The bitter taste of the fruit pulp and seeds can be removed using calcium carbonate (Nabhan, 1980). Adding water reduces the presence of bitter cucurbitacins in the fruit, foliage, and roots for consumption. Fruits must be boiled repeatedly before consumption (Cutler & Whitaker, 1961).

This plant has potential biomedical properties that are currently being investigated at a medicinal plant research institute in Mexico (Nabhan, 1980). Roots of the buffalo gourd have been used for medicinal purposes. Juice extracted from roots that have been pounded and boiled is used as a disinfectant and to treat toothaches (Niethammer, 1974). Pieces of the root are still sold in the southwestern United States.

This specie has the potential of a cultivated plant with the purpose of food crop, having seeds rich in oil and protein. It also produces an extensive storage root system rich in starch. It has evolved in the arid regions of North America and is adapted to growing on arid to semi-arid lands, which constitute much of the world's land mass – lands now marginal for crop production.

2.4.4 Ecology and Distribution

It is a native perennial plant considered a weed, widely distributed in northern Mexico and the southern United States (Wallace & Quesada-Ocampo, 2015) (Figure 2.6).

2.5 CURRENTLY ACCEPTED SCIENTIFIC NAME *DYSPHANIA AMBROSIOIDES* (L.) MOSYAKIN & CLEMANTS

2.5.1 Biology

The common name of this species is *Epazote* (from the Nahuatl *epatl*, meaning "skunk," alluding to its unpleasant smell). In Mexico there are 24 species that belong to the genus *Chenopodium* (Blanckaert et al.,2012).

It is a summer annual plant that flowers from May to November and reproduces only by seed (Dembitsky et al., 2008). It is characterized by being an upright, glandular plant, 20–30 cm tall. Its stem is simple or branched, sometimes reddish and striped. Its leaves are yellow on the underside and are oval or oblong, 2–6 cm long by 1–3 cm wide, covered with glands, sinuate-pinnatifid, with a thin petiole and oblong or deltoid lobes, without hairs or sticky covering (viscid) on the upper side. It has a loosely dichotomous inflorescence, with numerous axillary cymes arranged in long panicles with sessile flowers and pedicels. The fruits are surrounded by the perianth (a utricle), which is small with a thin wall, about 0.10 cm in diameter; horizontally, the seeds are 0.05 mm in diameter, dark brown or black and have an adherent pericarp. The seedlings are cylindrical hypocotyl, up to 22 mm; cotyledons with ovate to elliptic blades that are 2–4 mm long and 1.3–2 mm wide, without hairs; epicotyl that are 1.5–9 mm long and quadrangular; and have opposite leaves (Espinosa & Sarukhán, 1997).

2.5.2 Active Principles or Bioactive Compounds

The Amaranthaceae have a largely worldwide distribution. Members of this family are rich in bioactivity, such as *D. ambrosoides* which possesses antifungal, antibacterial and anti-inflammatory properties (Khan & Javaid, 2019; Khan & Javaid, 2020). *Epazote* oil has been reported to contain four hydroperoxides monoterpenes in addition to ascaridole (Kiuchi et al., 2002). Ascaridole is known to have an analgesic effect at doses of 100 mg/kg and at doses of 300 mg/kg to produce convulsions and lethal toxicity in mice (Okuyama et al., 1993). Likewise, fatal cases are recorded to have happened because of the misuse of this plant. *D. ambriosoides'* toxic amount of ascaridole is the main reason for poisoning in children in registered fatal cases (Marinoff & Urbina, 2009). In addition, *epazote* contains limonene (an important antioxidant) and camphor, among other compounds (Sagrero-Nieves & Bartley, 1995).

The chemistry has been little studied, and, in fact, there is only one study carried out in Mexico. It describes the presence in the aerial parts of the sesquiterpenes cryptomeridiol, its alpha-acetoxylated derivative and hydroxy-elemol; the flavonoids chrysin, pinocembrin, and

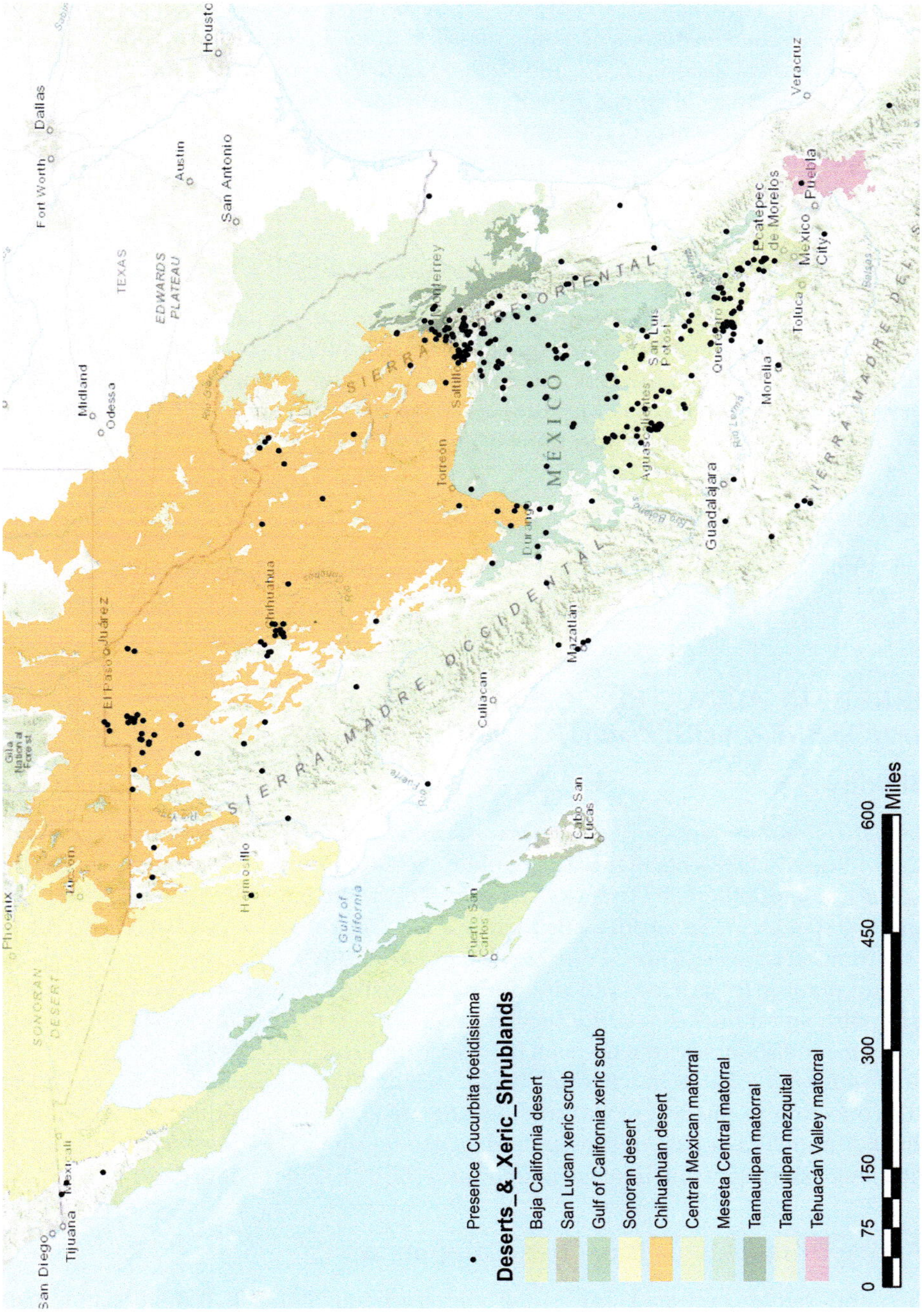

Figure 2.6 Map of the presence of *Cucurbita foetidissima* in the central highlands of Mexico.

pinostrobin; the sterols daucosterol, stigmasterol, and stigmas-2-en-3-beta-ol and the monoterpene geraniol acetate. The *in vitro* anthelmintic activity of *D. ambriosoides* leaf, stem, flower and seed extracts (fresh and dry) against freshly excysted *Fasciola hepatica* was verified, which were lethal to parasites at concentrations of 2.5mg plant/ml and 5.0plant/ml. Anthelmintic activity against *Fasciola hepatica*, *Ascaridia galli* and *Stomoxys calcitrans* larvae of the active compound pinocembrine (5,7-dihydroxiflavanone), isolated by fractionation in thin layer chromatography (TLC) of an extract of aerial parts of the plant, obtained with acetone and its essential oils of *D. ambrosioides*, are rich in monoterpenes, hydrocarbons and tannins (Camacho, 1991; Zarate & Xolocotzi, 1991; Calzada et al., 2003; Castillo-Juárez et al., 2009).

2.5.3 Ethnobotanical uses or Traditional Uses

It is classified as a medicinal plant because it has anthelmintic properties, and it turns out to soothe some stomach pains, intestinal worms (hookworm, roundworm), cramps, healing action inflammation in the belly, cough, cold, pneumonia, regulation of menstrual cycle, and fever (Gheno-Heredia et al., 2011; Grassi et al., 2013). It is also used for ceremonial and religious purposes as a condiment and to relieve infections in animals (Vibrans, 2009).

2.5.4 Ecology and Distribution

It is native to Mexico and grows in a weedy and ruderal habitat. It has been recorded in most of the states in Mexico (Villaseñor & Espinosa, 1998) (Figure 2.7).

2.6 CURRENTLY ACCEPTED SCIENTIFIC NAME *EUPHORBIA ANTISYPHILITICA* ZUCC

2.6.1 Biology

Candelilla is the common name of *Euphorbia antisyphilitica* Zucc, which means "little candle" (Barsch, 2004). The candelilla belongs to the Euphorbiaceae family, which is characterized by the presence of latex metabolite in its tissues (Hua et al., 2017). The *Euphorbia* genus is composed of more than 1,000 species in the world; while there are 138 accounted for in Mexico (Gordillo et al., 2002), it is estimated that there are 241 species (Steinmann, 2002).

It is a small plant less than 1-m tall that forms extensive colonies with multiple photosynthetic stems with small pink and white flowers. Vegetative reproduction is the most common form of underground shoots or root division (IBUNAM, 2019).

Euphorbia antisyphilitica is laticiferous, which means that extended cells found in tissues such as the roots and stems, where wax and hexane are extracted, produce latex (Mehrotra & Ansari, 2000). Candelilla is part of the Euphorbiaceae family and is considered toxic and can cause irritation of skin, eyes and mucous membranes (Barsch, 2004) (Figure 2.8).

2.6.2 Active Principles or Bioactive Compounds

The extracts and wax are an important source of phytochemicals, such as lipids, saponins, quinones and phenolic compounds, including ellagic acid, hydrolysable and condensed tannins, gallic acid and candelitannin (Rojas et al., 2021).

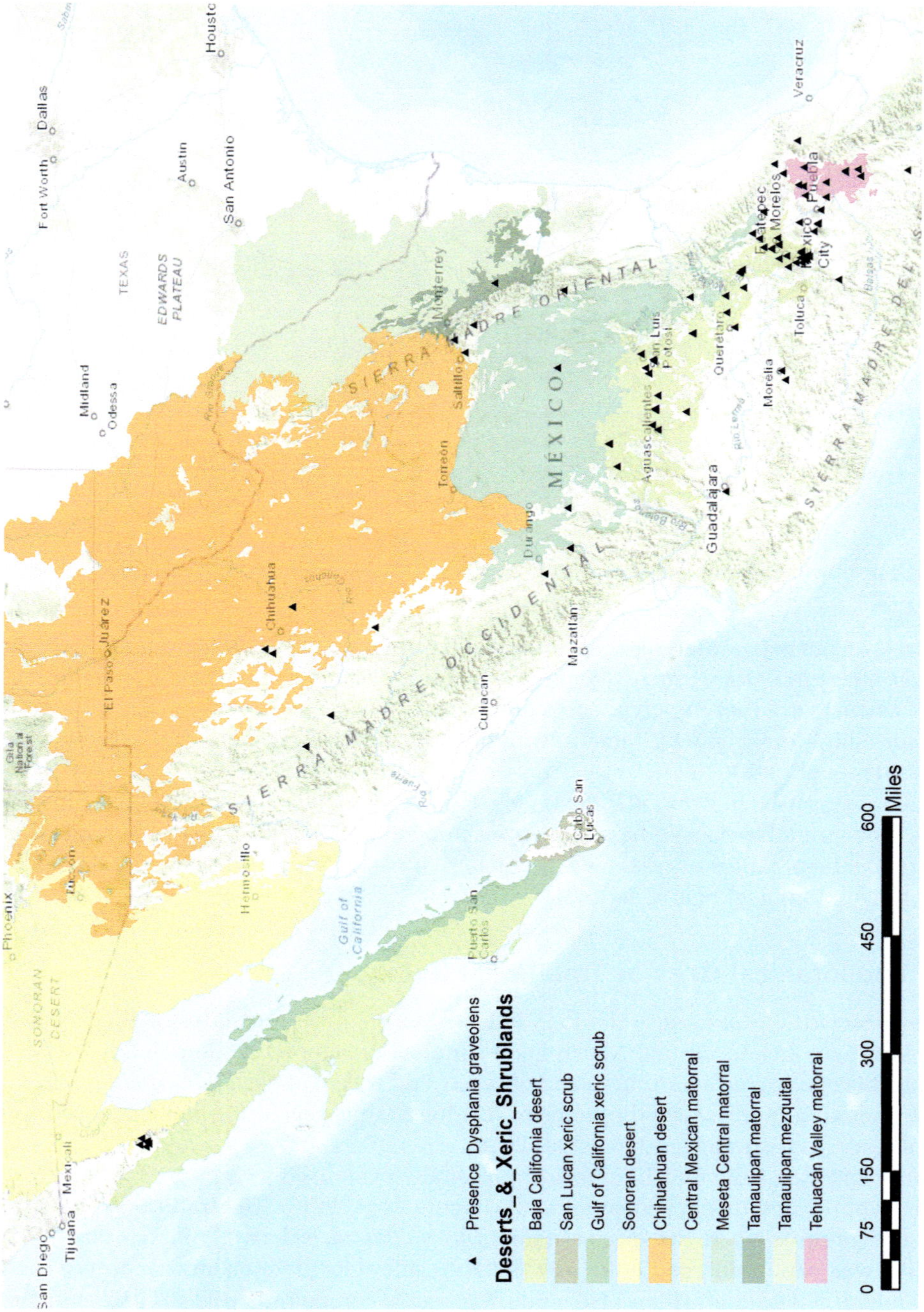

Figure 2.7 Map of the distribution of *Dysphania ambrosioide.s*

Figure 2.8 Specimens of *Euphorbia antisyphilitica* collected to extract the wax.

The extracts from dehydrated residues of candelilla stems act against phytopathogenic fungi such as *Alternaria alternata*, *Fusarium oxyzporum*, *Colletotrichum gloeosporoides* and *Rhizoctnia solani.* The extracts are rich in polyphenol components because of its ellagic acid content and its precursor ellagitannin, giving them antimutagenic, antiviral and anticarcinogenic effects (Ascacio-Valdés et al., 2013).

Candelilla wax is used in the food industry, as it is an eco-friendly input that is used in microemulsions, biodegradable packaging and oleogels and edible films for fruits (Aranda-Ledesma et al., 2022). The metabolites of residues have bioactivity as antioxidants such as flavonoid and polyphenolic compounds (Rojas et al., 2021)

2.6.3 Ethnobotanical Uses or Traditional Uses

This plant extracts the wax used in the food industry in the northern states of Mexico, mainly Coahuila, Durango, Zacatecas, and Nuevo León (Brailovsky Signoret & Hernández, 2010). This endemic plant is considered a natural source of cerote and wax in Mexico (Steinmann, 2002). The cerote is extracted from the sap of the stems by heating a large steel frying pan and applying sulfuric and other acids (Estrada-Castillón et al., 2018).

The leaves and stems of candelilla have traditionally been used as a purgative and to treat tooth pain, syphilis and urinary infections (Mariacute et al., 2013). The traditional use for the treatment of syphilis is where the species' name, *antisyphilitica*, is derived (Barsch, 2004).

Candelilla wax can inhibit bacterial growth as a result of its phytochemical content such as saponins and quinones. Plant stems also contain phenolic compounds with antibacterial activity, associated with the presence of antioxidants such as ellagic acid and gallic acid (Rojas et al., 2021).

2.6.4 Ecology and Distribution

The candelilla is a plant of the arid zones and belongs to the type of vegetation named desert shrub, which is associated with *Agave lechuguilla*, *Opuntia* spp., *Prosopis* spp., and *Larrea tridentate* (Miller et al., 2018).

The *Euphorbia antisyphilitica* Zucc is the more important plant of the scrub rosette vegetation, characteristic of regions with little rainfall (500 mm) in a semi-arid climate (Barrientos-Lozano & Rocha-Sánchez, 2013). It is a species that is also found in soils with the presence of gypsum, as in Cuatrocienegas, Coahuila (Ochoterena et al., 2020).

Its habitat is thin soils less than 50 cm deep, with abundant stoniness and good drainage, where the geological material is calcium carbonate (Hernández-Herrera et al., 2019). This plant prefers the soils leptosol, regosol and xerosol, with a medium texture (Hernandez-Centeno et al., 2020) (Figure 2.9).

2.7 CURRENTLY ACCEPTED SCIENTIFIC NAME *JATROPHA DIOICA* SESSÉ EX CERV.

2.7.1 Biology

Generally known as "Sangre de drago," "sangre de grado" or "sangregado," it is a species endemic to Mexico (Martínez-Calderas et al., 2019). This plant grows naturally in arid and semi-arid regions. It is a shrub of 50–150 cm high and owes its common name to the fact that it has a colorless astringent juice that changes on contact with air, turning reddish and resembling blood (Araujo-Espino et al., 2017). They generally form colonies because their rhizomes have thick roots and semi-woody stems that grow outward. The root can measure up to 5 m around the plant; it also has healing properties, which is why it was used in traditional medicine. Its branches are flexible and dark reddish in color. The leaves are deciduous and are grouped in fascicles that only appear during the rainy season. Its flowers develop on one side of the leaves; they are small and white with pink, grouped in fascicles. Its fruits are globular, 1.5 cm long and contain a brown seed inside (Flores-Torres et al., 2021). The flowering season is during the months of April and May. The plant is obtained by harvesting and is available throughout the year (Martínez-Calderas et al., 2019) (Figure 2.10).

2.7.2 Active Principles or Bioactive Compounds

Jatropha dioica has been described as a rich source of various phytochemicals of interest, which exhibit antitumor, gastroprotective, antibiotic, antiviral, antifungal and chemoprotective properties (Pérez-Pérez et al., 2020a; Gutiérrez-Tlahque et al., 2019), in addition to its antioxidant, chemopreventive and hyperglycemic properties. It is widely used in traditional medicine for the control of hair loss, kidney pain, digestion problems, eye irritation, caries, periodontal disease, toothache and skin cancer (Castro-Ríos et al., 2020). Its medicinal effects are associated with its phytochemical components; it has been identified as containing three biologically active diterpenes (rhyolozatrione, citlalitrione and jatrophone), a sterol, ellagic acid and oxalic acid, in addition to secondary metabolites, such as flavonoids and polyphenols (Araujo-Espino et al., 2017).

2.7.3 Ethnobotanical Uses or Traditional Uses

This species' main uses are medicinal, such as the use of latex in small doses to control cold sores and relieve dental pain. The seed oil extract is used as a purgative. To prevent hair loss,

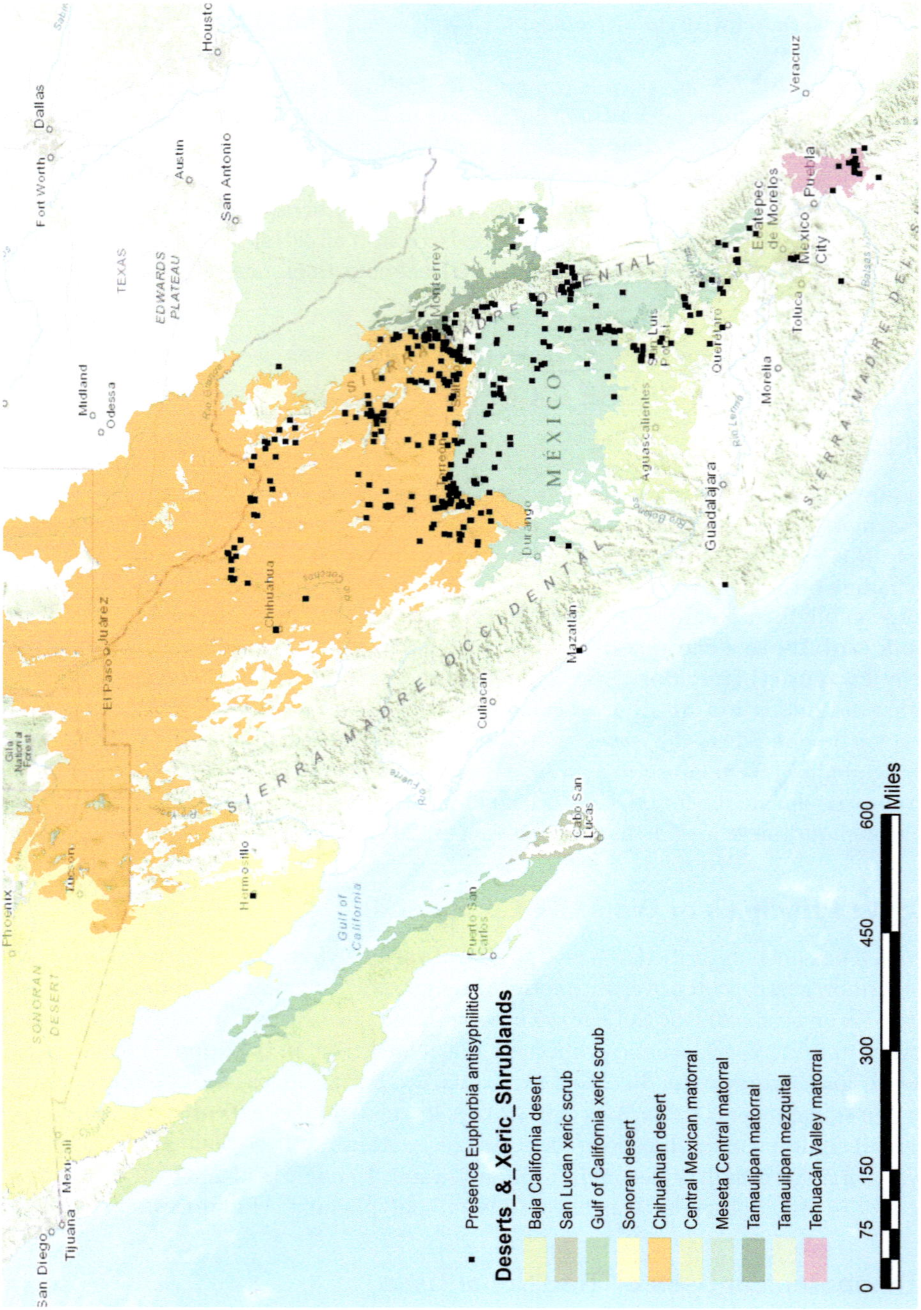

Figure 2.9 Map of the presence of Euphorbia antisyphilitica in the Chihuahuan Desert Region.

Figure 2.10 Set of individuals of *Jatropha dioica*.

dandruff and styes, the stems are boiled or the whole plant or root is crushed and applied as a hair tonic; it is also used as an infusion for the treatment of varicose veins and bruises, for which the plant is boiled and applied to the affected area. The fruit is used to treat irritated eyes by placing a drop of the juice in the fruit; for gingival bleeding and tooth mobility, the roots and stems are chewed. The aqueous extract of the root has also been found to have antibiotic activity against *Staphylococcus aureus* (Can-Aké et al., 2004). It is also commonly used in the agricultural sector for crop protection, as a live hedge and as food and feed. *J. dioica* has been used for biodiesel as fuel and biopesticide production (Pérez-Pérez et al., 2020b). Its potential use as a phytoaccumulator of heavy metals in soil has also been reported (Flores-Torres et al., 2021).

2.7.4 Ecology and Distribution

The genus *Jatropha* of the Euphorbiaceae family has 175 species, 45 of which are found in Mexico, of which 77% are endemic. They are native to Central America and the Caribbean and are also distributed in North America, Asia and Africa in tropical and subtropical areas (Pérez-Pérez, 2020a, b). They are shrubs characterized by an exudate that emanates when the plant suffers damage. These plants are used in traditional medicine in several countries such as India, Africa and Mexico as the presence of active principles such as carbohydrates, alkaloids, terpenes, lignans and cyclic peptides, phenols, saponins, camarins, tannins, flavonoids and phytosterols have been described (Thomas et al., 2008; Can-Ake et al., 2004). *J. dioica* is distributed from northern to central Mexico. It predominates in dry, warm and semi-warm climates, from 1,100 to 2,500 m above sea level (Martínez-Calderas, 2019). It can be found in dry climates such as hillsides, grassland, tropical deciduous forest and stony soil (Cordova-Tellez et al., 2015). The months in which the plant is found in greatest abundance are March, July and October and in smaller quantities during May and November (Barba et al., 2003) (Figure 2.11).

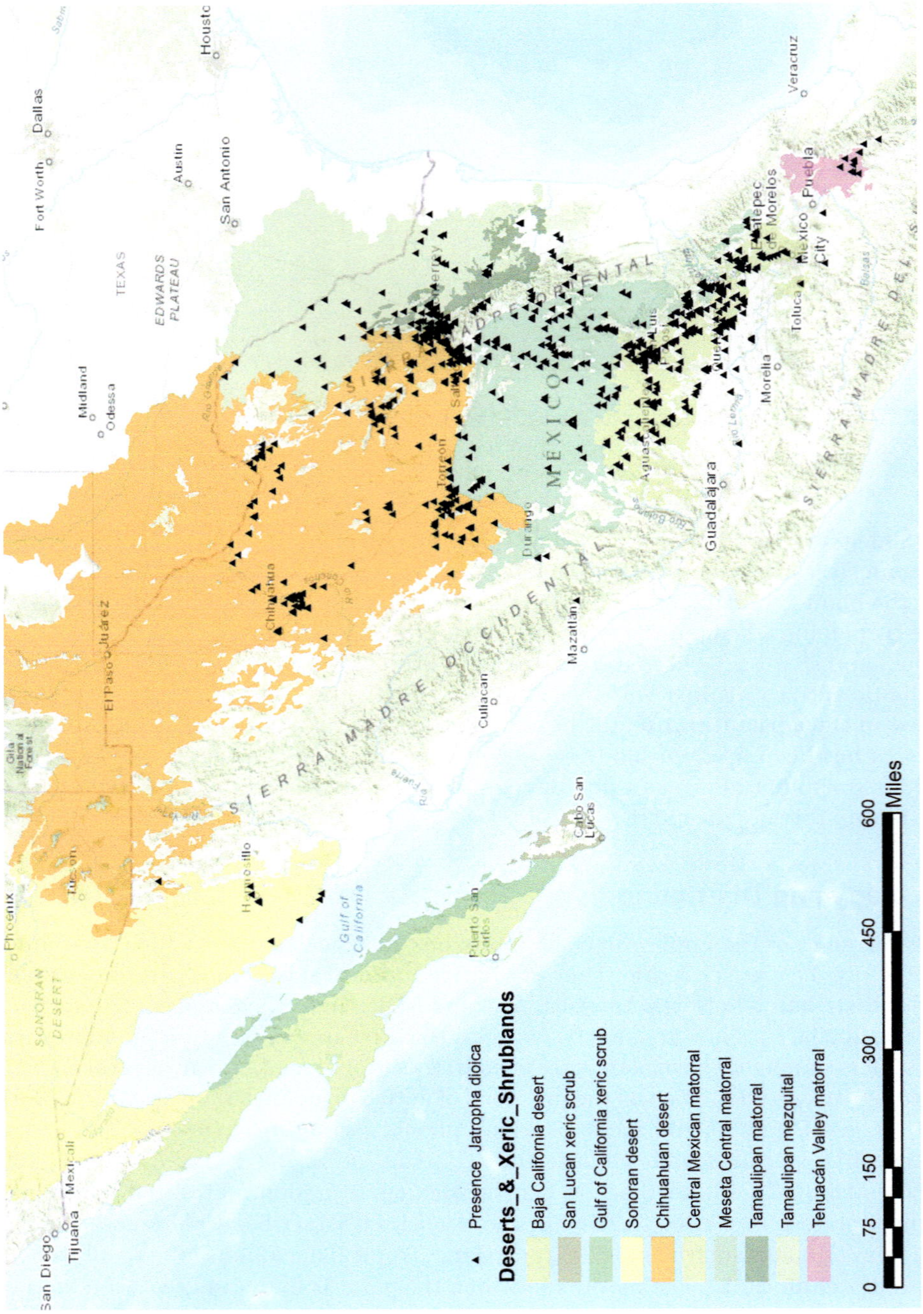

Figure 2.11 Map of the presence of *Jatropha dioica*.

2.8 CURRENTLY ACCEPTED SCIENTIFIC NAME *HETEROTHECA INULOIDES* CASS.

2.8.1 Biology

The *Heterotheca* genus is widely distributed in North America, being found from Canada to central Mexico, for which four species are recognized: *H. grandiflora* Nutt. *H. subaxilaris* (Lam.) Britton & Rusby, *H. inuloides* Cass. and *H. leptoglossa* DC., the latter two being endemic in Mexico (Nesom, 1990; Semple, 2008).

In Mexico, *H. inuloides* is generally known by the folk name of "*árnica*" or "*árnica mexicana*." This herb is one of the most commonly used native species in Mexico, it is a perennial or annual plant, often flowering in its first year (McVaugh, 1984; Rzedowski, 2001; Monroy-Ortíz & Monroy, 2004). Depending on its stage of development, it measures 25–30 cm and in its flowering stage, 70–100 cm tall. During its nonflowering stage, the plant is made up of only leaves and a tender stem, while in its flowering stage, it has an erect and woody stem, little branches with short villi, and its leaves have an ovate limb shape that are widened at the base, 3–10 cm long and 1–3.5 cm wide. As for its inflorescence, the flowers are grouped in a head (150 flowers), these measure 8–15 mm long (Maldonado, 2004; Maldonado-López et al., 2008). *H. inuloides* grows wild as a weed or secondary succession pioneer mainly in the cooler and temperate regions of Mexico in highlands at altitudes between 2,000 and 3,000 m above sea level, as well as, it has been found, in tropical forests, natural grasslands and xerophilous shrublands (Velázquez et al., 2003; Rojas et al., 2013; CONANP, 2014; Rodríguez-Chavez et al., 2017).

2.8.2 Active Principles or Bioactive Compounds

The bioactive compounds identified in both species of *arnica* have anti-inflammatory properties and belong to the family of sesquiterpene lactones such as chamissonolide, helenalin and hydrohelenalin (Lyss et al., 1998; Kos et al., 2005; Klaas et al., 2002; García-Pérez et al., 2016). The antimicrobial and antioxidant activity of Mexican *arnica* is due to the presence of phenolic compounds and flavonoids (quercetin and kamferol) (García-Pérez et al., 2016) that act synergistically with chamissonolide, helenalin and hydrohelenalin (Rodríguez Chávez et al., 2015). The presence of fatty acids in *arnica* oil favors the penetration through the skin of bioactive and anti-inflammatory compounds (Bilia et al., 2006).

The plant was used in homeopathic formulations to treat inflammatory conditions in humans with good results (Iannitti et al., 2016; Seeley et al., 2006; Fioranelli et al., 2016). On the other hand, this plant has been shown to activate cells of the immune system including neutrophils, which are cells that attack harmful bacteria (Olioso et al., 2016). Studies on the effect of *arnica* on edema (excess fluid in some tissue) induced by blood were also carried out, and it was shown that they were significantly reduced in rats (Conforti et al., 2007).

2.8.3 Ethnobotanical Uses or Traditional Uses

Arnica is used to treat inflammatory conditions or as antimicrobials (*Staphylococcus aureus*, *Eschericha coli*) and antioxidants (Lyss et al., 1998; García-Pérez et al., 2016) as well as for the treatment of bruises, sprains and rheumatic problems (Kamatani et al., 2014), wounds, bruises, pain and angina (Lannitti et al., 2016). Topical methods of use require the plant to be boiled in water and used to reduce wounds, pain, rheumatism, contusions, and ailments related to inflammatory processes and are more efficiently applied in cataplasm; orally, the plant is made into a tea or infusion to help ulcers, gastritis, diarrhea, and stomach-ache (Rodriguez-Chávez et al.,2017). All parts of the plant are used for medicinal purposes, from the stems, flowers, leaves, branches and roots. Preparation methods vary widely, including infusion, decoction, maceration and soaking, while the preparations are usually taken orally and topically. The most common form of preparation is with flowers, leaves and parts of the stems which are immersed in boiling water and ingested as a tea (González-Stuart, 2010)

2.8.4 Ecology and Distribution

H. inuloides is usually distributed in pine-oak forests, grasslands, and tropical deciduous forests (McVaugh, 1984; Rzedowski & Rzedowski, 2001). Its habitat is ruderal and adverse, disturbed grasslands, and clearings in forests as well as a weed in crop fields, pastures fields and disturbed areas (Rodríguez-Chávez et al., 2017).

The main distribution of *H. inuloides* is in the center of Mexico; it´s found in the states of Colima, Federal District, Durango, Guanajuato, Hidalgo, State of Mexico, Michoacán, Morelos, Nuevo León, Oaxaca, Puebla, Querétaro, San Luis Potosí, Tlaxcala and Veracruz (Villaseñor & Espinosa, 1998) (Figure 2.12).

2.9 CURRENTLY ACCEPTED SCIENTIFIC NAME *LIPPIA GRAVEOLENS* KUNTH

2.9.1 Biology

This plant is of the aromatic type and a non-timber forest species that grows wild in the arid and semi-arid zones of Mexico (Villavicencio-Gutiérrez et al., 2018).

It is also cultivated in several regions of the world and is harvested commercially for its characteristics as a spice, condiment and medicinal properties. There are numerous varieties of oregano for cultivation, mainly varying by the intensity of its aroma, the quality of the essential oil and the growing areas (Ocampo et al., 2009).

Lippia graveolens is a short-cycle shrubby perennial plant that is slender and aromatic, with a height that varies from 0.2 to 2 m, although it can reach up to 3 m in height. It has woody stems that are highly branched from the base and, therefore, it looks like a small shrub. The stems are reddish in color and can reach heights of up to 40 cm. Its leaves are oblong or elliptical, finely crenulated, very tomentose and hairy, are placed oppositely, show an oval shape and are very small, usually measuring between 5 and 15 mm. The leaves of this plant are hairy on the underside (Granados et al., 2013) (Figure 2.13).

2.9.2 Active Principles or Bioactive Compounds

Lippia graveolens contain an essential oil fraction that generates most of its economic value as a flavoring and additive for food products. Furthermore, fresh leaves and wasted plant material have been suggested as a bioactive fraction and alternative source of bioactive compounds such as polyphenols (flavonoids) and terpenes (Bautista-Hernández et al., 2021).

Flavonoids such as apigenin, luteolin, aglycones, aliphatic alcohols, terpene compounds and phenylpropane derivatives have been identified (Rassem et al., 2018).

Essential oils of the *Lippia* species contain limonene, b-caryophyllene, p-cymene, camphor, linalool, apinene and thymol, which may vary according to the chemotype (Rehman et al., 2016).

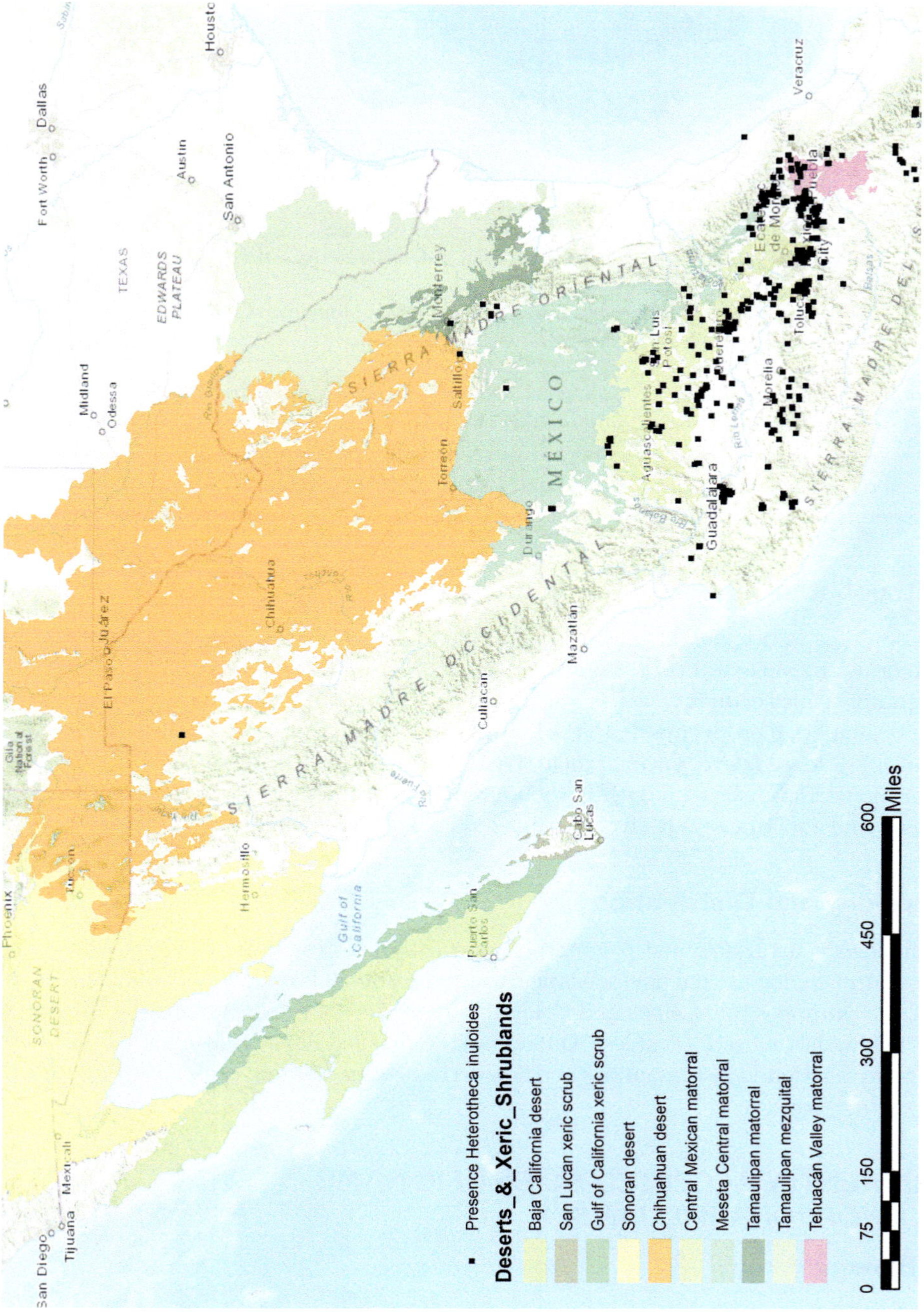

Figure 2.12 Map of the presence of *Heterotheca inuloides*.

Figure 2.13 Specimens of *Lippia graveolens* in Rancho San Juan, Monclova, Coahuila.

In methanolic extracts of *L. graveolens* leaves, seven minority iridoids, known as loganin, secologanin, secoxyloganin, dimethyl secologanoside, loganic acid, 8-epi-loganic acid and caryoptoside, and three majority iridoids, such as caryoptosidic acid and its derivatives 6'-O-p-coumaroyl and 6'-Ocaffeoyl (Arias et al., 2020), have been found. It also contains flavonoids such as naringenin, pinocembrin, lapachenol and icterogenin (Bautista-Hernández et al., 2021).

2.9.3 Ethnobotanical Uses or Traditional Uses

The collection of wild oregano is an essential activity for most of the rural communities where this shrub grows because it represents an opportunity for additional income and development for these groups whose economic activity is for survival (García-Pérez et al., 2012).

Due to its medicinal properties, both the leaf and the oil are used as an analgesic, anti-inflammatory, antipyretic, sedative, antidiarrheal, treatment of skin infections, antifungal, diuretic, remedy for menstrual disorders, antimicrobial, repellent, antispasmodic, treatment of respiratory diseases and abortifacient (Calvo-Irabién et al., 2014; Soto-Armenta et al., 2017).

2.9.4 Ecology and Distribution

Lippia graveolens is a variable and polymorphic species, consisting of populations with different morphological, phenological and phytochemical characteristics. It is mainly distributed in the states of Chihuahua, Durango, Tamaulipas and Coahuila, where 50% of the harvesting permits are concentrated, followed by Jalisco, Zacatecas, Querétaro, Hidalgo and Baja California. However, natural populations are also found in Tamaulipas and Sinaloa (Martínez-Natarén et al., 2018) (Figure 2.14).

2.10 CURRENTLY ACCEPTED SCIENTIFIC NAME *LOPHOPHORA WILLIAMSII*

2.10.1 Biology

The peyote plant is in the Cactaceae family, which has its natural distribution area in the Chihuahuan Desert of Mexico and the United States (Ermakova et al., 2021).

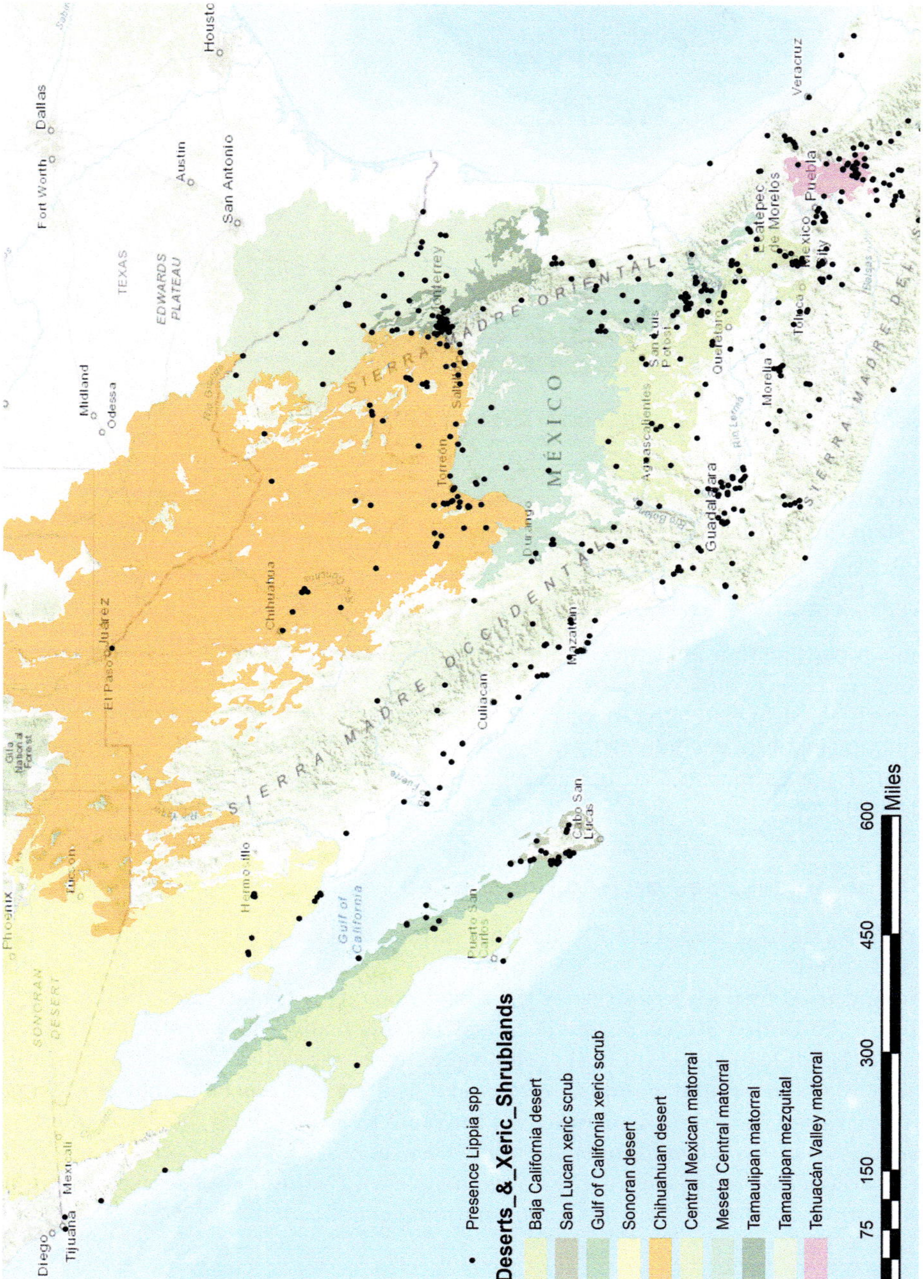

Figure 2.14 Map of the presence of *Lippia* spp.

Figure 2.15 Specimens of *Lophophora williamsii in* Marte Estacion, Coahuila.

Lophophora williamsii is a small, spineless cactus approximately 2–12 cm in diameter with a flat to dome shape and is distributed singly or in colonies (Newbold et al., 2020) (Figure 2.15).

2.10.2 Active Principles or Bioactive Compounds

Peyote contains dopamine that helps to tolerate the stress of the desert environment and provides an immune response against diseases. Its content of phenylalanine is hydroxylated to tyrosine, which is further hydroxylated to L-dopa or decarboxylated to tyramine.22 (Liu et al., 2020).

Peyote is a plant capable of accumulating high levels of alkaloids (Ibarra-Laclette et al., 2015), the best known is mescaline 3,4,5-trimethoxyphenethylamine, which has analgesic effects (LeBlanc et al., 2021).

2.10.3 Ethnobotanical Uses or Traditional Uses

Lophophora williamsii is a medicinal plant, but its harvest or possession of live specimens is not legal (Castle et al., 2014).

Peyote is a plant used by native peoples for their traditional ceremonies and rituals (Jones, 2007). Actually, in the United States, peyote is considered a Schedule 1 Controlled Substance (Terry & Trout, 2017), while in Mexico, it is a psychotropic substance with little or no therapeutic value, has a high addictive potential and is considered a health risk (Genet Guzmán & Labate, 2019). Peyote is harvested from the green crowns of the plant to avoid killing the whole plant to be used in medicine as *pomada* containing mescaline (Klein et al., 2015).

According to archaeological evidence, the native peoples of Coahuila and Texas have used peyote for the treatment of arthritis or joint inflammation (Feeney, 2018).

2.10.4 Ecology and Distribution

Peyote is a species that is at risk or vulnerable because of illegal extraction for its use in medicine, coupled with scarce existing populations found only in calcareous soil (Castle et al., 2014).

Its presence is limited to the Chihuahuan Desert. Its populations have an aggregate pattern in distribution and require nurse plants to survive, like *Larrea tridentata*, *Acacia sp.* and *Cordia parvifolia* (Naranjo et al., 2010) (Figure 2.16).

2.11 CURRENTLY ACCEPTED SCIENTIFIC NAME *OLNEYA TESOTA* A. GRAY

2.11.1 Biology

It is a monotypic genus of phanerogamous plants belonging to the Fabaceae family; its only species, *Olneya tesota*, is native to Mexico (Zuñiga-Tovar & Suzán-Azpiri, 2010). This tree is endemic to the Sonoran Desert and is considered one of the largest trees in the sarcocaulescent and xerophytic scrub communities; it is possible to find individuals over 15 m tall (Suzán et al., 1996). Mature trees may have a simple or branched trunk that produces shoots from the roots. It rarely loses its leaves, as these are continually replenished, and when they fall, they form layers of organic matter in varying degrees of decomposition that settle at the base of the trees. Flowering of *O. tesota* begins in March, and the fruits ripen in early summer; it has elliptical green leaves and white to pink flowers (SEMARNAT, 2014). The pods contain one to four large seeds that germinate after the first rains. Most seedlings die after germination due to lack of moisture, so the development of new individuals is sporadic and can take several decades. Ironwood is a slow-growing, long-lived species that can exceed 800 years of age (Suzán et al., 1996) (Figure 2.17).

2.11.2 Active Principles or Bioactive Compounds

The *O. tesota* seeds contain three lectins named PF1, PF2 and PF3; particularly, PF2 is a tetramer formed by 33 kDa subunits and is the most abundant molecule in the seeds with 3% of the total protein. This lectin is inhibited only by glycoprotein complex carbohydrates (Urbano-Hernández et al., 2015). The sequence of the amino-terminal end and some internal sequences of PF2 show high homology with the *Phaseolus vulgaris* lectin, which recognizes complex oligosaccharide structures of the bisected bistranded type, present in the cell membrane of erythrocytes (Lagarda-Diaz et al., 2017). Other authors point out that *O. tesota* contains high levels of methanol and hexane, substances with antibacterial properties, with derived uses for gastrointestinal diseases (Moreno-Salazar et al., 2008), as well as tannins, which have astringent properties (Waizel-Bucay & Martínez-Rico, 2011).

2.11.3 Ethnobotanical Uses or Traditional Uses

The plant parts of *O. tesota* have been put to many uses. Its hard, dark-colored wood is used as a raw material for some communities in the Sonoran Desert (Suzan et al., 1996). The flowers are used in infusions to alleviate various ailments associated with stomach and kidney ailments (Felker, 1981). The active compounds in the inflorescence have been reported to respond favorably to gastrointestinal disorders or infectious diseases (Rosas-Piñón et al., 2012). Traditionally, flower infusions in aqueous media with ethanol have been used (Sharma et al., 2017). Other studies indicate that *O. tesota* is very important for the diet of cattle, especially in times when the Gramineae family is scarce for foraging (Toyes-Vargas et al., 2013).

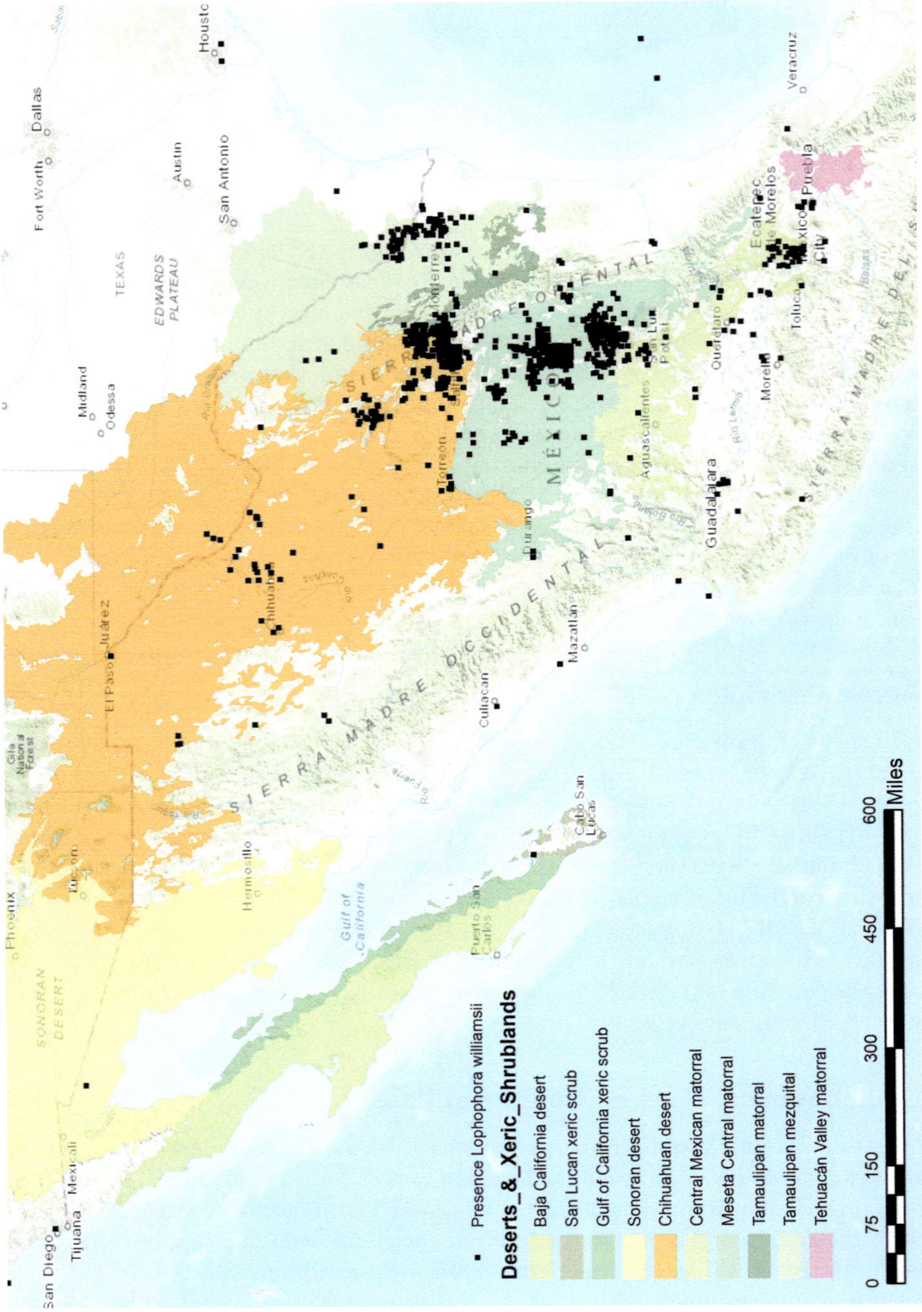

Figure 2.16 Map of the presence of *Lophophora williamsii* with a population in Coahuila and San Luis Potosi.

Figure 2.17 Specimens of *Olneya tesota* in Bermejillo, Durango.

2.11.4 Ecology and Distribution

The vegetation types of *O. tesota* are scattered in areas with warm and arid or semi-warm and very arid climates. In shrublands, known as arbosufrutescent, arborescent and arbocrasicaulescent, it is found mainly in plains and on low and medium-size hillsides (SEMARNAT, 2014); it grows in soils of moderate depth, with sandy, sandy-gravelly and sandy-loamy texture and in desert soils that are poor in organic matter. It is also a species that is very resistant to frost and high temperatures (Saduño et al., 2009).

Although there are areas where its density and percentage in the floristic composition is very important, such compact stands of trees are not found. Thanks to its great longevity and the perennial nature of its leaves, the microhabitat formed under its shade is stable. All these characteristics make ironwood a nurse species, i.e. it gives protection to other plants (Suzán et al., 1996) and provides shade, food and shelter for small wildlife mammals (Coronel-Arellano & López-González, 2010); in particular, some authors report that species such as the desert mule deer (*Odocoileus hemionus eremicus*) and the desert bighorn sheep (*Ovis canadensis weemsi*) depend on the branches and foliage provided by this legume (Marshal et al., 2004; Guerrero-Cárdenas et al., 2018).

It is considered a unique species within the genus *Olneya* and is distributed in the Sonoran Desert in the states of Arizona and California in the United States and in Mexico on the Baja California peninsula (Suzán et al., 1997). In the species present in the Mexican state of Sonora, it is mainly associated with mesquite and sarcocaulescent scrub, occupying a distribution gradient from 0 to 80 m above sea level. It is abundantly distributed in the center of the state of Sonora as well as near the coast and plains; it has been reported that the minimum winter temperature is an important factor for its distribution (SEMARNAT, 2014) (Figure 2.18).

2.12 CURRENTLY ACCEPTED SCIENTIFIC NAME *OPUNTIA FICUS-INDICA* (L.) MILL., 1768

2.12.1 Biology

Opuntia specimens are a very common genus of the Cactaceae family on the American continent (J. A. Reyes-Agüero et al., 2006). The specimens are bushy plants, with a maximum height of 6 meters, with erect growth made up of stems or cladodes (Porras Flórez et al., 2018). A cactus pear has a cladode as a base, whereas lanceolate cladodes branch out with a length of 39 cm and a thickness of 1.5 cm (IBUNAM, 2019). Vegetative reproduction of *Opuntia ficus-indica* is the most common form, with cladodes or stems, but it is also sexually propagated by seeds (J. Antonio Reyes-Agüero et al., 2005). The fruit is known as cactus pear or tuna in Mexico and has a thick shell with very fine spines, pulp, and abundant seeds (Ramírez-Rodríguez et al., 2020).

The *Opuntia ficus-indica* is a cacti plant with CAM that has the capacity to store water in its cladodes (Scalisi et al., 2016). Plants with CAM close their stomata in the day when the air temperature is high and open at night to take in CO_2, allowing them to manage scarce water in the arid regions (Chen & Blankenship, 2021) (Figure 2.19).

2.12.2 Active Principles or Bioactive Compounds

In nutrition, it has been found that cactus pear cladodes, processed, dehydrated and ground, have a high antioxidant potential and were an important source of flavonoids, phenolics, alkaloids and terpenoids, which have anticarcinogenic functions, such as the flavonoids protect cells against the consequences of oxidative stress (Msaddak et al., 2017).

The prickly pear or tuna gives fiber, vitamins and minerals and is also an important source of bioactive nutraceutical substances, such as some fatty acids, amino acids, sterols, carotenoids, phenolics and betalains, that have antioxidant, antibacterial, anti-inflammatory, antiproliferative and neuroprotective effects (Silva et al., 2021). Also, the seeds of the prickly pear are an excellent source of nutrients and have high amounts of phenolic compounds that have strong antioxidant properties. Generally, the seeds are eliminated after extraction of pulp or when consumed by people (Kolniak-Ostek et al., 2020).

2.12.3 Ethnobotanical Uses or Traditional Uses

The *Oputia ficus-indica* has been used by indigenous communities since before the pre-Hispanic period and has been used in many ways, such as nutrition from the fruit and cladodes (Ramírez-Rodríguez et al., 2020) and traditional medicine for patients with diabetes in rural communities (Estrada-Castillón et al., 2021).

In Mexico, adding the cladodes or fruit into daily diets has traditionally helped diabetic people (Silva et al., 2021). Consuming dehydrated cladodes helps patients with type 2 diabetes to reduce blood glucose, serum insulin and plasma glucose-dependent insulinotropic peptide peaks (López-Romero et al., 2014).

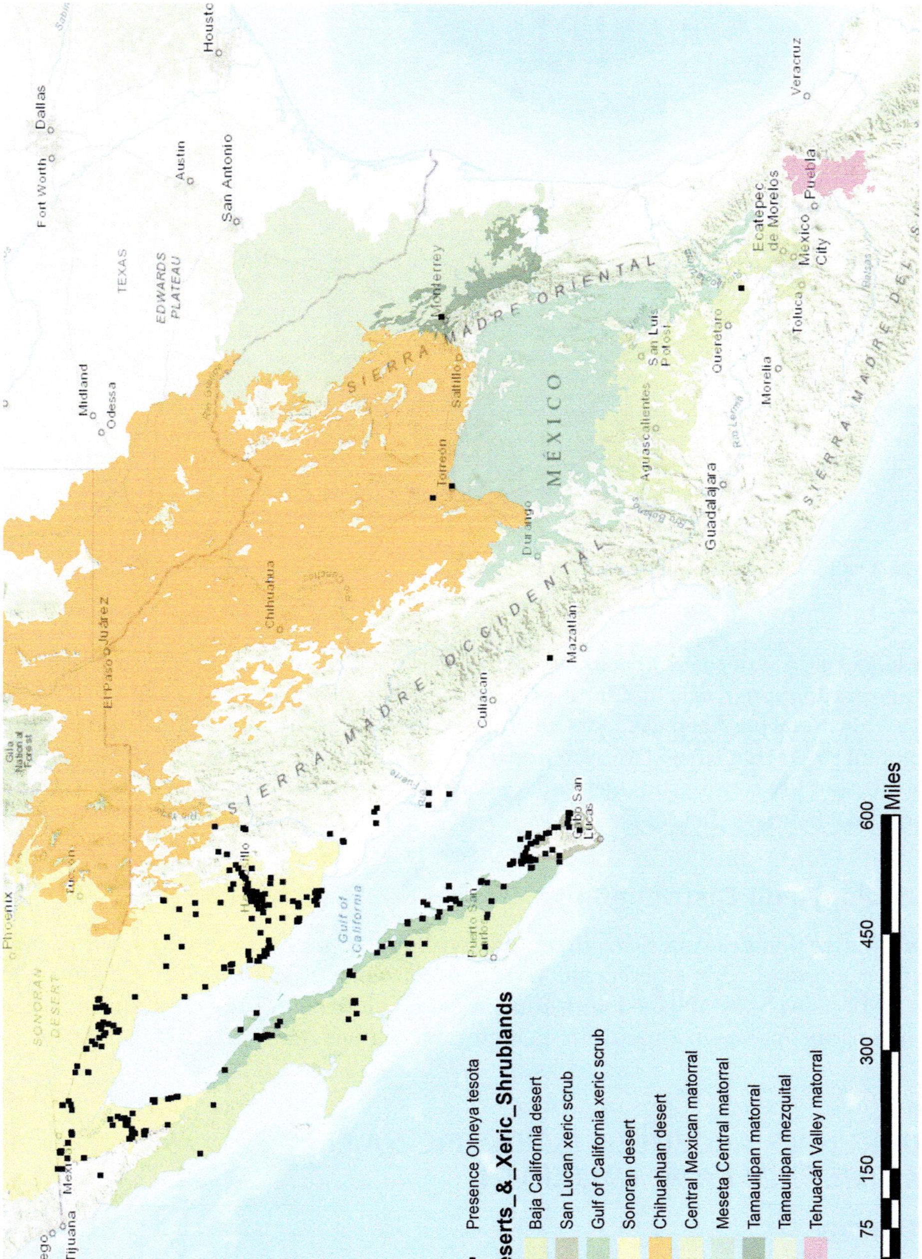

Figure 2.18 Map of the presence of *Olneya tesota*, a native of the Sonoran Desert.

Figure 2.19 Fruit and cladodies of *Opuntia ficus-indica.*

The presence of CAM in *Opuntia ficus-indica* makes it a widely used species with a high potential for forage production in arid and semi-arid regions, as in the case of India and Brazil (Kumar et al., 2022). This nopal has been used for the production of grana cochineal, a natural dye widely used in colonial times that turned Mexico into a producer of the textile dye (Ervin, 2012).

Nopals' efficient use of water makes cladodes and fruit a good option for feeding animals in the production of biomass (Reis et al., 2018).

2.12.4 Ecology and Distribution

Opuntia is a native plant of America, currently distributed in many parts of the world, such as Europe, Africa and Asia, where it is considered an invasive plant (Shackleton et al., 2011).

The plantations were established with this native plant since pre-Hispanic times. It is currently an important species in agriculture in many arid regions (Ervin, 2012) (Figure 2.20).

2.13 CURRENTLY ACCEPTED SCIENTIFIC NAME *PARTHENIUM INCANUM* KUNTH

2.13.1 Biology

The genus *Parthenium* contains approximately 16 species of shrubs, herbaceous perennials and annuals (Ebadi, 2006). The *mariola* is a greyish-green shrub that ranges from 30–120 cm tall and has a strong scent, many branches, and striated bark. It has alternate leaves with irregular lobes that are 5–10 cm long and the lower surface of the leaves have dense, pale green tomentum (McVaugh, 1984).

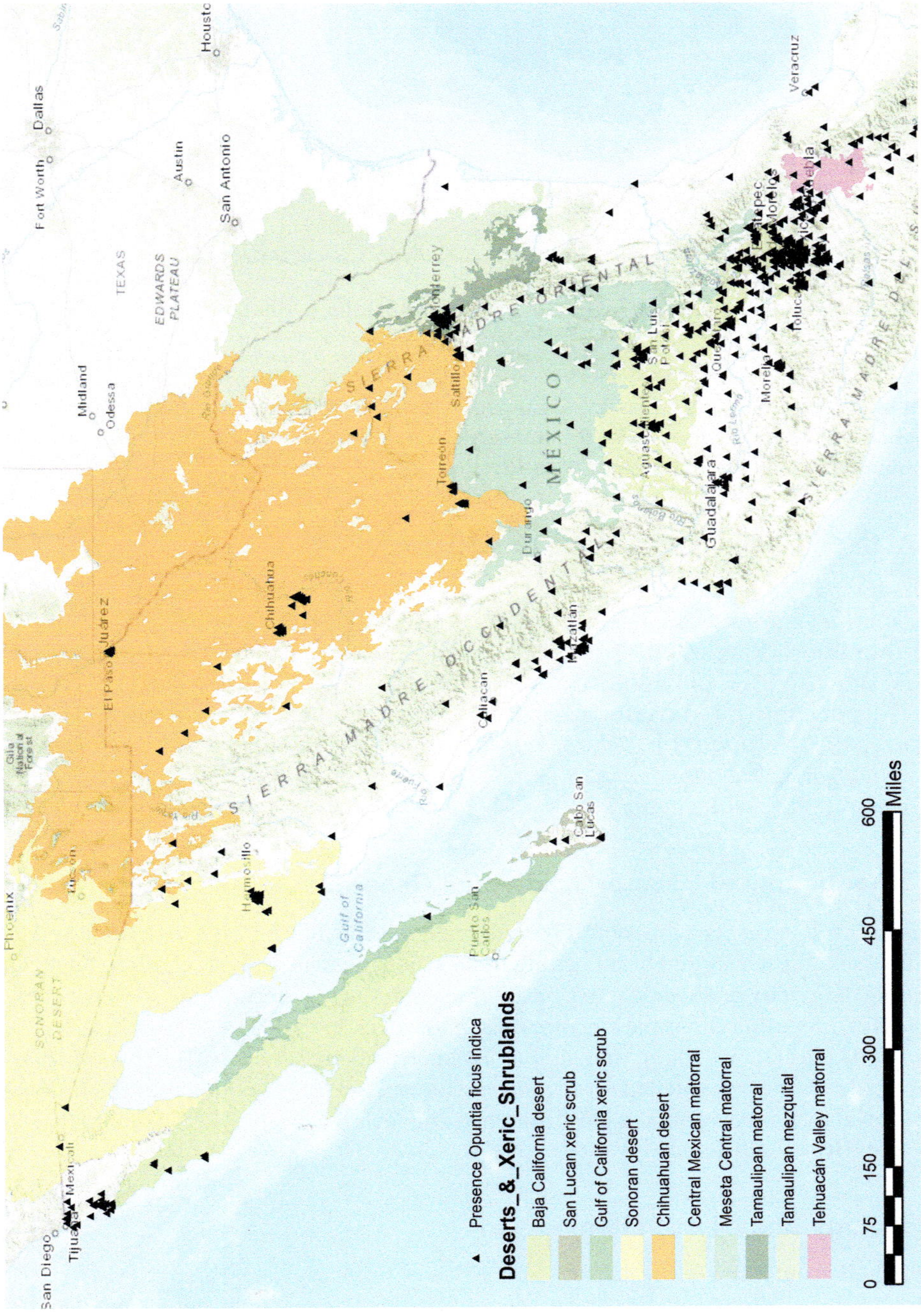

Figure 2.20 Map of presence of *Opuntia ficus-indica*.

Figure 2.21 *Parthenium incanum* in the town of Buenavista, Saltillo, Coahuila.

Heads of small creamy-white flowers are arranged in terminal cymose panicles above the leaves, and the achene fruits are 2 mm long. It is closely related to guayule (*Parthenium argentatum*) with which it can hybridize (Allred & Ivey, 2012) (Figure 2.21).

2.13.2 Active Principles or Bioactive Compounds

Various biological activities from related species (*P. hysterophorus*) have already been reported, such as antimicrobial, antioxidant, antihemolytic, cytotoxicity and lipid peroxidation inhibition (Kumar, 2013). The medicinal uses of this plant include treatment of indigestion, sluggish liver and mild constipation (Kane, 2006). The antimicrobial activity of parthenin was demonstrated against the growth of enteric bacterial pathogens, including *Salmonella typhi*, *S. typhimurium*, *Enterobacter aerogenes*, *Klebsiella pneumoniae*, *Escherichia coli*, *Proteus vulgaris*, *Pseudomonas aeruginosa*, *Staphylococcus epidermidis*, *S. aureus*, and *Shigella flexneri* (Siddhardha et al., 2012).

2.13.3 Ethnobotanical Uses or Traditional Uses

The whole plant is used except the root. It is consumed in infusions for bile, infections and other stomach diseases. The leaves and branches are boiled then taken as a tea (McVaugh, 1984). There are no records of toxicity or adverse reactions. It is used in the northern states of Mexico to treat gastric diseases, including constipation, diarrhea, poor digestion and stomach pain as well as to treat liver conditions (Kane, 2006). González (1988) mentions that, in the field, it is taken as tea for stomach pain, while it is smelled for dizziness. González (1998) explains that the infusion of this plant is used to combat stomach pain, diarrhea and fever. In Coahuila, this plant is reported as a remedy for liver conditions.

2.13.4 Ecology and Distribution

It is distributed from the southwestern United States from Nevada to southwestern Texas (McVaugh, 1984). In Mexico, it has been recorded from Sonora to Hidalgo and Querétaro (Villaseñor & Espinosa, 1998). *Mariola* is an important source of food for grazing animals in desert shrub lands (Villalobos, 2007).

The shrub occurs on low stony limestone slopes, deep soils with xerophytic scrub and grasslands. Its altitudinal distribution in western Mexico is from 800 to 2,000 m (Allred & Ivey, 2012). It flowers and fruits from May to September (McVaugh, 1984) (Figure 2.22).

2.14 CURRENTLY ACCEPTED SCIENTIFIC NAME *PINUS CEMBROIDES*

2.14.1 Biology

The trees are of a medium size, branching at low height, slow growing, and their wood is mainly used as fuel (Wolf, 1985; Bailey & Hawksworth, 1992; Romero et al., 1996). *P. cembroides* is distributed in Arizona, Texas and New Mexico in the United States and in 14 states of central and northern Mexico above 20° N, and as far south as the states of Querétaro and Hidalgo (Farjon & Styles, 1997). The importance of these species is ecological, economic and cultural; their adaptive potential is high, as it grows in places with little rainfall. This specie is useful for reforestation in degraded ecosystems in arid and semi-arid zones with altitudes of 800 to 2,800 m (Mohedano-Caballero et al., 1999), but birds and rodents consume the seeds (Romero et al., 1996). The seed (piñon) is collected for commercialization and self-consumption. It is a source of income for the owners of these forests and represents a source of pine nuts for the preparation of typical dishes and sweets (Fonseca, 2003; Hernández et al., 2011) (Figure 2.23).

2.14.2 Active Principles or Bioactive Compounds

Various investigations highlight the nutritional importance of this product, as it is an excellent source of protein, antioxidants, monounsaturated fatty acids, vitamin B1 (thiamin) and minerals, especially potassium and phosphorus. Both oleic and linolenic acid make up more than 85% of the total fatty acids contained in this seed, which are directly involved in the regulation of blood lipid levels, including total cholesterol and low-density lipoproteins (LDL), as well as the reduction of blood pressure (Sagrero-Nieves, 1992; Sen et al., 2016).

The Mexican pinion is a product highly appreciated by the cosmetic industry because of its high content of unsaturated fatty acids. Pine seeds are rich in oil (31%–68%) and contain polymethylene-interrupted unsaturated fatty acids with acis-5 ethylenic bond: cis-5, cis-9 18:2, cis-5, cis-9, cis-12 8:3, cis-5, cis-11 20:2 and cis-5, cis-11, cis-14 20:3 acids, with a trace of cis-5, cis-9, cis-12, cis-15 18:4 acid. Their percentage relative to total fatty acids varies from of 3.1% to 30.3%. The major cis-5 double bond-containing acid is generally the cis-5, cis-9, cis-12 18:3 acid (pinolenic acid). inoleic acid represents approximately one-half of the total fatty acids, whereas the content of oleic acid varies in the range of 14% -36% inversely to the sum of fatty acids containing acis-5 ethylenic bond and pinolenic represents about 15% of total fatty acids (Wolf & Bayard, 1995).

2.14.3 Ethnobotanical Uses or Traditional Uses

Pine nuts were very important in the diet of the indigenous tribes in northeastern Mexico and southern United States (Menniger, 1977). Actually, pine nuts are used in the production of candy, as ingredients in cooking lamb, pork, chicken and fish and combined in cakes, puddings and sauces (Rosengarten, 1984). *P. cembroides* is one of the smaller pines with edible nuts. Pine nuts have a nutritional value similar to that of other nuts sold worldwide (López Mata, 2001). Once collected and husked, it is generally used as an ingredient in sweets and salads.

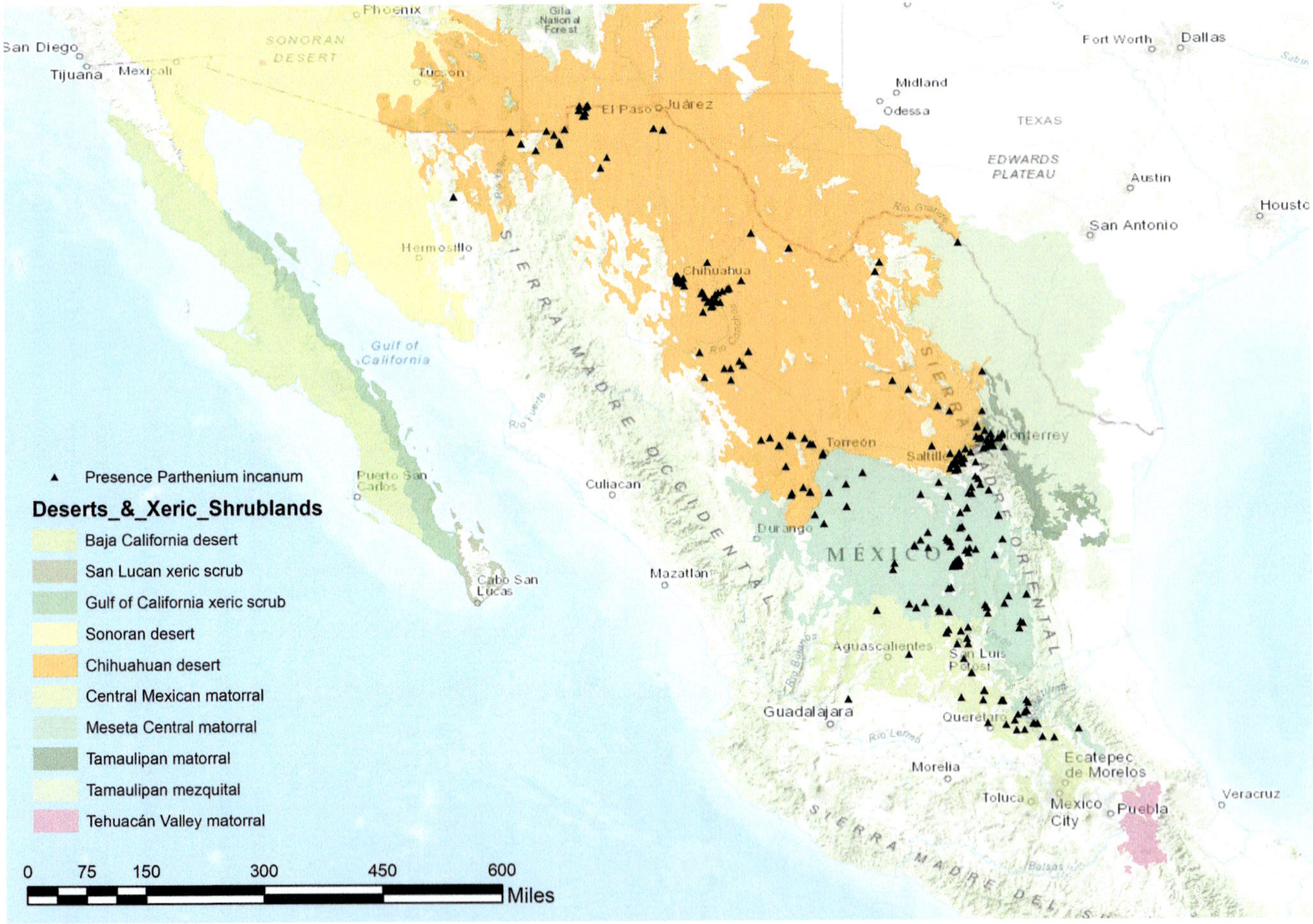

Figure 2.22 Map of the presence of *Parthenium incanum* only in the Chihuahuan Desert.

Figure 2.23 Forest of *Pinus cembroides* in the Sierra El Astillero, Durango

2.14.4 Ecology and Distribution

The distribution of *P. cembroides* is restricted to the low-lying mountains that surround the arid highlands of Mexico. Pines of this specie are found in Arizona, southwestern New Mexico, Texas, northern and central Mexico, up to Puebla (Romero Manzanares et al., 1996). *P. cembroides* prefers the dry climate of the Altiplano (BS of the Köppen climatic classification). Passini (1982, 1983) and Malusa (1992) indicate that area where the *P. cembroides* grows has a mean annual temperature from 12 °C to 18 °C, in which the hottest month is June and the coldest is December or January, and the annual rainfall is between 300 and 700 mm. Passini (1983) indicates that *P. cembroides* tolerates a warmer climate. It is drier and warmer in the Sierra Madre Oriental than in the Sierra Madre Occidental. According to Passini (1996) and Malusa (1992), this specie is found on the threshold of "cold winter" areas, and cold winters are necessary for the pines to produce, and there must be periods of moderate drought to favor the setting of the cones (Flores Díaz, 1989) (Figure 2.24).

2.15 CURRENTLY ACCEPTED SCIENTIFIC NAME *PROSOPIS* SPP.

2.15.1 Biology

The genus *Prosopis* constitutes important natural resources in arid and semi-arid regions of the world. Species of the genus *Prosopis* are shrubs or small trees in the family Fabaceae. They grow to a height of up to 12 m with a stem diameter of up to 1.2 m. It has deciduous, bipinnate,

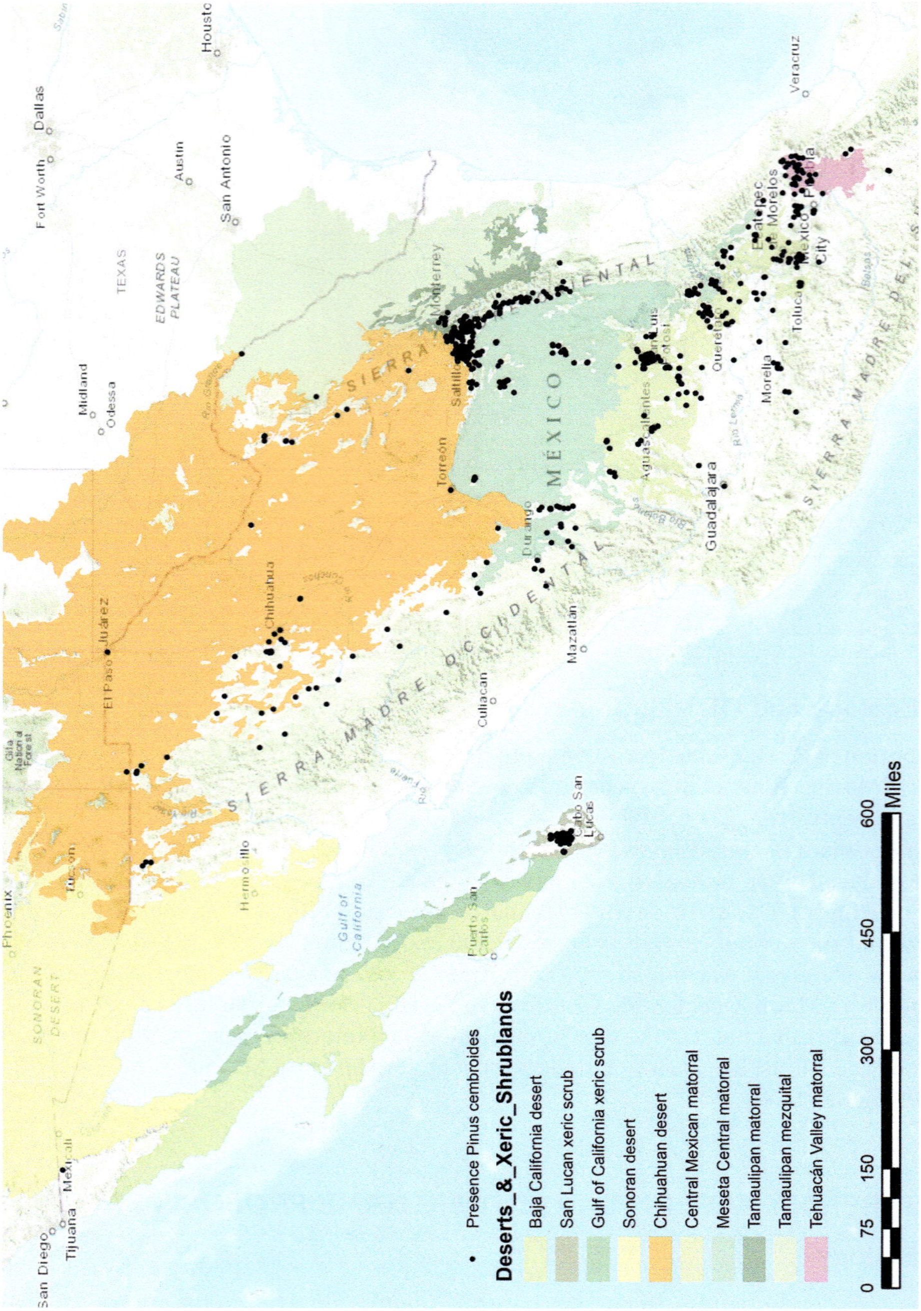

Figure 2.24 Map of the presence of *Pinus cembroides*. It is located on the borders of the Sierra Madre Oriental and the Occidental.

light green, compounded leaves with 12 to 20 leaflets. The flowers are in 5–10 cm long clusters of green-yellow cylindrical spikes at the ends of branches. Pods are 20 to 30 cm long and contain between 10 and 30 seeds per pod (Patel et al., 2018) (Figure 2.25).

2.15.2 Active Principles or Bioactive Compounds

Chemicals extracted from the wood include sugars, resins, volatile oils, fatty acids, tannins, alcohols and phenols. The main alkaloid juliflorine in *P. juliflora* was isolated and reported by Ahmad and Mahmood (1979), and Ott Longoni et al. (1980) reported the complete structure of juliflorine (juliprosopine) from *P. juliflora*. Juliflorine has been reported to possess antidermatophytic and antibacterial activities (Khursheed et al., 1986; Ahmad, 1986). Leaves contains alkaloids such as tryptamine, piperidine, phenethylamine and juliprosopine, which have antifungal and plant-growth inhibiting properties and might induce neuronal damage in animals (Tapia et al., 2000). Seeds contains fatty acids and a pentacyclic triterpenes (hexadecanoic, octadecanoic acids, glucopyranose, hydroquinone, glucopyranosides and galactose sugars) with antibacterial and antifungal activities (Sirmah, 2000).

Prosopis trees contain alkaloids that can intercalate with DNA: prosopinine, prosopine, tyramine and harman (Aqeel, 1989). Also, caffeic acid derivatives and quercetin 3-o-glucoside and quercetin 3-o-galactoside have been found, which bind with extracellular and soluble proteins and complexes to bacterial cell walls (Harzallah-Shhiri & Jannet, 2005). Caffeic acids have antibiotic properties against viruses, bacteria and fungi (Cowan, 1999).

Furthermore, Sirmah (2000) found gallocatechins, catechin, methylgallo-catechins, epicatechin, free sugars and fatty acids in *P. juliflora*'s bark. Pods contain galactomannans, mannoses, saturated and unsaturated fatty acids and free sugars that are used as food and medicine for both animals and humans. *P. juliflora* contains mesquitol (located in the heartwood), which was able to slow down the oxidation of methyl linoleate induced by 2, 2'-azobis 2-methylpropionitrile.

2.15.3 Ethnobotanical Uses or Traditional Uses

The growth and survival of man has depended upon the use of natural resources. *Prosopis* tree products are economically important sources of food, fodder, firewood, timber and soil fertility enrichment. Smoke from the leaves is used to tread eye troubles. Pods are eaten as a fresh fruit by humans in some areas for their sweet pulp and pleasant taste. The bark is used in leather tanning and yields an edible gum. The flowers are valuable for honey production. Bark and flowers are used medicinally, and, in some cases, the powdered bark is mixed with flour to make biscuits. Dried pods are a rich animal feed, which is liked by all cattle in dry seasons. Some extracts of the stem bark have anti-inflammatory properties. Aqueous and alcoholic extracts are markedly antibacterial. The bark is used as an anthelmintic, a tonic and as a cure for leukoderma, bronchitis, asthma, leprosy, dysentery, piles and tremors of the muscles. Any part from the tree is recommended for the treatment of snakebites and scorpion strings. Fuel and charcoal from the tree have high calorific value and burns well even when freshly cut.

2.15.4 Ecology and Distribution

Natural distribution of the 44 species of *Prosopis* (*Leguminosae*) occurs in arid zones in North and South America (European and Mediterranean Plant Protection Organization, 2019), northern Africa, southwestern Asia and the Indian subcontinent (Leakey & Last, 1980).

Figure 2.25 *Prosopis* spp. fruits in Rancho Los Ángeles, Saltillo, Coahuila.

The *Prosopis* genus is very important to arid environments because of their survival capacity and multipurpose uses; furthermore, they tolerate drought situations and improve soil fertility and are an important constituent of the vegetation system. The trees are well adapted to arid conditions and stand up well to the adverse vagaries of climate and browsing by animals, as cattle and wild animals readily browse it. In areas open to cattle browsing, the young trees assume a cauliflower-shaped bushy appearance. The adult trees demand light, and dense shade kills the seedlings.

Prosopis trees have a very deep tap root system (up to 50 m or more) (Mahoney, 1990) and does not generally compete with crops in agroforestry systems. Trees are not believed to compete for moisture or nutrients with crops grown close to the trunk. The improved physical soil conditions as well as the higher availability of nutrients under *Prosopis* canopy explain the better growth of the crops associated with it (Figure 2.26).

2.16 CURRENTLY ACCEPTED SCIENTIFIC NAME *QUERCUS* SPP.

2.16.1 Biology

The oaks belong to the Fagaceae family, which includes six to nine genera and approximately 600 to 900 species of plants and belong to the genus known as *Quercus*. The species is among the most widespread and species-rich tree genera in the northern hemisphere (Kremer & Hipp, 2019). The highest diversity is exhibited in Mexico and East Asia (Nixon, 2010; Galicia et al., 2015).

All oaks share a number of common biological characteristics: woody stems, leathery or tough leaves and the presence of acorns. Its growth form is commonly as a tree (with a height of 3–40 m) and some as shrubs (with heights of 10–60 cm). They are slow growing, so they have considerable longevity and grow mainly in temperate forests, although they can also be found in scrublands, grasslands and, in an intercalated form, in some dry forests (Scareli-Santos et al., 2013) (Figure 2.27).

2.16.2 Active Principles or Bioactive Compounds

The *Quercus* genus contains various classes of compounds such as glycosides, terpenoids, flavonoids (particularly flavan-3-ol), phenolic acids, fatty acids, sterols, and tannins (Vihna et al., 2016).

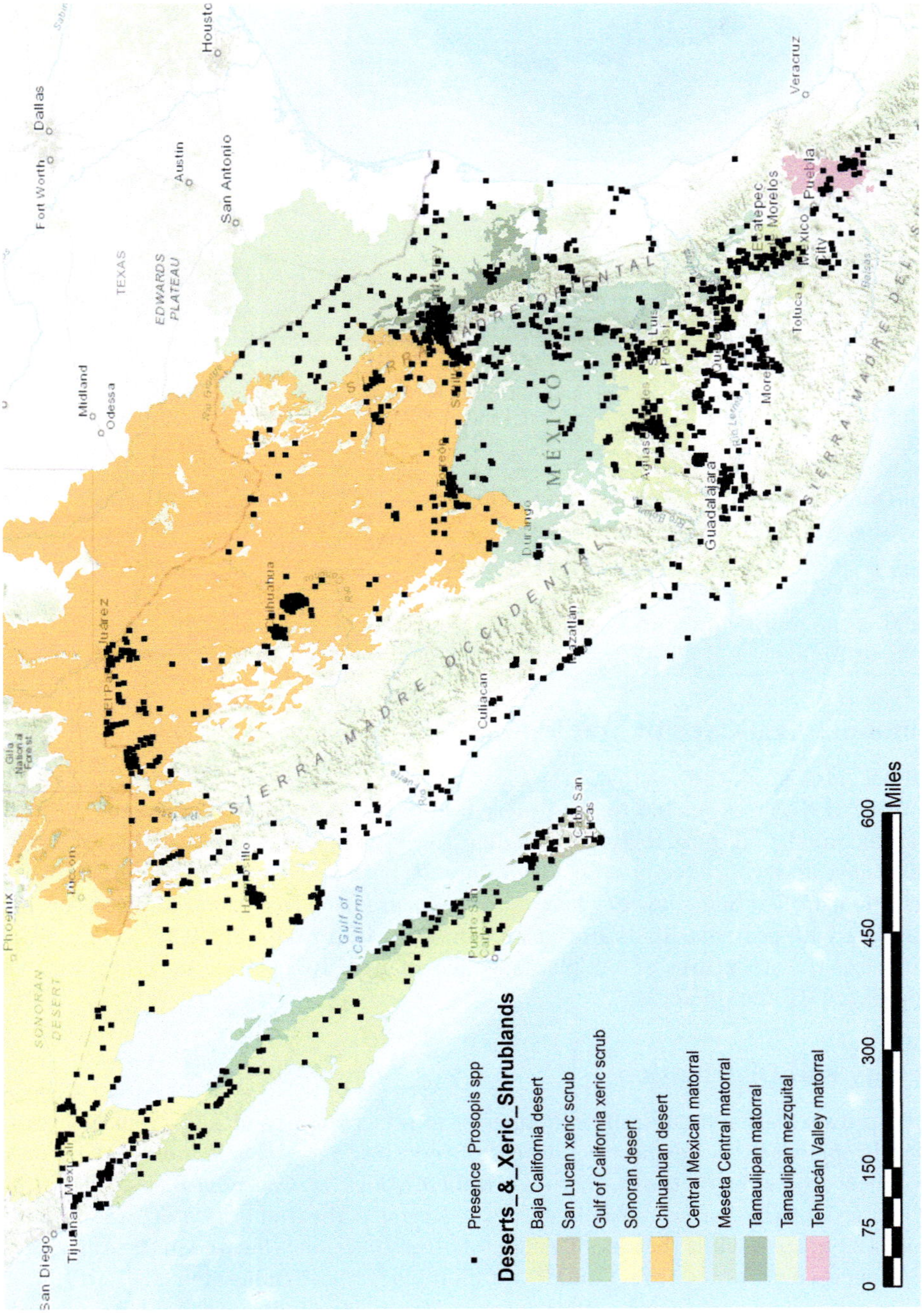

Figure 2.26 Map of *Prosopis* spp. It is the most distributed plant in Mexico.

Figure 2.27 Forest of *Quercus* spp. in the Sierra El Astillero, Durango.

The bark of Quercus bark has the potential to combat and control bacteria and fungi, from which they have been isolated compounds such as quercuschin, with six other compounds which were identified as quercetin, methyl gallate, gallic acid, betulinic acid, (Z)-9-octadecenoic acid methyl ester, and β-sitosterol glucoside (Gul et al., 2017).

Gentisic acid is the main compound found in the methanol leaves extract of *Quercus suber* (Custódio et al., 2015).

2.16.3 Ethnobotanical Uses or Traditional Uses

For medicinal use, almost all the organs of the plant are used: bark, leaves, flowers, roots and gills. Uses to treat 31 diseases related to the different apparatuses and systems of the human body have been recorded, and those related to the digestive system and the skin are use the largest number of species and whose recipes include mainly the bark and leaves (Taib et al., 2020).

For food use, acorns, leaf buds, flowers, leaves and galls are used to enrich the diet of indigenous communities with proteins, lipids and carbohydrates (Uddim & Rauf, 2012).

Forage comprises the structures of oak plants or acorns used to feed livestock, specifically swine or goats (García-Gómez et al., 2017).

2.16.4 Ecology and Distribution

Worldwide, oaks grow widely and naturally in temperate forests, tropical and semi-tropical forests and dry climate shrublands in the northern hemisphere. In contrast, very few oaks are found in tropical and semi-tropical ecosystems in the southern hemisphere (Aldrich & Cavender-Bares, 2011).

Oaks have evolved into two large areas or centers of diversity. The first is located in southeast Asia, with about 125 species, while the second center of diversification is located in the Americas, with about 250 species distributed from Canada to Colombia (Fried & Sulla-Menashe, 2019). The *Quercus* genus is very diverse in Mexico with more than 160 species, many of them endemic and distributed in different climates and soil types. They prefer temperate and subhumid environments at elevations from 1200 to 2600 m (Galicia et al., 2015) (Figure 2.28).

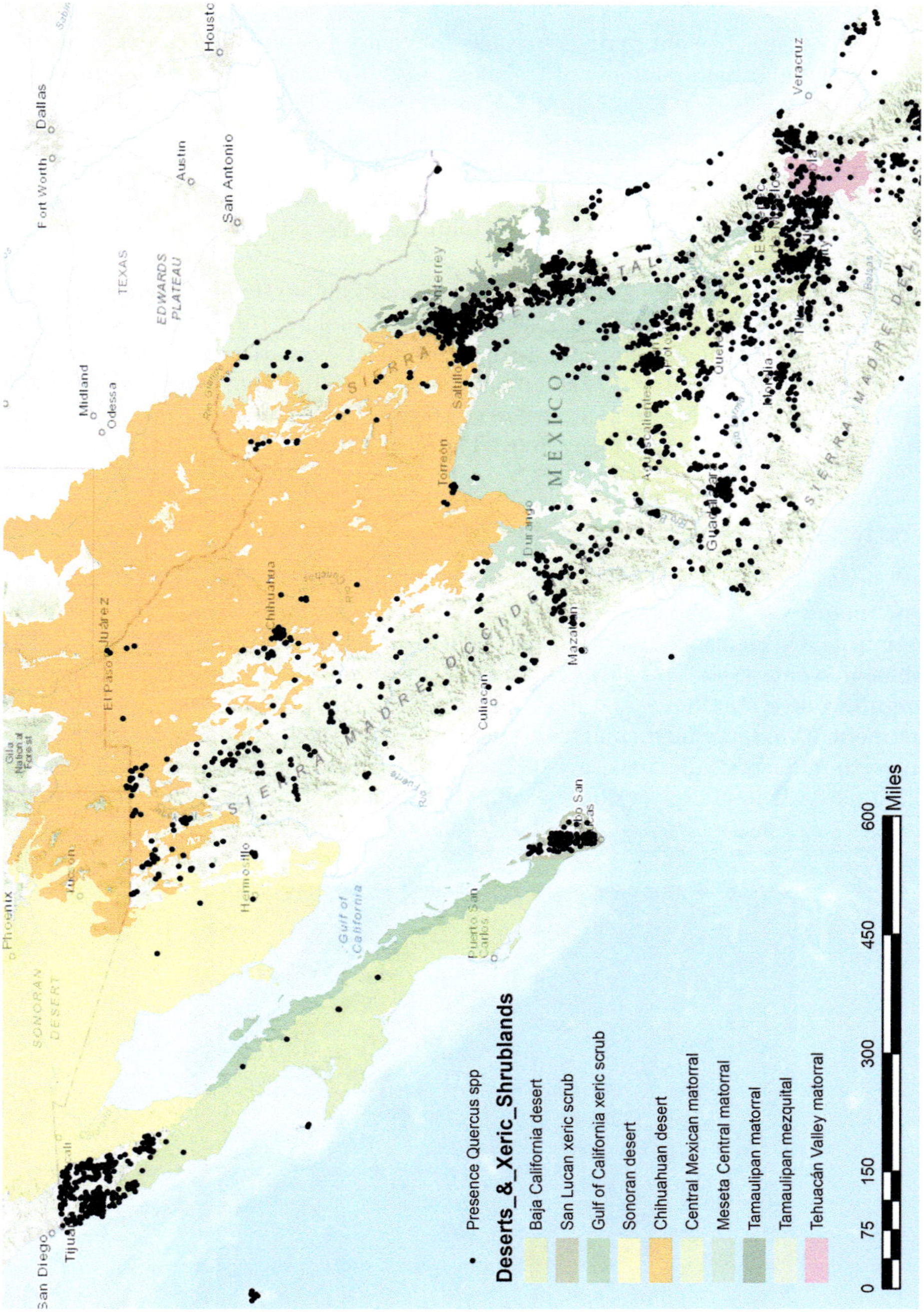

Figure 2.28 Map of *Quercus* spp. They are located on the borders of the Sierra Madre Oriental and Occidental.

2.17 CURRENTLY ACCEPTED SCIENTIFIC NAME *SELAGINELLA* SPP.

2.17.1 Biology

Selaginella species resist drought periods because they have poikilohydric properties (Mickel & Valdespino, 1992) and high contents of trehalose, which provides a protection mechanism against drought (Richards et al., 2001). In the dry season, their branches curl inward and look like golden balls 10 cm in diameter; in the humid season, their branches extend into a green rosette-like appearance (Korall et al., 1999; Korall & Kenrick, 2002). Several species of *Selaginella* grow in various Mexican states. They are commonly named "doradillas," "resurrection ferns," "resurrection plants," "much-kok," "texochitl yamanqui" and "flor de piedra" (Argueta et al., 1994; Martínez, 1989).

Plants are herbaceous, terrestrial, epilithic or occasionally epiphytic, evergreen or sometimes seasonally green. Rhizome erect, creeping, ascending or scandent, rhizophores are present at the lower part of rhizome, which bear roots. Two different leaf sizes are present on their dichotomous branch, which are simple and very small in size (~10 mm). The stem of the small *Selaginella* species grows up to approximately 3 cm, whereas stems of large species grow up to approximately 50 cm to 1 m long (Mukhopadhyay, 2001) (Figure 2.29).

2.17.2 Active Principles or Bioactive Compounds

Various species of *Selaginella* have medicinal potential because they are sources of bioactive compounds, mainly bioflavonoids (Queiroz et al., 2016; Zhao et al., 2017). Several reported chemical studies have identified bioflavonoids, such as amentoflavone, robustaflavone, (S)-2,3-dihydrorobustaflavone, (S)-2,3-dihydro-5-methoxy- robustaflavone, isocriptomerin, heveaflavone, hinoquiflavone trimetil eter and 5',5"- dihydroxy-7,7",4',4'"-tetramethoxy-amentoflavone, as main components of these medicinal plants (Quasim et al., 1985; Aguilar et al., 2008).

An immense number of studies have revealed that the *Selaginella* genus is rich with biflavonoids, steroids, alkaloids, alkaloidal glycosides, secolignans, lignans, neolignans, phenylpropanones

Figure 2.29 Specimens of *Selaginella* spp. in a dry period.

and caffeoyl derivatives (Sá et al., 2012). Most bioactive compounds isolated from various species of *Selaginella* belong to the class of flavonoids, alkaloids, lignans, pigments, phenylpropanoids, steroids, benzenoids, quinoids, carbohydrates, coumarins, chromones and oxygen heterocycle (Almeida et al., 2013)

2.17.3 Ethnobotanical Uses or Traditional Uses

The *Selaginella* genus is used in folk medicine as an antidepressant, aphrodisiac, diuretic, analgesic and anti-inflammatory agent as well as a female fertility enhancer (Agra et al., 2008). Traditional treatments with Decoction of the whole plant are traditionally used to treat urinary obstruction, cystitis, renal calculus, kidney inflammation and waist and back pains (Vázquez et al., 2005, Aguilar et al., 2013). Treatments also include digestive problems, cough, bronchitis and parasitic infections (Argueta et al., 1994; Márquez et al., 1999; Martínez, 1989).

Different species of the genus *Selaginella* are exploited for various ethnomedicinal purposes, mainly to cure fever, jaundice, hepatic disorders, cardiac diseases, cirrhosis, diarrhea, cholecystitis, sore throat and cough as well as to promote blood circulation, remove blood stasis and stop external bleeding after trauma and separation of the umbilical cord (Singh & Singh, 2015; Adnan et al., 2021). Different species of *Selaginella* have been used as food, medicine, in handicrafts and as ornaments from time immemorial. As the distribution of *Selaginella* species is seen worldwide, usage of these plants has been observed in traditional ways by people around the world for various purposes (Adnan, 2021).

2.17.4 Ecology and Distribution

The *Selaginella* genus comprises about 700–800 species (Hirai & Prado, 2000). Approximately 270 species of *Selaginella* are found in America, of which 61 occur in Brazil, and the genus is widely distributed throughout America, Africa and Europe (Tryon & Tryon, 1982).

Tropical and subtropical areas have maximum diversity in which plants can grow in various types of soil and climate. Some of the species are also found in extreme climates such as dry desert, cool alpine and tundra (Judd et al., 1999; Tryon & Tryon, 1982) (Figure 2.30).

2.18 CURRENTLY ACCEPTED SCIENTIFIC NAME *SIMMONDSIA CHINENSIS* (LINK) C.K. SCHNEID

2.18.1 Biology

This is a shrubby plant native to the Sonoran Desert, mainly in the southeastern United States and adjacent areas in northwestern Mexico. It is a dioecious perennial shrub that reaches 1–3.5 meters tall and is stiff, leafy and many-branched. It is usually a long-lived plant, and its wood is hard and pale yellow in color (Benzioni, 2010). It occurs in different soil types, geology, climate, vegetation and altitude, which influences seed production, phenology and population dynamics. In particular, wild populations of *S. chinensis* are subject to the influence of biotic and abiotic environments, which can be determined in the seed production, which is the species' main product of importance for the inhabitants of the abovementioned regions (Maldonado et al., 1995). It is an obligate cross-pollinated plant; therefore, it exhibits a high variability in its phenology, morphological parameters and yield (Hassanein et al., 2012).

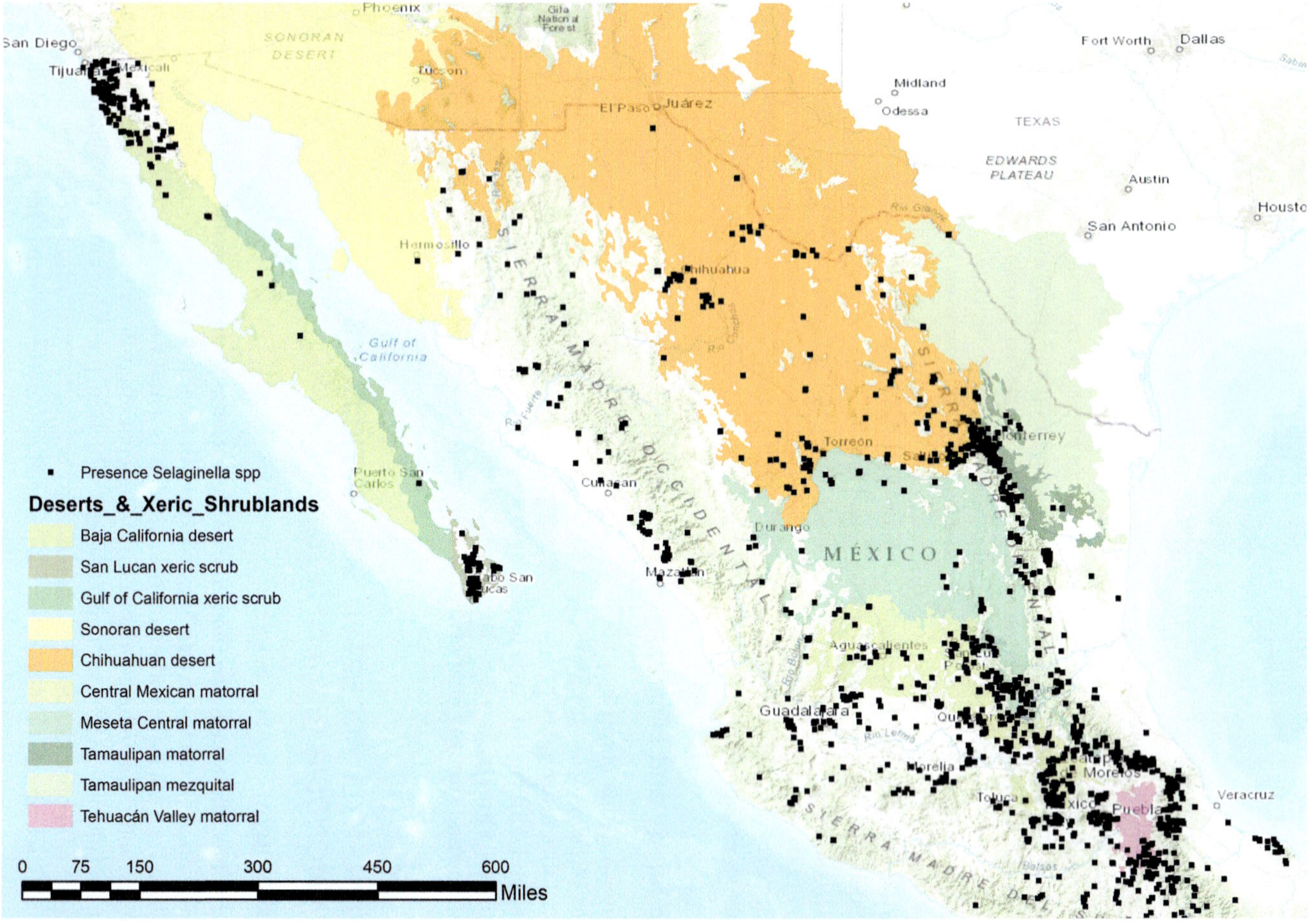

Figure 2.30 Map of the presence of *Selaginella* spp., which is abundant in the mountains.

Furthermore, there are records indicating that less than 1% of plants from the seed of wild populations have the potential to produce economically acceptable yields (Reedy & Chikara, 2010).

2.18.2 Active Principles or Bioactive Compounds

The uniqueness of the jojoba plant stems from the unusual presence, quantity and chemical structure of liquid wax in its seeds, which consists of approximately 50% of the seed weight (Al-Widyan & Al-Muhtaseb, 2010). The wax is composed of long-chain esters. These can be exploited in various industrial applications, the most prominent being pharmaceuticals and cosmetics, because of the similarity of their structures and properties to human skin sebum (Tietel et al., 2021). Jojoba oil is a mixture of long-chain esters (C36–C46) of fatty acid and fatty alcohol, distinguishing it from other vegetable oils, which are based on triglycerides (Mokhtari et al., 2019; Agarwal et al., 2018).

In addition, tocopherols and phytosterols are two other groups of bioactive molecules that have been observed in *S. chinensis* oil. Tocopherols are common in oil crops and serve as lipophilic antioxidants and oil stabilizers, which also possess health-promoting properties. In the oil obtained from this plant, relatively high concentrations (417 ppm) of antioxidants were recorded, with gamma-tocopherol as the main component at 79.2%, and alpha-, beta- and delta-tocopherol at lower concentrations (El-Mallah & El-Shami, 2009).

2.18.3 Ethnobotanical Uses or Traditional Uses

It is widely used in pharmaceutical and cosmetic formulations because of its unique structural characteristics and health benefits, and a large number of studies have been reported on the beneficial effects of the antioxidant activity of the oil obtained from this plant on human health (Tietel et al., 2021). Likewise, the uptake capacity of methanolic and ethanolic extracts of both jojoba seed and leaf by the 2,2′-diphenyl-1-picrylhydrazyl (DPPH) method has been demonstrated; under this methodology it was found that *S. chilensis* extracts possess similar scavenging efficacy (approximately 40% at 500 µg ml-1) (Kara, 2017). Other studies (Abdel-Wahhab et al., 2016) reported the hepatoprotective impact of an ethanolic seed extract with positive results at doses of 0.5 and 1 mg kg-1, showing that hepatic oxidative stress was attenuated in the induced mycotoxin damage model.

Also, the plant has been used as an antimicrobial agent. *S. chinensis* extracts have been shown to have antibacterial properties through a formulation with 80% methanol, ethanol, acetone, isopropanol and ethyl acetate (Wagdy & Taha, 2012). Finally, jojoba oil has long been used in dermocosmetological products. Among its main applications, jojoba is a key component of the oil phase used in numerous topical formulations, as well as serving as a carrier and enhancer of active compounds (Di Berardino et al., 2006).

2.18.4 Ecology and Distribution

It grows in areas of lower rainfall, on mountain slopes and valleys, and much of its distribution is confined to the Sonoran Desert (Reddy & Chikara, 2010). The largest and fastest growing populations are in frost-free areas with an annual rainfall of 200–400 mm. It grows in a diversity of preferably well-drained and aerated desert soils, from porous rock to clay, from slightly acidic to alkaline (Maldonado et al., 1995). It is distributed in semi-dry temperate, warm dry and warm semi-dry climate types, all with summer rainfall and a winter rainfall of more than 10.2% (Figure 2.31).

2.19 CURRENTLY ACCEPTED SCIENTIFIC NAME *TAXODIUM MUCRONATUN* TEN

2.19.1 Biology

Taxodium mucronatum, "ahuehuete" or "sabino" has a straight shaft that sometimes presents irregular bulges and is frequently divided into two or three at the base. Its bark is smooth, grayish or reddish brown, from whose fibrous structures are detached in longitudinal shapes (Enríquez-Peña & Suzán-Azpiri, 2011).

The tree has a wide and irregular crown, with twisted, extended and robust branches, with dark green, simple, alternate, scaly leaves that measure 10– 22 mm long and 0.5 to 1 mm wide. It is a monoecious species with male and female flowers. Its aromatic fruit is oval to globose and green in color with internal resin glands (Suzán-Azpiri et al., 2007) (Figure 2.32).

2.19.2 Active Principles or Bioactive Compounds

The flavonoids cryptomerin A and B, isocryptomerin, hinokiflavone, hyperoside, podocarpus flavone A, quercetin glycoside and sciadoptin have been isolated in the leaves of *T. mucronatum*. The diterpene 8-beta-hydroxy-pimar-15-en-19-oic acid was found in a sample of leaves and fruit. The essential oil of *T. mucronatum* contains 72.21% monoterpenes, 10.70% sesquiterpenes and 16.80% other compounds (Luján-Hidalgo et al., 2012).

2.19.3 Ethnobotanical Uses or Traditional Uses

The resin produced by *T. mucronatum* was used in pre-Hispanic times to cure wounds, ulcers, skin diseases, toothache and gout. The resin obtained by burning wood chips relieves bronchitis and chest ailments. The bark is used as an emmenagogue (which causes or promotes menstruation) and a diuretic. For the treatment of sores and circulatory problems, a decoction of the bark, leaves, fruit and shoots is taken on an empty stomach (Cortés-Arroyo et al., 2011).

This tree has great beauty, favors the diversity of flora and fauna and has recreational benefits (Enríquez-Peña & Suzán-Azpiri, 2011). The ahuehuete is also of great historical and cultural importance in Mexico (Suzán-Azpiri et al., 2007).

2.19.4 Ecology and Distribution

The *T. mucronatum* has high water requirements and grows in riparian habitats located in permanent or semi-permanent streams and occasionally in sites without direct contact with water, but with a superficial water level to which it has direct access (Cortés-Barrera et al., 2010). The species has a wide distribution in Mexico except for the peninsulas of Baja California and Yucatan (Villanueva-Diaz et al., 2020).

It has been estimated to grow in an area of 48,958.5 ha in Mexico. The state of Oaxaca has the largest area (14,341.8 ha, 29.2 %), followed by Chiapas (4,791.9 ha, 9.7 %), and Michoacán (4,711.2 ha, 9.6 %) (Martínez-Sifuentes et al., 2021).

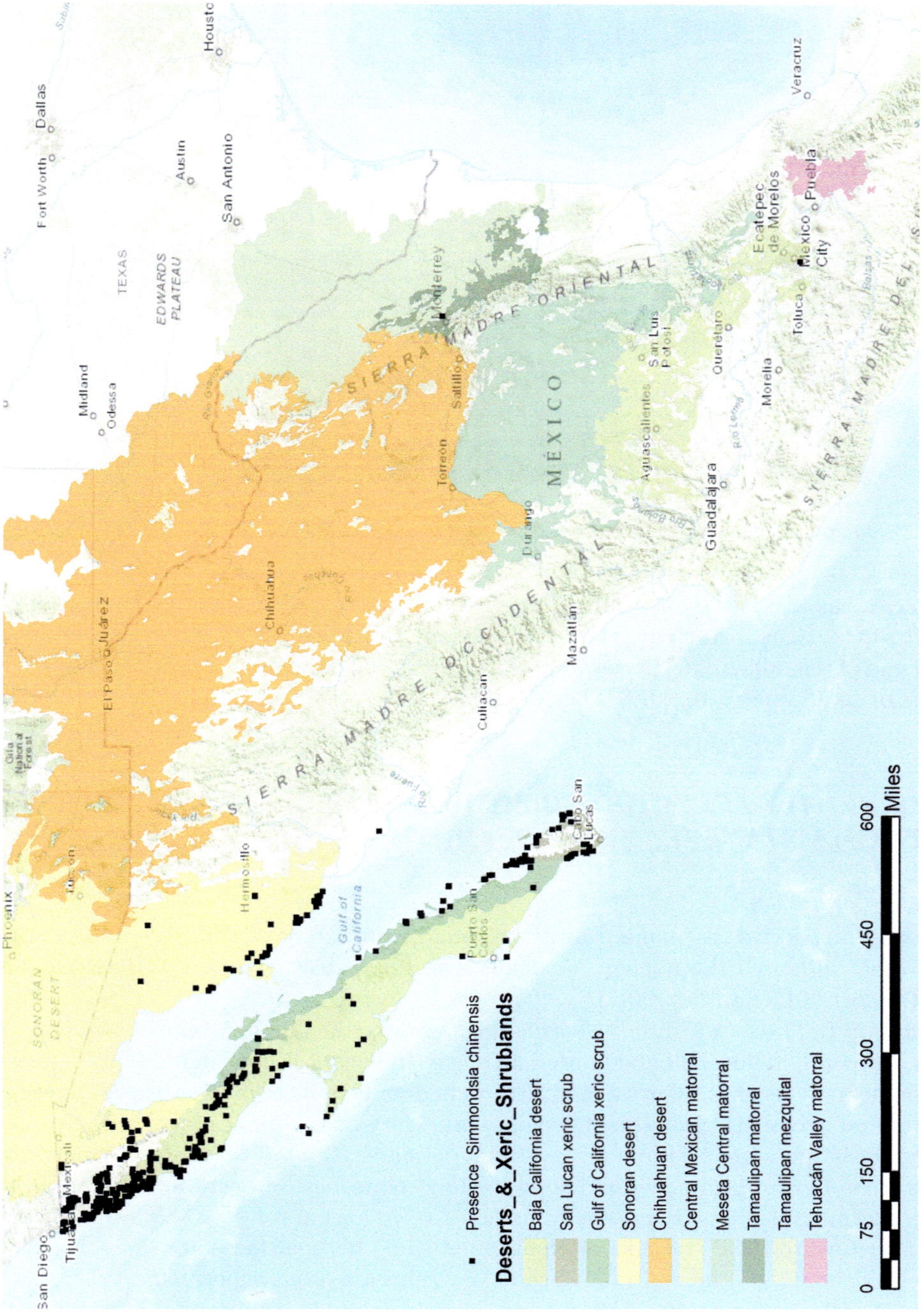

Figure 2.31 Map of the presence of *Simmondsia chinensis* in the Sonoran Desert.

Figure 2.32 *Taxodium mucronatun* in the Cañón de Fernández, Lerdo, Durango.

The species has been used for dendroclimatic and hydrological studies, as it is the longest-lived species in Mexico (Correa-Díaz et al., 2014; Osorio-Osorio et al., 2020; Villanueva-Díaz et al., 2020) and one of the millennial species in the eastern United States where it reaches more than 2,600 years of age (Stahle et al., 2019) (Figure 2.33).

2.20 CURRENTLY ACCEPTED SCIENTIFIC NAME *TECOMA STANS* (L.) JUSS EX KUNTH

2.20.1 Biology

The *Tecoma* genus derives its name from the Nahuatl word *tecomaxochitl* which means "cup-shaped flower," although the ancient Mexicans also knew it as *nixtamaxochitl* (Sánchez de Lorenzo-Cáceres, 2015; Ramírez, 2017).

Tecoma stans (L.) Juss. ex Kunth is a species belonging to the Bignoniaceae family, popularly known as yellowing, yellow bell and garden little ipe. This plant is considered to be a shrub or low tree and usually reaches 3.0 m tall and 20.0 cm in diameter. Its trunk has brown, teretiform, ribbed bark and irregularly puberulent, lepidote branches. It has imparipinnate leaves, 1.0–9.0 cm long or absent petioles, 2.0 mm long or absent petiolules, 3-9-foliolate or sometimes simple 1-foliolate leaves in young branches and opposite, lanceolate, base cuneate, apex acute, toothed or serrated, terminal leaflets that range from 2.0--5.0 cm long and 1.0–6.0 cm wide. The laterals are longer toward the distal end, often attenuate, and both surfaces are sparsely lepidote, with simple trichomes on the main vein. Secondary veins are often puberulous at the base and sometimes over the entire surface. Inflorescences are terminal or subterminal in racemes of up to 20 yellow flowers. Capsules are 7.0–21.0 cm long, 5.0–7.0 mm wide, linear, tapering toward the ends, teretiform when immature, surface lenticelate and almost glabrous to slightly lepidote.

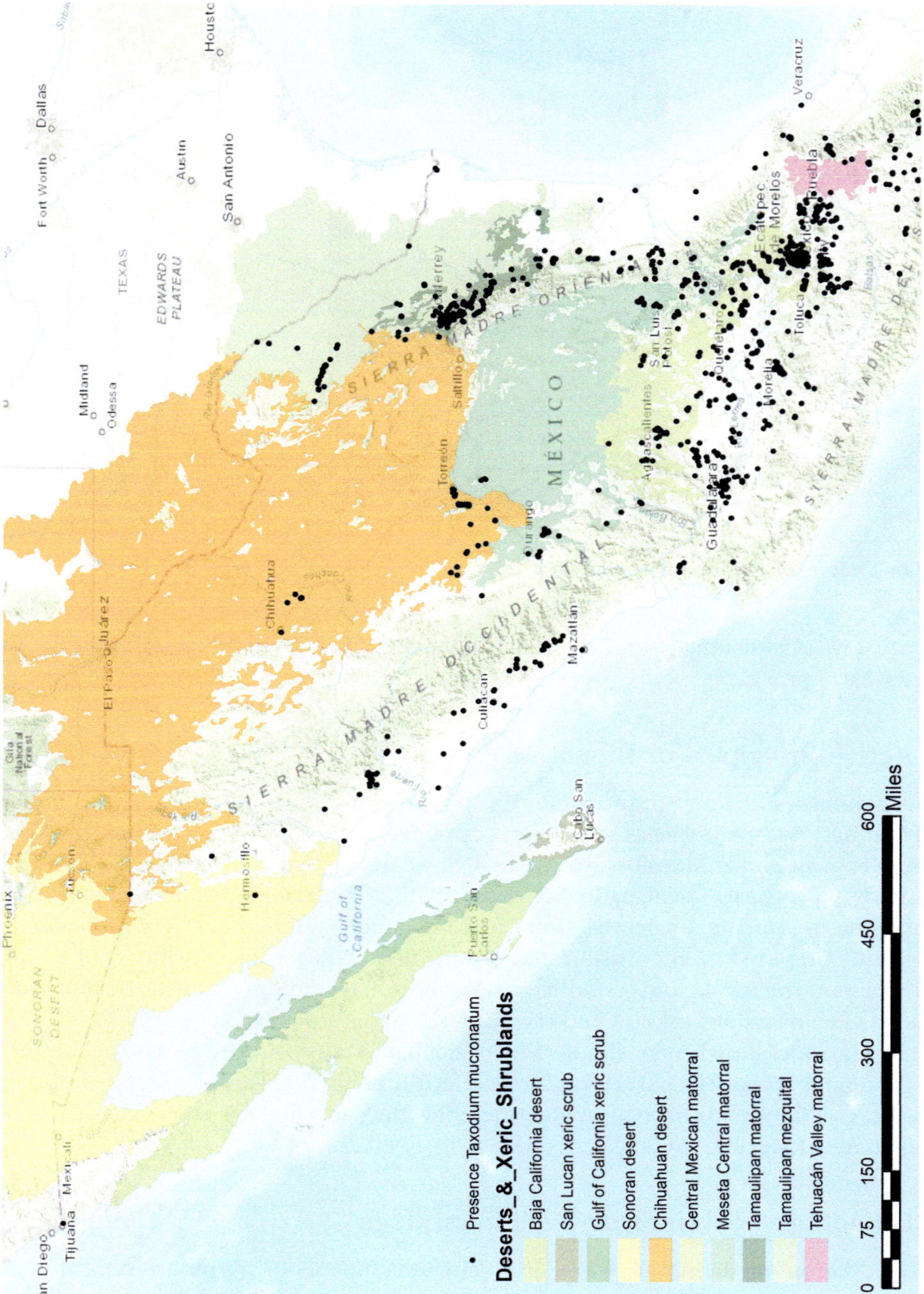

Figure 2.33 Map of the presence of *Taxodium mucronatum.*

Figure 2.34 Flowers of *Tecoma stans* in Lerdo, Durango.

The seeds are 3.0–5.0 mm long, 2.4–2.7 cm wide and have hyaline-membranous wings (Martínez & Ramos, 2012) (Figure 2.34).

2.20.2 Active Principles or Bioactive Compounds

This species exhibits antibacterial, antioxidant, antinociceptive, anti-inflammatory, antidiabetic and larvicidal activities, and these effects are correlated to presence of alkaloids, anthraquinones, phenolic compounds, steroids, glycosides, hydrocarbons, essential oils, tannins, terpenes and saponins (Alonso-Castro et al., 2010; Prasanna et al., 2013; Salem et al., 2013).

In most of the treatments, the leaves, stems and/or branches, bark, flowers or roots are used. They are usually prepared in an infusion that is administered as a tea. The leaves of the plant contain the monoterpene alkaloids actidin, boschniakinin, tecomanin, tecostatin, and tecostidin; alkaloids such as indole, eskatol and tryptamine; monoterpenes, including aucubin, plantarenalóside, stanside, stansioside, alpha-stansioside, beta-stansioside and 5-deoxystansioside; the benzyl components caffeic, paracoumaric and ferulic acids; and the flavonoid syringic acid. In the flowers, two flavonoids have been identified: the glucoside and the rutinoside of cyanidin (Medicinal Plants of Mexico, 2011).

2.20.3 Ethnobotanical Uses or Traditional Uses

A total of 54 different medicinal uses and 56 chemical components are reported for this plant (Ramírez, 2017), so it has a wide use in the herbal medicine of Mexico and Central America, mainly for diabetic, urinary and digestive disorders (Anburaj et al., 2016; Dewangan et al., 2017; Kumar, 2017). Likewise, it is used in Mexico to treat different digestive ailments such as stomach pain, dysentery, gastritis, poor digestion, indigestion, heartburn (gastroesophageal reflux), intestinal atony and liver problems. To treat empacho (indigestion), an infusion to which rose de Castilla and *viuxita* (spp. n/r) are added is recommended to make, and against dysentery only tapacola (*Waltheria americana* L.) is added (BDMTM, 2009a). The flowers and bark are used to treat various types of cancer (Anburaj et al., 2016).

2.20.4 Ecology and Distribution

Tecoma spp. belongs to the botanical family Bignoniaceae. The members of this family are characterized as being rare woody trees, shrubs or climbers times herbaceous, they are rarely herbaceous with sarmentose growth. It has a mostly pantropical geographic distribution, although some species can be found in temperate zones. However, its center of diversity is tropical America (Mokche et al., 2008; Castillo & Rossini, 2010; Laboratory of Vascular Plant Systematics, 2017).

It is native to Mexico and is currently found from the southern part of Florida, Texas and Arizona (United States), through all of Mexico and Central America, reaching northern Venezuela and along the Andes to northern Argentina. It is widely distributed throughout the Mexican Republic (CONABIO, 2018).

It is native to North America, although it is an invasive species in various regions of Africa. It is the official flower of the Virgin Islands and a symbol of the Bahamas (Gallardo-Alba, 2019). It lives in low and medium subdeciduous forests, preferably in secondary vegetation (Secretaria de Desarrollo y Medio Ambiente [SEDUMA], 2012). It is found in open and sunny areas as well as in secondary vegetation. It is frequently found on the edge of paths and highways or on walls (Zamora, 2011). Its altitudinal distribution varies from 0 to 2,000 meters above sea level in places with rainfall of 600–2,000 mm, but it grows well in dry sites with poor but well-drained soils and is intolerant to frost (Salazar & Soihet, 2001). It occurs in warm, semi-warm, dry, very dry and temperate climates (Olivas, 1999). Flowering occurs during the months of July to August in Mexico and Nicaragua, from November to May in El Salvador and throughout the year in Costa Rica and Puerto Rico. Pollination is carried out by honey bees (Salazar & Soihet, 2001). The fruits have been observed in February and March (Zamora, 2011). It is a species widely used as an ornamental in parks and gardens. In traditional medicine, the leaves, root and bark are used to control diabetes (SEDUMA, 2012).

Of a wide distribution in tropical America (Rsedowski & Calderón de Rzedowski, 1993), it thrives mainly in tropical forests, oak forests, xerophytic scrub and in disturbed areas.

Its distribution extends from the southernmost parts of Florida, Texas and Arizona (United States) and is widely distributed in almost all the states in Mexico and in Central America to the north of Venezuela, along the Andes, to the north of Argentina, it is also widespread in the Antilles (Rzedowski & Calderón de Rzedowki, 1993) (Figure 2.35).

2.21 CURRENTLY ACCEPTED SCIENTIFIC NAME *TURNERA DIFFUSA* WILLD. EX SCHULT

2.21.1 Biology

Turnera diffusa is a plant used in folk medicine as a natural stimulant (Zhao et al., 2007). The genus *Turnera* belongs to the family Passifloraceae and comprises about 135 species from the warm and tropical parts of the Americas and two from Africa (one from the southwest and one from the east) (Arbo & Espert, 2009; Thulin et al., 2012).

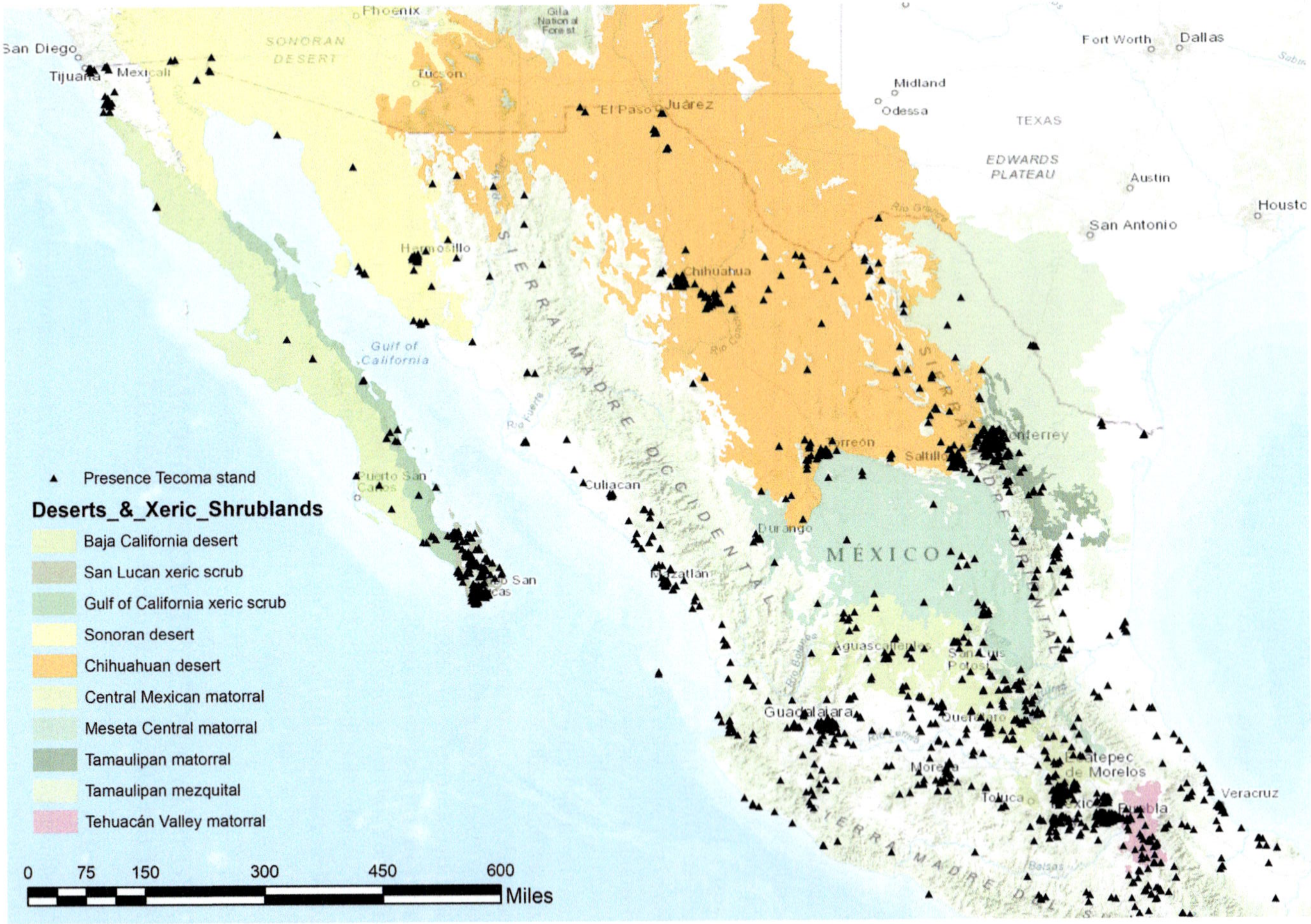

Figure 2.35 Map of the wide presence of *Tecoma stans* in Mexico.

Plants vary from sub-shrubs, shrubs, or, rarely, to trees (Alvarado-Cárdenas, 2006). *T. diffusa* is generally known as "damiana" and "yerba del venado." This species is a robustly growing and spreading plant that consists of herbaceous stems and is indigenous to southwestern Texas and Mexico (Perez et al., 1984; Hoffman, 1991). This species is a small shrub, between 0.30 to 2.00 m, deciduous, highly branched, reddish-brown and woody. Its alternate leaves are aromatic, pale green or yellow green in color, broadly lanceolate, serrate margin, sweet smelling and displays showy flowers with 5 yellowish petals in 5 styles that appear during the summer (Alcaraz-Meléndez et al., 2007, Zhao et al., 2007) and fruits in small capsules (Alvarado-Cárdenas, 2006).

2.21.2 Active Principles or Bioactive Compounds

The genus *Turnera* is known for the presence of important secondary metabolites (Szewczyk & Zidorn, 2014). The economic importance lies in its industrial use as an infusion beverage, flavoring, liqueur ingredient and use in traditional Mexican herbal medicine, for which it is attributed to stimulating nervous, aphrodisiac and diuretic effects (Osuna-Leal & Meza-Sánchez, 2000). Likewise, antimicrobial effects and possible hypoglycemic properties have also been reported (Zhao et al., 2007).

A phytochemical investigation identified 35 compounds in *Turnera diffusa*, including flavonoids, terpenoids, saccharides, phenolics and cyanogenic derivatives (Zhao et al., 2007) and 24 isolated structures to investigate the antiaromatase activity of *Turnera diffusa* (Zhao et al., 2008). It is also used to improve the flavor of desserts, ice cream, sweets and beverages (Garza-Juárez et al., 2011). In addition, it has antioxidant activity similar to quercetin (Salazar, 2008).

Essential oils are commercially employed for the preparation of liquors, cosmetic products and food flavoring (Alcaraz-Meléndez et al., 2004). Its chemical composition includes flavonoids, phenolic compounds and terpenoids among others (Zhao et al., 2007), most of which are known as antioxidants.

2.21.3 Ethnobotanical Uses or Traditional Uses

The effects of *Turnera diffusa* as a tonic, a stimulant, an aphrodisiac, and generally as a tonic in neurasthenia and impotency were already scientifically recorded in the early twentieth century by Manuel Pio Correa (Otsuka et al., 2010). Recently, in the (Argueta, 1994) *Atlas de las Plantas de la Medicina Tradicional Mexicana*, damiana is attributed the same set of ethnopharmacological applications. Moreover, usages as a remedy against stomachache, lung diseases related to tobacco abuse, bladder and kidney infections, rheumatism, diabetes, and scorpion stings are recorded. According to Martinez (1969), damiana was used, in particular, by the native people of northern Mexico as a tonic and aphrodisiac.

In traditional Mexican medicine, damiana is used as an aphrodisiac, for liver diseases, depression, anxiety, neurosis and as expectorant stimulant (Alcaraz-Meléndez et al., 2004). *Turnera diffusa* is used extensively as an anticough, diuretic, and aphrodisiac agent. It has antibacterial activity against the most common gastrointestinal diseases in Mexico (Osuna-Leal & Meza-Sánchez, 2000).

Popular knowledge recognizes the aphrodisiac properties of *Turnera diffusa* (Estrada-Castillón, 2014; Domínguez & Hinojosa, 1976). According to Estrada-Castillón et al. (2021), it is used for circulatory ailments in the Cuatrocienegas valley in Coahuila, Mexico, and most people mentioned that is an excellent remedy against body weakness. Some people use it daily to obtain better physical performance at work in the fields. Both virtues of this plant have been detailed in studies in which at least 20 different chemical compounds have been detected (Domínguez & Hinojosa, 1976); however, it is still unknown which compound is responsible for the aphrodisiac activity (Spencer & Seigler, 1981), although the aphrodisiac effect has been demonstrated in rats (Arletti, 1999).

The Guarijios de Sonora (Chihuahua, northern Mexico) use *Turnera diffusa* against influenza and pain and as an aphrodisiac (Aguilar et al., 1994). In the southern part of Nuevo León (northern Mexico), it is used as an aphrodisiac (poor man's Viagra) and to treat gastrointestinal and skin disorders (Estrada-Castillón et al., 2012). In Monterrey (Nuevo León, Mexico), preparations made from this plant are used as a remedy against impotence, as an aphrodisiac and to promote ovulation (González-Stuart, 2010).

2.21.4 Ecology and Distribution

According to Alvarado-Cárdenas (2006) and Villaseñor (2016), *T. diffusa* is distributed from the southern United States to northern South America (northeast Brazil), including the Antilles. In Mexico, this plant is found in the northern and northeastern states, from Baja California to the southeastern Yucatan and Quintana Roo. It grows in arid and semi-arid areas in the northwestern region of Mexico in brushy hills, sandy soil (Alcaraz-Meléndez et al., 2011), in xeric scrub, tropical deciduous forest and secondary vegetation at elevations of 150–1,500 m. Flowering occurs from February to November and fruiting from May to November (Figure 2.36).

2.22 CURRENTLY ACCEPTED SCIENTIFIC NAME *YUCCA FILIFERA* CHABAUD

2.22.1 Biology

The species *Yucca filifera* belongs to the Agavaceae family, generally known as "palma china" or "palma corriente" and is native to North America, northern Mexico and central Mexico (Cambrón-Sandoval et al., 2013). The flowers of this species are edible and highly nutritious (Pellmyr, 2003). It can reach more than 10 m in height; the stem is monopodic and robust in its initial part, later branching, covering and protecting itself largely with dead leaves. It has the capacity to reproduce sexually and vegetatively by budding on the same tree (Knox, 2010; Granados-Sánchez & López-Ríos, 1998) (Figure 2.37).

2.22.2 Active Principles or Bioactive Compounds

Y. filifera seeds contain oils, linoleic acid and steroidal sapogenin in the leaves (Castillo et al., 2012). The crude protein content of the flower is 113–275 g kg^{-1} DM^{-1} and crude fiber 104–177 g kg^{-1} (Sotelo et al., 2007). Castillo-Reyes et al. (2015) reported a total polyphenol concentration in *Y. filifera* of 5.28×10^4 ppm.

2.22.3 Ethnobotanical Uses or Traditional Uses

The main uses of *Y. filifera* are in human food from the flower and fruit and in livestock fodder using the inflorescence (Flores, 2001; Pacheco-Pantoja et al., 2021). Saponins, steroids important for the agrochemical industry, have also been reported (Arce-Montoya et al., 2006). In the pharmaceutical industry, edible oil is also produced from the seed or used for the pulp industry in the manufacture of Kraft paper, a paper resistant to wear and tear (Garcia-Berfon, 2021). The fibers of the leaves are used in rigging and cordage, its roots contain saponins, and its fruits, called "dates," are edible. In addition, this plant has drought-resistant characteristics, which is why it is used in plantations and reforestation in arid areas (Cambrón-Sandoval et al., 2013).

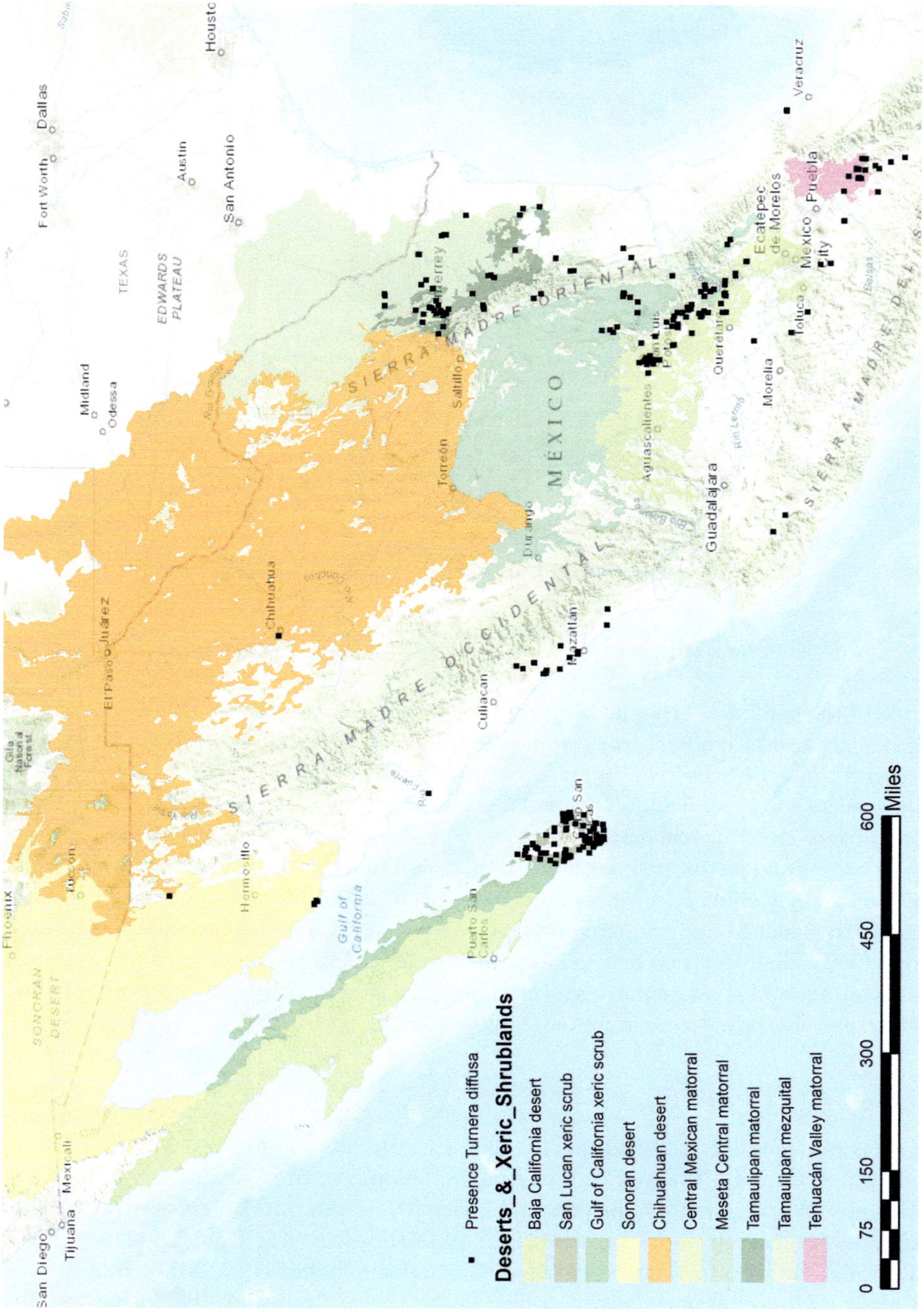

Figure 2.36 Map of the presence of *Turnera diffusa* showing little distribution.

Figure 2.37 Specimen of *Yucca filifera* in the town of Gómez Farías, Saltillo, Coahuila.

Yucca are one of the genera of ornamental plants and representative flora of the arid zones of Mexico. The leaves are used to make thread, ropes and cloth, and the flowers in buds and flower stalks, both raw and cooked, are used for human consumption (Pellmyr, 2003). The fruits also serve as food for some insects and other wildlife (Granados-Sánchez & López-Ríos, 1998). Freshly cut *Y. filifera* flowers can be purchased in local markets and, before being used in a variety of dishes, the flowers detached from stems and pistils are cooked in salted water. In order to avoid bitterness, in some places, the water is usually changed several times during the cooking process (Sánchez-Trinidad, 2017). When cooled, the cooked flowers are combined with tomatoes, lemon juice and salt to create a simple but nutritious salad that can be used as a garnish for thick stews. Other traditional uses include various stuffings and stews, sometimes combining *Y. filifera* flowers with egg, pork, shrimp, or vegetables. They can also be battered and fried (Mulík & Ozuna, 2020).

2.22.4 Ecology and Distribution

The genus *Yucca* is represented by 35–40 species that are cultivated in Central America, North America and southern Canada. The species *Yucca filifera* belongs to the Agavaceae family and is native to North America and northern and central Mexico (Oumama, 2014). It grows in arid areas in generally shallow, stony soils of calcareous origin at altitudes varying from 1,500 to 1,800 m where rainfall totals from 200 to 400 mm per year (Granados-Sánchez et al., 2011). This species grows in the upper and middle parts of the slopes. Under these conditions, the species entails interspecific competition with some species of different sizes; in addition, the habitat where it thrives is in xeric sites (Rentería & Cantú, 2003; Briones & Villarreal, 2001) (Figure 2.38).

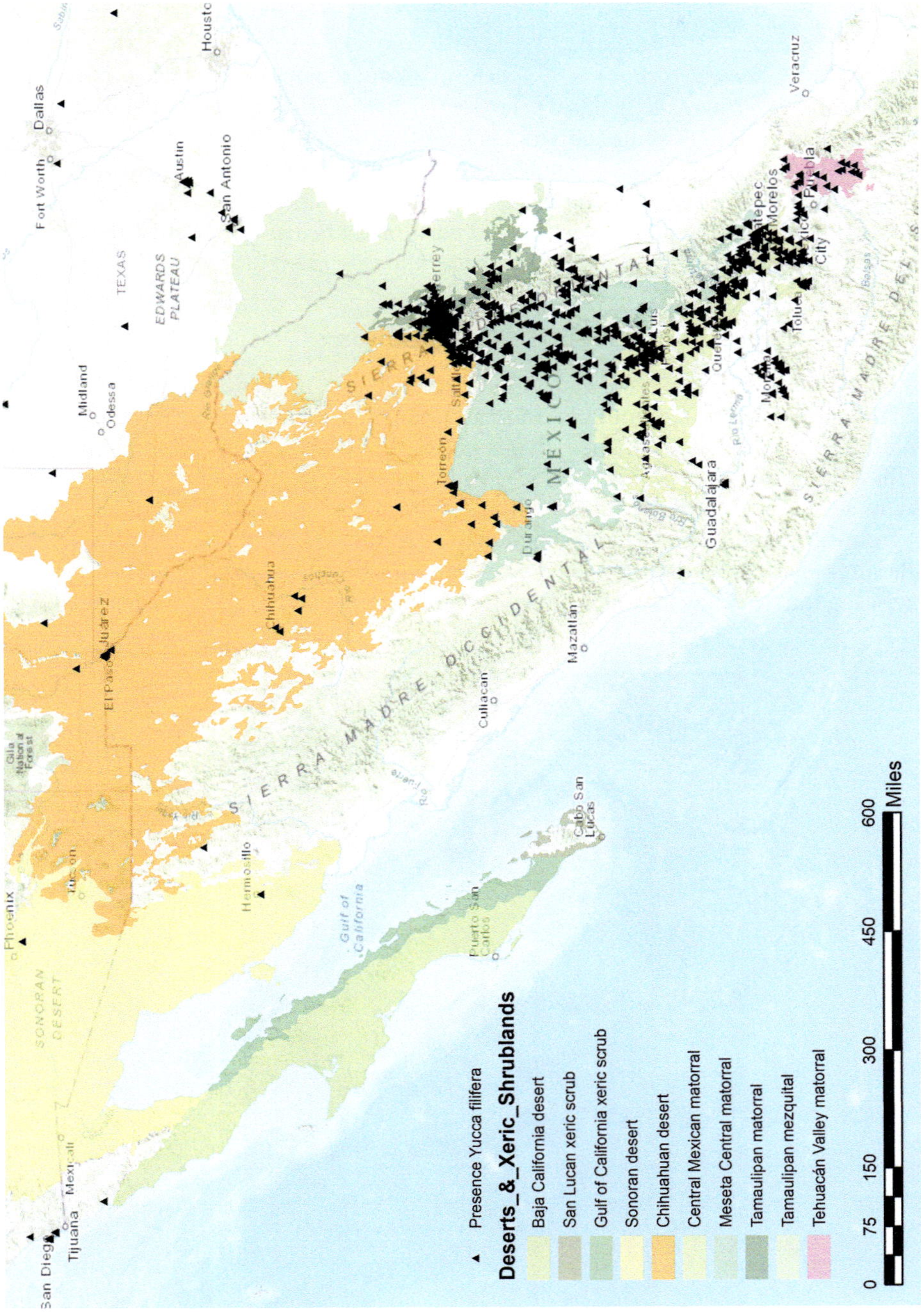

Figure 2.38 Map of the presence of *Yucca filifera*.

2.23 CURRENTLY ACCEPTED SCIENTIFIC NAME *YUCCA CARNEROSANA* (TREL) MCKELVEY

2.23.1 Biology

The samadoca palm (*Y. carnerosana*) is a specie of phanerogamous plant belonging to the Asparagaceae family (Pellmyr, 2003); it is a very slow-growing, shrubby perennial plant with an average height of 3 m, although some individuals reach up to 10 m. The trunk is thick and scaly with a diameter of 15–40 cm; it rarely branches at the top. The flowering stem grows in the central part of the rosette, completely overhanging the foliage. The floral stem branches with 15–30 pedicels with white bracts. The flower is white and perfumed; the fruit is indehiscent with thick, flat, half-round black seeds (Sheldon, 1980) (Figure 2.39).

2.23.2 Active Principles or Bioactive Compounds

Y. carnerosana contains good quality sapogenin in adult fruits. The presence of cholesterol was identified in the seed of unripe fruits, indicating the possible preceding nature of these substances in the formation of steroidal saponin. An aqueous extract has been found to have inhibitory activity against the development of *R. stolonifer*, *C. gloeosporioides*, and *P. digitatum* (Jasso de Rodriguez et al., 2011). In addition, the fibers contain cellulose, lignin and hemicellulose that have various functional groups that increase the bioadsorption of heavy metals (Medellín-Castillo, 2017).

Lopez-Ramirez et al. (2021) reported the presence of 27 compounds in cultivated *Y. carnerosana* plants, suggesting that this species is a promising source for the study and production of biomass

Figure 2.39 *Yucca carnerosana* on the campus of UAAAN in Buenavista, Saltillo, Mexico.

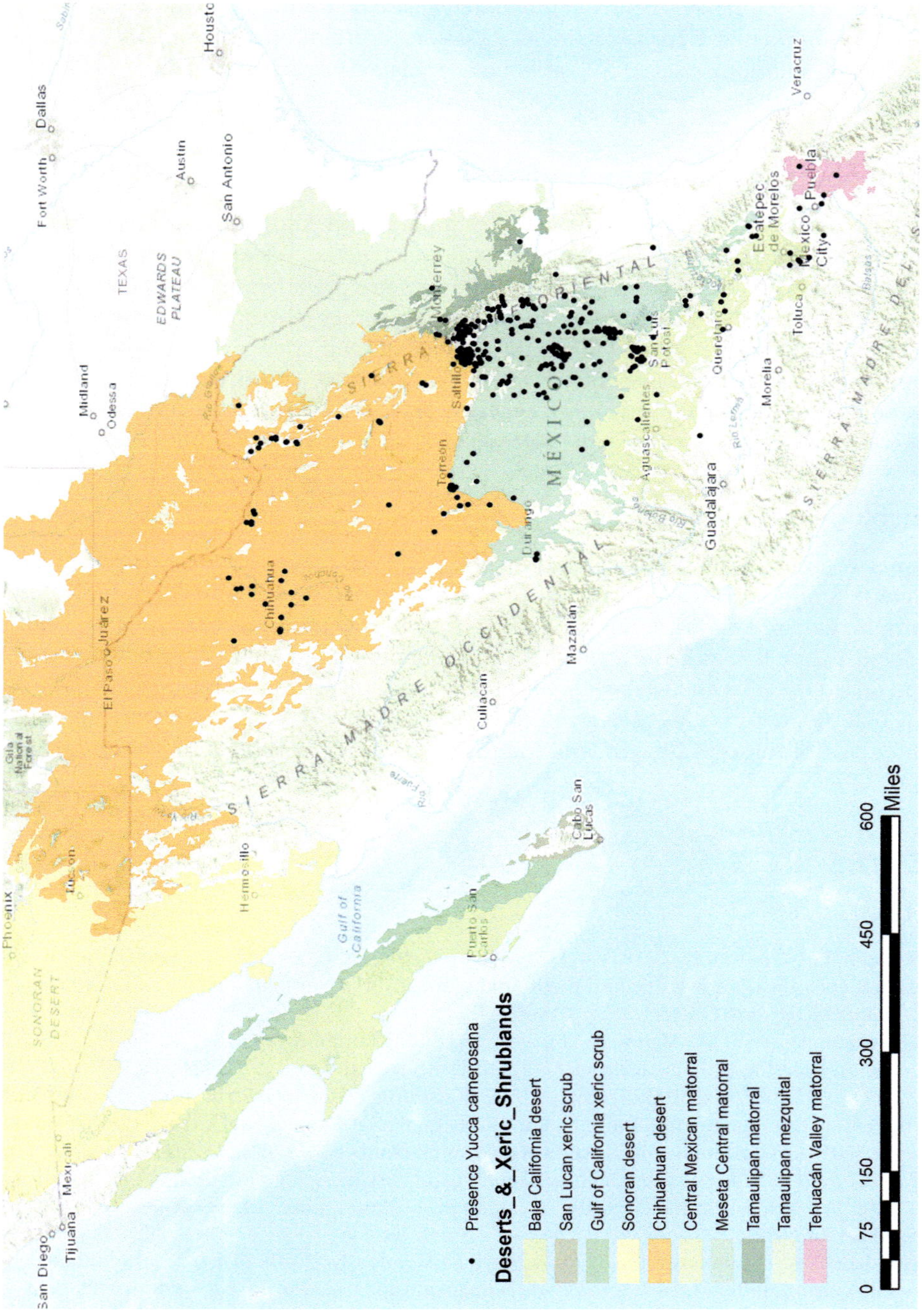

Figure 2.40 Map of the presence of *Yucca carnerosana*.

and plant metabolites. On the other hand, the antifungal activity of *Y. carnerosana* against the development of fruit postharvest fungi, such as *Rhizopus stolonifer*, *Colletotrichum gloeosporioides* and *Penicillium digitatum*, has been demonstrated (Jasso de Rodríguez et al., 2011). Similarly, evidence has been found that *Y. carnerosana* possesses biosorption capacity to remove Pb(II) ions present in aqueous solutions related to its phenolic acids and lignans (Medellín-Castillo et al., 2017).

2.23.3 Ethnobotanical Uses or Traditional Uses

The samandoca palm is used for the extraction of a hard fiber known as ixtle, which is used to make ropes and brushes. It is also used for local construction, cleaning and dietary needs (Sheldon, 1980). From an edible point of view, its flowers are prepared as stewed food and can also be made into flour and used to prepare an alcoholic beverage (Estrada et al., 2013). Likewise, inflorescences have been used as feed for goats because of their chemical composition, digestibility of organic matter and energy content (Mellado et al., 2009; Trenti-Very et al., 2021). It has also been reported that the dried leaves of *Y. carnerosana* serve as shelter for the flat-headed bat (*Myotis planiceps*), which is distributed in northwestern Mexico (Sil-Berra et al., 2022).

2.23.4 Ecology and Distribution

Y. carnerosana is a common succulent plant in the arid and semi-arid zones of north-central Mexico (zona ixtlera) and the southwestern United States (Pellmyr, 2003). It is a typical plant of the rosetophytic-izotal desert scrub, adapted to survive in conditions of low water availability (Granados-Sánchez & López-Ríos, 1998). Natural populations develop in alluvial soils, commonly found on slopes, hills and well-drained shallow soils. They are distributed in an altitudinal range of 1,000–2,200 meters above sea level. At lower elevations, it can be mixed with *Yucca filifera*, forming part of the rosetophytic and microphilous desert scrub (Maksimov et al., 2019) (Figure 2.40).

REFERENCES

Introduction

Caballero-Serrano, V., McLaren, B., Carrasco, J.C., Alday, J.G., Fiallos, L., Amigo, J., & Onaindia, M. (2019). Traditional ecological knowledge and medicinal plant diversity in Ecuadorian Amazon home gardens. *Global Ecology and Conservation*, *17*, e00524.

Dinerstein, E., Olson, D., Joshi, A., Vynne, C., Burgess, N.D., Wikramanayake, E., … & Saleem, M. (2017). An ecoregion-based approach to protecting half the terrestrial realm. *BioScience*, *67*(6), 534–545. https://doi.org/10.1093/biosci/bix014. https://www.gislounge.com/terrestrial-ecoregions-gis-data/

Lara Reimers, E.A., Fernández Cusimamani, E., Lara Rodríguez, E.A., Zepeda del Valle, J.M., Polesny, Z., & Pawera, L. (2018). An ethnobotanical study of medicinal plants used in Zacatecas state, Mexico. *Acta Soc Bot Pol.*, *87*(2), 3581. https://doi.org/10.5586/asbp. https://www.researchgate.net/publication/326072495_An_ethnobotanical_study_of_medicinal_plants_used_in_Zacatecas_state_Mexico [accessed Feb 01 2022].

Magaña Alejandro, M. A., Gama Campillo, L. M., & Mariaca Méndez, R. (2010). El uso de las plantas medicinales en las comunidades Maya-Chontales de Nacajuca, Tabasco, México. *Polibotánica*, (29), 213–262.

Pascual Casamayor, D., Pérez Campos, Y. E., Morales Guerrero, I., Castellanos Coloma, I., & González Heredia, E. (2014). Algunas consideraciones sobre el surgimiento y la evolución de la medicina natural y tradicional. *Medisan*, *18*(10), 1467–1474.

Petrovska, B.B. (2012). Historical review of medicinal plants' usage. *Pharmacognosy Reviews*, *6*(11), 1.
Reimers, E., Cusimamani, E., Rodriguez, E., Zepeda del Valle, J., Polesny, Z., & Pawera, L. (2018). An ethnobotanical study of medicinal plants used in Zacatecas state, Mexico. *Acta Societatis Botanicorum Poloniae*, *87*(2), 3581. https://doi .org /10 .5586 /asbp
Rzedowski, J. & Huerta M L. (1978). *Vegetación de Mexico*. (1a ed.). Editorial Limusa.
Rzedowski, J. (1991). Diversidad y orígenes de la flora fanerogámica de Mexico. *Acta botánica mexicana*. *14*, 3–21.
Rzedowski, J. y Calderón de Rzedowski, G. (1993). Flora del Bajio y de regiones adyacentes. *Familia Bignoniaceae*, *22*, 1–44. Instituto de Ecología. http://inecolbajio.inecol.mx/floradelbajio/documentos /fasciculos/ordinarios/Bignoniaceae%2022.pdf
Rzedowski, J. (2005). Mexico como área de origen y diversificación de linajes vegetales. In: Llorente, J. & J.J. Morrone (eds.). *Regionalización biogeográfica en Iberoamérica y tópicos afines. Las Prensas de Ciencias* (pp. 375–382). Mexico, D.F., Mexico: Universidad Nacional Autónoma de Mexico.
Tamang, J. P. (2010). Diversity of fermented beverages and alcoholic drinks. In J. P. Tamang & K. Kailasapathy (Eds.), *Fermented Foods and Beverages of the World* (pp. 85–125). CRC Press.
Villavicencio-Gutiérrez, E. E., Cano-Pineda, A., Castillo-Quiroz, D., Hernández-Ramos, A., & Martínez-Burciaga, O. U. (2021). Manejo forestal sustentable de los recursos no maderables en el semidesierto del norte de México. *Revista mexicana de ciencias forestales*, *12*(SPE1), 31–63.

Acacia spp.

Anderson, D.M.W., Farquhar, J.G.K., & McNab, C.G.A. (1984). The gum exudates from some *Acacia* subspecies of the series botryocephalae. *Phytochemistry*, *23*, 579–580.
Brown, W.L. (1960). Ant, acacias, and browsing animals. *Ecology*, *41*, 587–592.
Coe, M., & Coe, C.F. (1987). Large herbivores, *Acacia* trees and bruchid beetles. *South African Journal of Science*, *83*, 624–635.
Demole, E., Enggist, P., & Stoll, M. (1969). Sur les constituants de l'essence absolue de cassie (Acacia farnesiana will). *Helvetica Chimica Acta*, *52*, 24–32.
Gwynne, M.D. (1969). The nutritive value of *Acacia* pods in relation to Acacia seed distribution by ungulates. *East African Wildlife*, *7*, 176–178.
Joshi, K.C., Bansal, R.K., Sharma, T., Murray, R.D.H., Forbes, I.T., Cameron, A.F., & Maltz, A. (1979). Two novel cassane diterpenoids from *Acacia jacquemontii*. *Tetrahedron*, *35*, 1449–1453.
Kjaer, A., Knudsen, A., & Larsen, P.O. (1961). Amino acid studies. *Acta Chem. Scand.*, *15*, 1193–1195.
Lapornik, B., Prošek, M., & Wondra, A.G. (2005). Comparison of extracts prepared from plant by-products using different solvents and extraction time. *Journal of Food Engineering*, *71*, 214–222.
Letizia, C. S., Cocchiara, J., Wellington, G. A., Funk, C., & Api, A. M. (2000). Cassie absolute (Acacia farnesiana(L.) willd.). *Food and Chemical Toxicology*, *38*, 27–29. https://doi.org/10.1016/S0278 -6915(00)80011-3
McVaugh, R. (1987). *Leguminosae in W. R. Anderson* (ed. F. Novo-Galiciana, Vol. 5). Ann Arbor: The University of Michigan Press.
Mors, W.B., do Nascimento, M.C., Ruppelt Pereira, B.M., & Pereira, N.A. (2000). Plant natural products as active against snake-bite molecular approach. *Phytochemistry*, *55*, 627–642.
No Name (2000). Cassie absolute (*Acacia farnesiana* (L.) Willd). *Food and Chemical Toxicology*, *38*(Supplement 3), s27–s29.
Purata, S.E., Greenberg, R., Barrientos, V., López Portillo, J. (1999). Economic potential of the huizache, *Acacia pennatula* (Mimosoideae) in central Veracruz, Mexico. *Economic Botany*, *53*, 15–29.
Rico, A.M.deL. (1980). *El género Acacia (Leguminosae) en Oaxaca*. Tesis de Licenciatura. Facultad de Ciencias, UNAM.
Rico, A.M.deL. (1984). The genus Acacia in Mexico. *Bull. Int. Group. Study Mimosoid*, *12*, 50–59.
Rzedowski, J. & Huerta M L. (1978). *Vegetación de méxico*. (1a ed.). Editorial Limusa.
Siegler, D.S., & Ebinger, J.B. (1988). *Acacia macracantha*, *A pennatula*, and *A. cochliacantha (Fabacea: Mimosoideae)* species complexes in Mexico. *Systematic Botany*, *13*, 7–15.

Agave salmiana

Alonso-Castro, A.J., Villarreal, M.L., Salazar-Olivo, L.A., Gomez-Sanchez, M., Dominguez, F., & Garcia-Carranca, A. (2011). Mexican medicinal plants used for cancer treatment: Pharmacological, phytochemical and ethnobotanical studies. *Journal of Ethnopharmacology, 133*(3). https://doi.org/10.1016/j.jep.2010.11.055

Álvarez-Ríos, G.D., Pacheco-Torres, F., Figueredo-Urbina, C.J., & Casas, A. (2020). Management, morphological and genetic diversity of domesticated agaves in Michoacán, Mexico. *Journal of Ethnobiology and Ethnomedicine, 16*(1). https://doi.org/10.1186/s13002-020-0353-9

Castañeda-Nava, J.J., Rodríguez-Domínguez, J.M., Camacho-Ruiz, R.M., Gallardo-Valdez, J., Villegas-García, E., & Gutiérrez-Mora, A. (2019). Morphological comparison among populations of Agave salmiana Otto ex Salm-Dyck (Asparagaceae), a species used for mezcal production in Mexico. *Flora: Morphology, Distribution, Functional Ecology of Plants, 255*. https://doi.org/10.1016/j.flora.2019.03.019

Chávez-Güitrón, L.E., Florencia del, C.S.P., Pérez-Salinas, E.A., Caballero, J., Vallejo-Zamora, A., & Sandoval-Zapotitla, E. (2019). Variation of epidermal-foliar characters of Agave salmiana subsp. Salmiana (Asparagaceae) in the center of Mexico. *Botanical Sciences, 97*(4). https://doi.org/10.17129/botsci.2159

Díaz, E.V., García Nava, J.R., Peña Valdivia, C.B., Ramírez Tobías, H.M., & Ramos, V.M. (2011). Seed size, emergence and seedling development of maguey (Agave salmiana Otto ex Salm-Dyck). *Revista Fitotecnia Mexicana, 34*(3). https://doi.org/10.35196/rfm.2011.3.167

Escamilla-Treviño, L.L. (2012). Potential of plants from the genus agave as bioenergy crops. In *Bioenergy Research, 5*(1). https://doi.org/10.1007/s12155-011-9159-x

Estrada, E., Villarreal, J.A., Cantú, C., Cabral, I., Scott, L., & Yen, C. (2007). Ethnobotany in the Cumbres de Monterrey National Park, Nuevo León, Mexico. *Journal of Ethnobiology and Ethnomedicine, 3*. https://doi.org/10.1186/1746-4269-3-8

Lara-Ávila, J.P., & Alpuche-Solís, Á.G. (2016). Análisis de la diversidad genética de agaves mezcaleros del centro de Mexico. *Revista Fitotecnia Mexicana, 39*(3). https://doi.org/10.35196/rfm.2016.3.323-330

Leal-Díaz, A.M., Noriega, L.G., Torre-Villalvazo, I., Torres, N., Alemán-Escondrillas, G., López-Romero, P., Sánchez-Tapia, M., Aguilar-López, M., Furuzawa-Carballeda, J., Velázquez-Villegas, L.A., Avila-Nava, A., Ordáz, G., Gutiérrez-Uribe, J.A., Serna-Saldivar, S.O., & Tovar, A.R. (2016). Aguamiel concentrate from Agave salmiana and its extracted saponins attenuated obesity and hepatic steatosis and increased Akkermansia muciniphila in C57BL6 mice. *Scientific Reports, 6*. https://doi.org/10.1038/srep34242

Martínez Salvador, M., Rubio Arias, H., & Ortega Rubio, A. (2005). Population structure of maguey (Agave salmiana ssp. crassispina) in southeast Zacatecas, Mexico. *Arid Land Research and Management, 19*(2). https://doi.org/10.1080/15324980590916495

Muñiz-Márquez, D.B., Contreras, J.C., Rodríguez, R., Mussatto, S.I., Wong-Paz, J.E., Teixeira, J.A., & Aguilar, C.N. (2015). Influence of thermal effect on sugars composition of Mexican Agave syrup. *CYTA - Journal of Food, 13*(4). https://doi.org/10.1080/19476337.2015.1028452

Pinedo-Espinoza, J.M., Gutiérrez-Tlahque, J., Santiago-Saenz, Y.O., Aguirre-Mancilla, C.L., Reyes-Fuentes, M., & López-Palestina, C.U. (2020). Nutritional composition, bioactive compounds and antioxidant activity of wild edible flowers consumed in semiarid regions of Mexico. *Plant Foods for Human Nutrition, 75*(3). https://doi.org/10.1007/s11130-020-00822-2

Popoca, J., Aguilar, A., Alonso, D., & Villarreal, M.L. (1998). Cytotoxic activity of selected plants used as antitumorals in Mexican traditional medicine. *Journal of Ethnopharmacology, 59*(3). https://doi.org/10.1016/S0378-8741(97)00110-4

Santos-Zea, L., Fajardo-Ramírez, O.R., Romo-López, I., & Gutiérrez-Uribe, J.A. (2016). Fast centrifugal partition chromatography fractionation of concentrated agave (Agave salmiana) sap to obtain saponins with apoptotic effect on colon cancer cells. *Plant Foods for Human Nutrition, 71*(1). https://doi.org/10.1007/s11130-015-0525-2

Smith, G.F., & Figueiredo, E. (2012). A further species of Agave L., A. salmiana Otto ex Salm-Dyck (subsp. Salmiana) var. salmiana (Agavaceae), naturalised in the Eastern Cape Province of South Africa. *Bradleya, 30*. https://doi.org/10.25223/brad.n30.2012.a22

Stewart, J.R. (2015). Agave as a model CAM crop system for a warming and drying world. *Frontiers in Plant Science, 6*(September). https://doi.org/10.3389/fpls.2015.00684
Tello-Balderas, J.J., & García-Moya, E. (2017). El maguey (Agave, subgénero Agave) en el altiplano potosino-zacatecano. *Botanical Sciences, 48*. https://doi.org/10.17129/botsci.1350
Villarreal Morales, S.L., Enríquez Salazar, M.I., Michel Michel, M.R., Flores Gallegos, A.C., Montañez-Saens, J., Aguilar, C.N., & Herrera, R.R. (2019). Metagenomic microbial diversity in aguamiel from two agave species during 4-year seasons. *Food Biotechnology, 33*(1). https://doi.org/10.1080/08905436.2018.1547200

Cucurbita foetidissima

Bemis, W.P., Berry, J.W., Weber, C.W. (1978). The buffalo gourd, a potential crop for arid lands. *Arid Lands Newslett, 8*, 1–7.
Bemis, W.P., Berry, J.W., Weber, C.W. (1979). The buffalo gourd: A potential arid land crop. In G.A. Ritchie (ed.), *New Agricultural Crops, AAAS Selected Symposium 38*. Boulder, CO: Westview Press.
Berry, J. W., Bemis, W. P., Weber, C. W., & Philip, T. (1975). Cucurbit root starches. Isolation and some properties of starches from Cucurbita foetidissima and Cucurbita digitata. *Journal of Agricultural and Food Chemistry, 23*(4), 825–826.
Cutler, H. C., & Whitaker, T. W. (1961). History and distribution of the cultivated cucurbits in the Americas. *American Antiquity, 26*(4), 469–485.
Dittmer, H. J., & Talley, B. P. (1964). Gross morphology of tap roots of desert cucurbits. *Botanical Gazette, 125*(2), 121–126.
Dreher, M. L., Weber, C. W., Bemis, W. P., & Berry, J. W. (1980). Cucurbit seed coat composition. *Journal of Agricultural and Food Chemistry, 28*(2), 364–366.
Gómez González, A., Rangel Guerrero, J. M., Morales Flores, F., Aquino Pérez, G., Santana García, M. A., & Silos Espino, H. (2019). Diagnóstico de poblaciones silvestres de calabacilla loca en el altiplano central de México. *Revista mexicana de ciencias agrícolas, 10*(7), 1517–1528.
Hogan, L., & Bemis, W.P. (1983). Buffalo gourd and jojoba: Potential new crops for arid lands. *Advances in Agronomy, 36*, 317–349.
Lancaster, M., Storey, R., & Bower, N. W. (1983). Nutritional evaluation of buffalo gourd: Elemental analysis of seed. *Economic Botany, 37*, 306–309.
Nabhan, G. (1980). Ethnobotany of wild cucurbits in arid North America, An annotated bibliography [Unpub. ms, Dept, Plant Sciences, University of Arizona, Tucson, AZ].
Nelson, J. M., Scheerens, J. C., Berry, J. W., & Bemis, W. P. (1983). Effect of plant population and planting date on root and starch production of buffalo gourd grown as an annual. *Journal of the American Society for Horticultural Science, 108*(2), 198–201.
Niethammer, C. (1974). *American Indian Food and Lore*. New York: Macmillan.
Scheerens, J. C., Bemis, W. P., Dreher, M. L., & Berry, J. W. (1978). Phenotypic variation in fruit and seed characteristics of buffalo gourd. *Journal of the American Oil Chemists' Society, 55*, 523–525.
Wallace, E., Adams, M., & Quesada-Ocampo, L.M. (2015). First report of downy mildew on buffalo gourd (Cucurbita foetidissima) caused by Pseudoperonospora cubensis in North Carolina. *Plant Disease, 99*(12), 1861–1861.
Whitaker, T. W., & Bemis, W. P. (1975). Origin and evolution of the cultivated Cucurbita. *Bulletin of the Torrey Botanical Club, 102*, 362–368.

Dysphania ambrosioides

Blanckaert, I., Paredes-Flores, M., Espinosa-García, F. J., Pinero, D., & Lira, R. (2012). Ethnobotanical, morphological, phytochemical and molecular evidence for the incipient domestication of Epazote (Chenopodium ambrosioides L.: Chenopodiaceae) in a semi-arid region of Mexico. *Genetic Resources and Crop Evolution, 59*, 557–573.
Calzada, F., Velázquez, C., Cedillo-Rivera, R., & Esquivel, B. (2003). Antiprotozoal activity of the constituents of Teloxys graveolens. *Phytotherapy Research, 17*(7), 731–732.

Castillo-Juárez, I., González, V., Jaime-Aguilar, H., Martínez, G., Linares, E., Bye, R., & Romero, I. (2009). Anti-Helicobacter pylori activity of plants used in Mexican traditional medicine for gastrointestinal disorders. *Journal of Ethnopharmacology, 122*(2), 402–405.

Dembitsky, V., Shkrob, I., & Hanus, L. O. (2008). Ascaridole and related peroxides from the genus Chenopodium. *Biomedical Papers of the Medical Faculty of Palacky University in Olomouc, 152*(2). 29–33.

Espinosa-García, F.J., & Sarukhán, J. (1997). *Manual de Malezas del Valle de Mexico: Claves, Descripciones e Ilustraciones* (407 p). Mexico, D.F.: Fondo de Cultura Económica.

Gheno-Heredia, Y.A., Nava-Bernal, G., Martínez-Campos, Á.R., & Sánchez-Vera, E. (2011). Las plantas medicinales de la organización de parteras y médicos indígenas tradicionales de Ixhuatlancillo, *Veracruz*, Mexico y su significancia cultural. *Polibotánica, 31*, 199–251.

Grassi, T.L., Malheiros, A.M.S.C., da Silva Buss, Z., Monguilhott, E.D., Frode, T.S., de Souza, M.M. (2013). From popular use to pharmacological validation: a study of the anti-flammatory, anty- nociceptive and healing effects of Chenopodium ambrosioides extract. *J- Ethnopharmacol.*, 145–138.

Khan, I. H., & Javaid, A. (2019). Antifungal, antibacterial and antioxidant components of ethyl acetate extract of quinoa stem. *Plant Protection, 3*(3), 125–130.

Khan, I. H., & Javaid, A. (2020). Anticancer, antimicrobial and antioxidant compounds of quinoa inflorescence. *Advancements in Life Sciences, 8*(1), 68–72.

Kiuchi, F., Itano, Y., Uchiyama, N., Honda, G., Tsubouchi, A., Nakajima-Shimada, J., & Aoki, T. (2002). Monoterpene hydroperoxides with trypanocidal activity from Chenopodium ambrosioides. *Journal of Natural Products, 65*(4), 509–512.

Marinoff, M.A., Martínez, J.L., & Urbina, M.A. (2009). Precauciones en el empleo de plantas medicinales. *Boletín Latinoamericano y del Caribe de Plantas Medicinales y Aromáticas, 8*(3), 184–187.

Montoya-Cabrera, M.A., Escalante-Galindo, P., Meckes-Fisher, M., Sánchez-Vaca, G., Flores-Alvarez, E., & Reynoso-García, M. (1996). Envenenamiento mortal causado por el aceite de epazote, Chenopodium graveolens. *Gaceta Medica de Mexico, 132*, 433–438.

Palomares-Alonso, F., Rojas-Tomé, I.S., Rocha, V.J., Hernández, G.P., González-Maciel, A., Ramos-Morales, A., Jung-Cook, H. (2015). Cysticidal activity of extracts and isolated compounds from Teloxys graveolens: In vitro and in vivo studies. *Experimental Parasitology, 156*, 79–86.

Sagrero-Nieves, L., & Bartley, J. P. (1995). Volatile constituents from the leaves of Chenopodium ambrosioides L. *Journal of Essential Oil Research, 7*(2), 221–223.

Trivellato Grassi, L., Malheiros, A., Meyre-Silva, C., da Silva Buss, Z., Monguilhott, E. D., Fröde, T. S., ... de Souza, M. M. (2013). From popular use to pharmacological validation: A study of the anti-inflammatory, anti-nociceptive and healing effects of Chenopodium ambrosioides extract. *Journal of Ethnopharmacology, 145*(1), 127–138.

Vibrans, H., Hanan-Alipi, A.M., & Mondragón-Pichardo, J. (2009). Malezas de Mexico, Suaeda torreyana S.Watson. http://www.conabio.gob.mx/malezasdemexico/chenopodiaceae/chenopodium-graveolens/fichas/ficha.htm

Villaseñor, J.L., & Espinosa-García, F.J. (1998). Catálogo de Malezas de Mexico. In Universidad Nacional Autónoma de Mexico, Consejo Nacional Consultivo Fitosanitario y Fondo de Cultura Económica. Mexico, D.F., 449 p.

Zarate Aquino, M. A., & Hernandez Xolocotzi, E. (1991). Cultural practices of the epazote de zorrillo (Teloxys graveolens (Willd.) WA Weber) a kind of medicinal crop. *Serie Recursos Naturales Renovables*. Colegio de Postgraduados, Montecillo, Mexico.

Euphorbia antisyphilitica

Aranda-Ledesma, N. E., Bautista-Hernández, I., Rojas, R., Aguilar-Zárate, P., del Pilar Medina-Herrera, N., Castro-López, C., & Martínez-Ávila, G. C. G. (2022). Candelilla wax: Prospective suitable applications within the food field. *Food Science and Technology LWT, 159*(113170).

Ascacio-Valdés, J., Burboa, E., Aguilera-Carbo, A.F., Aparicio, M., Pérez-Schmidt, R., Rodríguez, R., & Aguilar, C.N. (2013). Antifungal ellagitannin isolated from Euphorbia antisyphilitica Zucc. *Asian Pacific Journal of Tropical Biomedicine, 3*(1). https://doi.org/10.1016/S2221-1691(13)60021-0

Barsch, F. (2004). Candelilla (Euphorbia antisyphilitica): Utilization in Mexico and international trade. *Medicinal Plant Conservation, 9*(10), 46–50.

Barrientos-Lozano, L., & Rocha-Sánchez, A.Y. (2013). A new species of the genus Pterodichopetala (Orthoptera: Tettigoniidae: Phaneropterinae) from northeastern Mexico. *Journal of Orthoptera Research, 22*(1). https://doi.org/10.1665/034.022.0102

Brailovsky Signoret, D., & Hernández, H.M. (2010). Mazapil, Zacatecas: Diversity and Conservation of Cacti in a Poorly-Known Arid Region in Northern Mexico. *Cactus and Succulent Journal, 82*(5). https://doi.org/10.2985/015.082.0502

Estrada-Castillón, E., Villarreal-Quintanilla, J.Á., Rodríguez-Salinas, M.M., Encinas-Domínguez, J.A., González-Rodríguez, H., Figueroa, G.R., & Arévalo, J.R. (2018). Ethnobotanical survey of useful species in Bustamante, Nuevo León, Mexico. *Human Ecology, 46*(1). https://doi.org/10.1007/s10745-017-9962-x

Gordillo, M.M., Ramírez, J.J., Durán, R.C., Arriaga, E.J., García, R., Cervantes, A., & Hernández, R.M. (2002). Los géneros de la familia Euphorbiaceae en Mexico (parte A). *Anales del Instituto de Biología. Serie Botánica, 73*(2), 155–196.

Hernandez-Centeno, F., Lõpez-De La Peña, H.Y., & Rojas, R. (2020). Candelilla (Euphorbia Antisiphylitica Zucc): Geographical Distribution Climate and Edaphology. In *Food Process Engineering and Quality Assurance*. https://doi.org/10.1201/9781315232966-16

Hernández-Herrera, J., Moreno-Reséndez, A., Manuel Valenzuela-Núñez, L., & Martínez-Salvador, M. (2019). Modelación de la presencia de Euphorbia antisyphilitica Zucc mediante propiedades físicas y químicas del suelo Modelation of the presence of Euphorbia antisyphilitica Zucc with physical and chemical properties of soil. *Ecosistemas y Recursos Agropecuarios, 6*(18), 499–511. https://doi.org/10.19136/era.a6n18.1910

Hua, J., Liu, Y., Xiao, C.J., Jing, S.X., Luo, S.H., & Li, S.H. (2017). Chemical profile and defensive function of the latex of Euphorbia peplus. *Phytochemistry, 136*, 56–64.

IBUNAM. (2019). Euphorbia antisyphilitica var. luxurians Miranda, ejemplar de: Herbario Nacional de Mexico (MEXU). In *Datos Abiertos UNAM (en línea)*. http://datosabiertos.unam.mx/IBUNAM:MEXU:T18770

Menchaca, M. D. C. V., Morales, C. R., Star, J. V., Cárdenas, A. O., Morales, M. E. R., González, M. A. N., & Gallardo, L. B. S. (2013). Antimicrobial activity of five plants from Northern Mexico on medically important bacteria. *African Journal of Microbiology Research, 7*(43), 5011–5017.

Mehrotra, N.K., & Ansari, S.R. (2000). Effect of iron on euphorbia antisyphilitica zucc. *Journal of Herbs, Spices and Medicinal Plants, 6*(4). https://doi.org/10.1300/J044v06n04_10

Miller, E.T., Mccormack, J.E., Levandoski, G., & Mckinney, B.R. (2018). Sixty years on: Birds of the Sierra del Carmen, Coahuila, Mexico, revisited. In *Bulletin of the British Ornithologists' Club, 138*, 4. https://doi.org/10.25226/bboc.v138i4.2018.a4

Ochoterena, H., Flores-Olvera, H., Gómez-Hinostrosa, C., & Moore, M.J. (2020). *Gypsum and Plant Species: A Marvel of Cuatro Ciénegas and the Chihuahuan Desert*. https://doi.org/10.1007/978-3-030-44963-6_9

Rojas, R., Tafolla-Arellano, J.C., & Martínez-ávila, G.C.G. (2021). Euphorbia antisyphilitica zucc: A source of phytochemicals with potential applications in industry. In *Plants, 10*(1). https://doi.org/10.3390/plants10010008

Steinmann, V.W. (2002). Diversidad y endemismo de la familia Euphorbiaceae en Mexico. *Acta Botanica Mexicana, 61*. https://doi.org/10.21829/abm61.2002.909

Jatropha dioica

Araujo-Espino, D.I., Zamora-Perez, A.L., Zúñiga-González, G.M., Gutiérrez-Hernández, R., Morales-Velazquez, G., & Lazalde-Ramos, B.P. (2017). Genotoxic and cytotoxic evaluation of *Jatropha dioica* Sessé ex Cerv. by the micronucleus test in mouse peripheral blood. *Regulatory Toxicology and Pharmacology, 86*, 260–264. https://doi.org/10.1016/j.yrtph.2017.03.017

Barba, A.M.D.L.D., Hernández, D.M.C., & De la Cerda, L.M. (2003). *Plantas útiles de la región semiárida de Aguascalientes*. Primera Edición. Universidad Autónoma de Aguascalientes. ISBN: 968 5073 61 9.

Can-Aké, R., Erosa-Rejón, G., May-Pat, F., Peña-Rodríguez, L.M., & Peraza-Sánchez, S.R. (2004). Bioactive terpenoids from roots and leaves of *Jatropha gaumeri*. *Revista de la Sociedad Química de Mexico*, *48*(1), 11–14. Retrieved February 18, 2022. http://www.scielo.org.mx/scielo.php?pid=S0583-76932004000100003&script=sci_arttext

Castro-Ríos, R., Melchor-Martínez, E.M., Solís-Cruz, G.Y., Rivas-Galindo, V.M., Silva-Mares, D.A., & Cavazos-Rocha, N.C. (2020). HPLC method validation for *Jatropha dioica* extracts analysis. *Journal of Chromatographic Science*, *58*(5), 445–453. https://doi.org/10.1093/chromsci/bmaa004

Córdova-Téllez, L., Bautista, L.E., Zamarripa, C., A., Rivera L., J.A., Pérez V., A., Sánchez S., O.M., Martínez, H.J & Cuevas S., J.A. (2015). Diagnóstico y plan estratégico de la Red *Jatropha* spp. en Mexico. In *SNICS* (116p). Mexico: SINAREFI.

Flores-Torres, G., Solís-Hernández, A.P., Vela-Correa, G., Rodríguez-Tovar, A.V., Cano-Flores, O., Castellanos-Moguel, J., Octavio-Pérez, N., Chimal-Hernández, A., Moreno-Espíndola, I.P., Salas-Luévano, M.A., Chávez-Vergara, B.M., Rivera-Becerril, F. (2021). Pioneer plant species and fungal root endophytes in metal-polluted tailings deposited near human populations and agricultural areas in Northern Mexico. *Environmental Science and Pollution Research*, *28*(39), 55072–55088. https://doi.org/10.1007/s11356-021-14716-6

Gutiérrez-Tlahque, J., Aguirre-Mancilla, C.L., López-Palestina, C., Sánchez-Fernández, R.E., Hernández-Fuentes, A.D., & Martin Torres-Valencia, J. (2019). Constituents, antioxidant and antifungal properties of *Jatropha dioica* var. *dioica*. *Natural Product Communications*, *14*(5), 1934578X19852433. https://doi.org/10.1177/1934578X19852433

Martínez-Calderas, J.M., Palacio-Núñez, J., Martínez-Montoya, J.F., Olmos-Oropeza, G., Clemente-Sánchez, F., & Sánchez-Rojas, G. (2019). Distribución y abundancia de *Jatropha dioica* en el centro-norte de Mexico. *Agrociencia*, *53*(3), 433–446. Retrieved February 18, 2022. https://www.agrociencia-colpos.mx/index.php/agrociencia/article/view/1794/1791

Perez-Perez, J. U., Guerra-Ramirez, D., Reyes-Trejo, B., Cuevas-Sanchez, J. A., & Guerra-Ramirez, P. (2020a). Actividad antimicrobiana in vitro de extractos de Jatropha dioica Sesee contra bacterias fitopatogenas de tomate. *Polibotánica*, *49*, 125–133. https://doi.org/10.18387/polibotanica.49.8.

Perez-Perez, J. U., Reyes-Trejo, B., Guerra-Ramirez, D., & Cuevas-Sanchez, J. A. (2020b). Seed oil Jatropha dioica Sesee as a biodiesel potencial resource. *Revista bio ciencias*, *7*, e728. https://doi.org/10.15741/revbio.07.e728.

Thomas, R., Sah, N.K., & Sharma, P.B. (2008). Therapeutic biology of *Jatropha curcas*: A mini review. *Current Pharmaceutical Biotechnology*, *9*(4), 315–324. https://doi.org/10.2174/138920108785161505

Heterotheca inuloides

Bilia, A. R., Bergonzi, M. C., Mazzi, G., & Vincieri, F. F. (2006). Development and stability of semisolid preparations based on a supercritical CO2 Arnica extract. *Journal of Pharmaceutical and Biomedical Analysis*, *41*(2), 449–454.

CONANP [Comisión Nacional de Áreas Naturales Protegidas] (2014). Estudio previo justificativo para el establecimiento del área natural protegida de competencia de la Federación con la categoría de Reserva de la Biosfera "Desierto Semiárido de Zacatecas", ubicada en el estado de Zacatecas, *Mexico*, 303 p.

Conforti, A., Bellavite, P., Bertani, S., Chiarotti, F., Menniti-Ippolito, F., & Raschetti, R. (2007). Rat models of acute inflammation: A randomized controlled study on the effects of homeopathic remedies. *BMC Complementary and Alternative Medicine*, *7*(1), 1–10.

Fioranelli, M., Bianchi, M., Roccia, M.G., & DI Nardo, V. (2016). Effects of Arnica comp. Heel on reducing cardiovascular events in patients with stable coronary disease. *Minerva Cardioangiol.*, *64*(1), 34–40.

García-Pérez, J. S., Cuéllar-Bermúdez, S. P., Arévalo-Gallegos, A., Rodríguez-Rodríguez, J., Iqbal, H. M., & Parra-Saldivar, R. (2016). Identification of bioactivity, volatile and fatty acid profile in supercritical fluid extracts of Mexican arnica. *International Journal of Molecular Sciences*, *17*(9), 1528.

González-Stuart, A.E. (2010). Use of medicinal plants in Monterrey, Mexico. *Notulae Scientia Biologicae*, *2*(4), 7–11.

Iannitti, T., Morales-Medina, J.C., Bellavite, P., Rottigni, V., and Palmieri, B. (2016). Effectiveness and safety of *Arnica montana* in post-surgical setting, pain and inflammation. *American Journal of Therapeutics, 23*(1), e184–e197.

Kamatani, J., Iwadate, T., Tajima, R., Kimoto, H., Yamada, Y., Masuoka, N., Kubo, I., & Nihei, K.I. (2014). Stereochemical investigation and total synthesis of inuloidin, a biologically active sesquiterpenoid from Heterotheca inuloides. *Tetrahedron, 70*, 3141–3145.

Klaas, C. A., Wagner, G., Laufer, S., Sosa, S., Della Loggia, R., Bomme, U., ... Merfort, I. (2002). Studies on the anti-inflammatory activity of phytopharmaceuticals prepared from Arnica flowers. *Planta Medica, 68*(5), 385–391.

Kos, O., Lindenmeyer, M.T., Tubaro, A., Sosa, S., & Merfort, I. (2005). New sesquiterpene lactones from Arnica tincture prepared from fresh flower heads of *Arnica montana*. *Planta medica. 71*(11), 1044–1052.

Lannitti, T., Morales-Medina, J. C., Bellavite, P., Rottigni, V., & Palmieri, B. (2016). Effectiveness and safety of Arnica montana in post-surgical setting, pain and inflammation. *American Journal of Therapeutics, 23*(1), e184–e197.

Lyß, G., Knorre, A., Schmidt, T. J., Pahl, H. L., & Merfort, I. (1998). The anti-inflammatory sesquiterpene lactone helenalin inhibits the transcription factor NF-κB by directly targeting p65. *Journal of Biological Chemistry, 273*(50), 33508–33516.

Maldonado, Y. (2004). *Cuantificación de Caladenos e Isocadalenos anti-inflamatorios de árnica (heterotheca inuloides cass) en plantas sometidas a fertilización y cortes sucesivos* (Doctoral dissertation, tesis doctoral. Mexico: Michoacán. Universidad Michoacana De San Nicolas De Hidalgo).

Maldonado–López, Y., Linares–Mazari, E., Bye, R., Delgado, G., and Espinosa–García, F.J. (2008). Mexican Arnica Anti–Inflammatory Action: Plant Age is Correlated with the Concentration of Anti–Inflammatory Sesquiterpenes in the Medicinal Plant Heterotheca inuloides Cass.(Asteraceae) 1. *Economic Botany, 62*(2), 161–170.

McVaugh, R., 1984. *Flora Novo-Galiciana. A Descriptive Account of the Vascular Plants of Western Mexico. Vol. 12. Compositae*. Ann Arbor: University of Michigan.

Monroy-Ortiz, C., & Monroy, R. (2004). Análisis preliminar de la dominancia cultural de las plantas útiles en el estado de Morelos. *Botanical Sciences*, (74), 77–95.

Nesom, G.L. (1990). Taxonomy of Heterotheca sect. Heterotheca (Asteraceae: Astereae) in Mexico, with comments on the taxa of the United States. *Phytologia*.

Olioso, D., Marzotto, M., Bonafini, C., Brizzi, M., & Bellavite, P. (2016). Arnica montana effects on gene expression in a human macrophage cell line. Evaluation by quantitative real-time PCR. *Homeopathy, 105*(2), 131–147.

Rojas, S., Castillejos-Cruz, C., & Solano, E. (2013). Floristic and phytogeographic relations of the xeric scrubland in Tecozautla Valley, Hidalgo, Mexico. *Botanical Sciences, 91*(3), 273–294.

Rodríguez-Chávez, J. L., Coballase-Urrutia, E., Nieto-Camacho, A., & Delgado-Lamas, G. (2015). Antioxidant capacity of "Mexican arnica" Heterotheca inuloides Cass natural products and some derivatives: Their anti-inflammatory evaluation and effect on C. elegans life span. *Oxidative Medicine and Cellular Longevity, 2015*.

Rodríguez-Chávez, J.L., Egas, V., Linares, E., Bye, R., Hernández, T., Espinosa-García, F.J., & Delgado, G. (2017). Mexican Arnica (Heterotheca inuloides Cass. Asteraceae: Astereae): Ethnomedical uses, chemical constituents and biological properties. *Journal of Ethnopharmacology, 195*, 39–63.

Rzedowski, G.C. de, & Rzedowski, J. (2001). *Flora fanerogámica del Valle de Mexico*. (2a ed.). Pátzcuaro, Michoacán, Mexico: Instituto de Ecología y Comisión Nacional para el Conocimiento y Uso de la Biodiversidad.

Semple, J.C. (2008). Cytotaxonomy and cytogeography of the goldenaster genus Heterotheca (Asteraceae: Astereae). *Botany, 86*(8), 886–900.

Villaseñor, R.J.L., & F.J. Espinosa, G. (1998). *Catálogo de malezas de Mexico*. Mexico, D.F: Universidad Nacional Autónoma de Mexico. Consejo Nacional Consultivo Fitosanitario. Fondo de Cultura Económica.

Velázquez, A., Fregoso, A., Bocco, G., & Cortez, G. (2003). The use of landscape approach in Mexican forest indigenous communities to strengthen long-term forest. *Interciencia, 28*(11), 632–638.

Lippia graveolens

Arias, J., Mejía, J., Córdoba, Y., Martínez, J. R., Stashenko, E., & del Valle, J. M. (2020). Optimization of flavonoids extraction from Lippia graveolens and Lippia origanoides chemotypes with ethanol-modified supercritical CO2 after steam distillation. *Industrial Crops and Products, 146*, 112170.

Bautista-Hernández, I., Aguilar, C.N., Martínez-Ávila, G.C.G., Torres-León, C., Ilina, A., Flores-Gallegos, A.C., Kumar Verma, D., & Chávez-González, M.L. (2021). Mexican Oregano (Lippia graveolens Kunth) as source of bioactive compounds: A review. *Molecules, 26*, 5156. https://doi.org/10.3390/molecules26175156

Calvo-Irabién, L.M., Parra-Tabla, V., Acosta-Arriola, V., Escalante-Erosa, F., Díaz-Vera, L., Dziba, G.R., & Peña-Rodríjuez, L.M. (2014). Essential Oils of Mexican Oregano (Lippia graveolens Kunth) Populations along an Edapho-Climatic Gradient. *Chemistry & biodiversity, 11*, 1010–1021.

García-Pérez, E., Francisco, F., Castro-Álvarez, G.-U., Alejandra, J., & García-Lara, S. (2012). Revisión de la producción, composición fitoquímica y propiedades nutracéuticas del orégano mexicano. *Revista mexicana de ciencias agrícolas, 3*(2), 339–353.

Granados Sánchez, D., Martínez Salvador, M., López Ríos, G.F., & Rodríguez Yam, G.A. (2013). Ecología, aprovechamiento y comercialización del orégano (Lippia graveolens HBK) en Mapimí, Durango. *Revista Chapingo. Serie ciencias forestales y del ambiente, 19*(2), 305–322.

Martínez-Natarén, D.A., Parra-Tabla, V., Ferrer-Ortega, M.M., & Calvo-Irabién, L.M. (2013). Genetic diversity and genetic structure in wild populations of Mexican oregano (Lippia graveolens H.B.K.) and its relationship with the chemical composition of the essential oil. *Plant Systematics and Evolution, 300*(3), 535–547. https://doi.org/10.1007/s00606-013-0902-y

Ocampo, V.R.V., Malda, B.G.X., & Suárez, R.G. (2009). Biología reproductiva de orégano mexicano (Lippia graveolens Kunth) en tres condiciones de aprovechamiento. *Agrociencia, 43*(5), 475–482.

Rassem, H., Nour, A., & Yunus, R. (2018). Biological activities of essential oils: A review. *Pacific International Journal, 2*, 63–76.

Rehman, R., Hanif, M., Mushtaq, Z., & Al-Sadi, A. (2016). Biosynthesis of essential oils in aromatic plants: A review. *Food Reviews Internationa, 32*, 117–160.

Soto-Armenta, L.C., Sacramento-Rivero, J.C., Acereto-Escoffié, P.O., Peraza-González, E.E., Reyes-Sosa, C.F., & Rocha-Uribe, J.A. (2017). Extraction yield of essential oil from Lippia graveolens leaves by steam distillation at laboratory and pilot scales. *Har Krishan Bhalla Sons, 20*, 610–621.

Villavicencio-Gutiérrez, E.E., Hernández-Ramos, A., Aguilar-González, C.N., & García-Cuevas X. (2018). Estimación de la biomasa foliar seca de Lippia graveolens Kunth del sureste de Coahuila. *Revista Mexicana de Ciencias Forestales, 9*(45). https://doi.org/10.29298/rmcf.v9i45.139

Lophophora williamsii

Castle, L.M., Leopold, S., Craft, R., & Kindscher, K. (2014). Ranking tool created for medicinal plants at risk of being overharvested in the wild. *Ethnobiology Letters, 5*(1). https://doi.org/10.14237/ebl.5.2014.169

Ermakova, A., Whiting, C.V., Trout, K., Clubbe, C., Terry, M.K., & Fowler, N. (2021). Densities, plant sizes, and spatial distributions of six wild populations of lophophora williamsii (cactaceae) in Texas, U.S.A. *Journal of the Botanical Research Institute of Texas, 15*(1). https://doi.org/10.17348/JBRIT.V15.I1.1057

Feeney, K. (2018). Texas peyote culture. *Cactus and Succulent Journal, 90*(1). https://doi.org/10.2985/015.090.0104

Genet Guzmán, M., & Labate, B. (2019). Reflexiones sobre la expansión y legalidad del campo peyotero en Mexico. *Frontera Norte, 31*. https://doi.org/10.33679/rfn.v1i1.2060

Ibarra-Laclette, E., Zamudio-Hernández, F., Pérez-Torres, C.A., Albert, V.A., Ramírez-Chávez, E., Molina-Torres, J., Fernández-Cortes, A., Calderón-Vázquez, C., Olivares-Romero, J.L., Herrera-Estrella, A., & Herrera-Estrella, L. (2015). De novo sequencing and analysis of Lophophora williamsii transcriptome, and searching for putative genes involved in mescaline biosynthesis. *BMC Genomics, 16*(1). https://doi.org/10.1186/s12864-015-1821-9

Jones, P.N. (2007). The Native American Church, Peyote, and Health: Expanding Consciousness for Healing Purposes. *Contemporary Justice Review, 10*(4). https://doi.org/10.1080/10282580701677477

Klein, M.T., Kalam, M., Trout, K., Fowler, N., & Terry, M. (2015). Mescaline concentrations in three principal tissues of Lophophora Williamsii (Cactaceae): Implications for sustainable harvesting practices. *Haseltonia, 20*. https://doi.org/10.2985/026.020.0107

LeBlanc, R., de Silva, S., & Terry, M. (2021). Analysis of over-the-counter analgesics purported to contain mescaline from the peyote cactus (lophophora williamsii: cactaceae). *Journal of the Botanical Research Institute of Texas, 15*(1). https://doi.org/10.17348/JBRIT.V15.I1.1055

Liu, Q., Gao, T., Liu, W., Liu, Y., Zhao, Y., Liu, Y., Li, W., Ding, K., Ma, F., & Li, C. (2020). Functions of dopamine in plants: A review. In *Plant Signaling and Behavior, 15*(12). https://doi.org/10.1080/15592324.2020.1827782

Naranjo, G., De, O., Alejandra, H., & María, M. (2010). Patrón de distribución espacial y nodricismo del peyote (Lophophora williamsii) en Cuatrociénegas , Mexico. *Cáctaceas y Suculentas Mexicanas, 55*.

Newbold, R., De Silva, S., & Terry, M. (2020). Correlation of mescaline concentrations in Lophophora williamsii (Cactaceae) with rib numbers and diameter of crown (U.S.A.). *Journal of the Botanical Research Institute of Texas, 14*(1). https://doi.org/10.17348/jbrit.v14.i1.901

Terry, M., & Trout, K. (2017). Regulation of peyote (Lophophora williamsii: Cactaceae) in the U.S.A.: A historical victory of religion and politics over science and medicine. *Journal of the Botanical Research Institute of Texas, 11*(1). https://doi.org/10.17348/jbrit.v11.i1.1146

Olneya tesota

Coronel-Arellano, H. & López-González, C.A. (2010). The small mammal community associated with ironwood (*Olneya tesota*). In: Halvorson, W.L., Schwalbe, C.R., & Van Riper, C. (Eds.). *Southwestern desert resources* (pp. 1–14). University of Arizona Press. Retrieved February 18, 2022. https://www.researchgate.net/publication/310477066_THE_SMALL_MAMMAL_COMMUNITY_ASSOCIATED_WITH_IRONWOOD_OLNEYA_TESOTA

Felker, P. (1981). Uses of tree legumes in semiarid regions. *Economic Botany, 35*(2), 174–186. Retrieved February 18, 2022. https://www.osti.gov/servlets/purl/5167206

Guerrero-Cárdenas, I., Álvarez-Cárdenas, S., Gallina, S., Corcuera, P., Ramírez-Orduña, R., & Tovar-Zamora, I. (2018). Variación estacional del contenido nutricional de la dieta del borrego cimarrón (*Ovis canadensis* weemsi), en Baja California Sur, Mexico. *Acta Zoológica Mexicana (nueva serie), 34*, 1–18. https://doi.org/10.21829/azm.2018.3412113

Lagarda-Diaz, I., Guzman-Partida, A.M., & Vazquez-Moreno, L. (2017). Legume lectins: Proteins with diverse applications. *International Journal of Molecular Sciences, 18*(6), 1242. https://doi.org/10.3390/ijms18061242

Marshal, J.P., Bleich, V.C., Andrew, N.G., & Krausman, P.R. (2004). Seasonal forage use by desert mule deer in southeastern California. *The Southwestern Naturalist, 49*(4), 501–505. Retrieved February 18, 2022. https://www.researchgate.net/profile/Jason-Marshal/publication/232667181_Seasonal_forage_use_by_desert_mule_deer_in_southeastern_California/links/55d039f708ae502646aa4aad/Seasonal-forage-use-by-desert-mule-deer-in-southeastern-California.pdf

Maldonado, C.M., Bermúdez, E.M., & Armendáriz, J.L.D. (1995). Caracterización ecológica de las poblaciones naturales de jojoba (*Simmonclsia chinensis* (Link) Schneider) en el estado de Sonora. *Revista Mexicana de Ciencias Forestales, 20*(78), 3–22. Retrieved February 18, 2022. http://cienciasforestales.inifap.gob.mx/index.php/forestales/article/download/1006/2312

Moreno-Salazar, S.F., Verdugo, A.E., López, C.C., Martínez, E.B., Candelas, T.M., & Robles-Zepeda, R.E. (2008). Activity of medicinal plants, used by native populations from Sonora, Mexico, against enteropathogenic bacteria. *Pharmaceutical Biology, 46*(10–11), 732–737. https://doi.org/10.1080/13880200802215800

Rosas-Piñón, Y., Mejía, A., Díaz-Ruiz, G., Aguilar, M.I., Sánchez-Nieto, S., & Rivero-Cruz, J.F. (2012). Ethnobotanical survey and antibacterial activity of plants used in the Altiplane region of Mexico for the treatment of oral cavity infections. *Journal of Ethnopharmacology, 141*(3), 860–865. https://doi.org/10.1016/j.jep.2012.03.020

Sañudo-Torres, R.R., Vázquez-Peñate P., Armenta-López, C., Azpiroz-Rivero, H.S., Campos-Beltrán, C., Ibarra-Ceceña, M.G., & Félix-Herrán, J.A. (2009). Tratamientos pregerminativos en semillas de palo fierro

(*Olneya tesota* A. Gray) y propagación en sustrato de composta de lirio acuático (*Eichhornia crassipes*). *Ra Ximhai: Revista científica de sociedad, cultura y desarrollo sostenible*, *5*(3), 329–334. Retrieved February 18, 2022. http://uaim.edu.mx/webraximhai/Ej-15articulosPDF/07%20Palo_Fierro.pdf

Secretaria del Medio Ambiente y Recursos Naturales (SEMARNAT). (2014). *Plan de Manejo Tipo para la Conservación y Aprovechamiento Sustentable de Olneya tesota, Gray (palo fierro) Manejo Extensivo*. Mexico, D. F. 41 pp. Retrieved February 18, 2022. http://www.semarnat.gob.mx/sites/default/files/documentos/vidasilvestre/planes/pmt_olneya_tesota_2014.pdf

Sharma, A., del Carmen Flores-Vallejo, R., Cardoso-Taketa, A., & Villarreal, M.L. (2017). Antibacterial activities of medicinal plants used in Mexican traditional medicine. *Journal of Ethnopharmacology*, *208*, 264–329. https://doi.org/10.1016/j.jep.2016.04.045

Suzán, H., Nabhan, G.P., & Patten, D.T. (1996). The importance of *Olneya tesota* as a nurse plant in the Sonoran Desert. *Journal of Vegetation Science*, *7*(5), 635–644. https://doi.org/10.2307/3236375

Suzán, H., Patten, D.T., & Nabhan, G.P. (1997). Exploitation and conservation of ironwood (*Olneya tesota*) in the Sonoran Desert. *Ecological Applications*, *7*(3), 948–957. https://doi.org/10.1890/1051-0761(1997)007[0948:EACOIO]2.0.CO;2

Toyes-Vargas, E.A., Murillo-Amador, B., Espinoza-Villavicencio, J.L., Carreón-Palau, L., & Palacios-Espinosa, A. (2013). Composición química y precursores de ácidos vaccénico y ruménico en especies forrajeras en Baja California Sur, Mexico. *Revista mexicana de ciencias pecuarias*, *4*(3), 373–386. Retrieved February 18, 2022. http://www.scielo.org.mx/pdf/rmcp/v4n3/v4n3a8.pdf

Urbano-Hernández, G., Guzmán-Partida, A.M., del Carmen Candia-Plata, M., López-Cervantes, G., & Vázquez-Moreno, L. (2015). Use of the *Olneya tesota* lectin (PF2) for purification and histochemical detection of a spleen heat shock protein. *Biotecnia*, *17*(1), 3–9. Retrieved February 18, 2022. https://biotecnia.unison.mx/index.php/biotecnia/article/download/5/4

Waizel-Bucay, J., & Martinez-Rico, I. (2011). Algunas plantas usadas en Mexico en padecimientos periodontales. *Revista aDM*, *58*(2), 73–88. Retrieved February 18, 2022. https://www.researchgate.net/profile/Jose-Waizel-2/publication/289131068_Plantas_empleadas_en_odontalgias/links/5689c19408ae1e63f1f90353/Plantas-empleadas-en-odontalgias.pdf

Zuñiga-Tovar, B., & Suzán-Azpiri, H. (2010). Comparative population analysis of desert ironwood (*Olneya tesota*) in the Sonoran Desert. *Journal of Arid Environments*, *74*(2), 173–178. https://doi.org/10.1016/j.jaridenv.2009.08.004

Opuntia ficus-indica

Chen, M., & Blankenship, R.E. (2021). Photosynthesis: Photosynthesis. In *Encyclopedia of Biological Chemistry* (3rd ed., Vol. 2). https://doi.org/10.1016/B978-0-12-819460-7.00081-5

Departamento de Botánica, Instituto de Biología (IBUNAM), *Opuntia ficus-indica* (L.) Mill., ejemplar de: Herbario Nacional de Mexico (MEXU), Plantas Vasculares. En *Portal de Datos Abiertos UNAM (en línea)*. Mexico: Universidad Nacional Autónoma de Mexico. Disponible en: http://datosabiertos.unam.mx/IBUNAM:MEXU:1183333. Fecha de actualización: 24/11/2019, 9:17:25 p.m. Fecha de consulta: 23/02/2022, 1:17:03 p.m.

Ervin, G.N. (2012). Indian fig cactus (opuntia ficus-indica (L.) miller) in the Americas: An uncertain history. *Haseltonia* (17). https://doi.org/10.2985/1070-0048-17.1.9

Estrada-Castillón, E., Villarreal-Quintanilla, J.Á., Encina-Domínguez, J.A., Jurado-Ybarra, E., Cuéllar-Rodríguez, L.G., Garza-Zambrano, P., ..., & Gutiérrez-Santillán, T.V. (2021). Ethnobotanical biocultural diversity by rural communities in the Cuatrociénegas Valley, Coahuila; Mexico. *Journal of Ethnobiology and Ethnomedicine*, *17*(1), 1–22.

Kolniak-Ostek, J., Kita, A., Miedzianka, J., Andreu-Coll, L., Legua, P., & Hernandez, F. (2020). Characterization of bioactive compounds of opuntia ficus-indica (L.) mill. seeds from spanish cultivars. *Molecules*, *25*(23). https://doi.org/10.3390/molecules25235734

Kumar, S., Palsaniya, D. R., Kumar, T. K., Misra, A. K., Ahmad, S., Rai, A. K., ... Bhargavi, H. A. (2022). Survival, morphological variability, and performance of Opuntia ficus-indica in a semi-arid region of India. *Archives of Agronomy and Soil Science*, 1–18.

López-Romero, P., Pichardo-Ontiveros, E., Avila-Nava, A., Vázquez-Manjarrez, N., Tovar, A.R., Pedraza-Chaverri, J., & Torres, N. (2014). The effect of nopal (Opuntia Ficus Indica) on postprandial blood glucose, incretins, and antioxidant activity in Mexican patients with type 2 diabetes after consumption of two different composition breakfasts. *Journal of the Academy of Nutrition and Dietetics, 114*(11). https://doi.org/10.1016/j.jand.2014.06.352

Msaddak, L., Abdelhedi, O., Kridene, A., Rateb, M., Belbahri, L., Ammar, E., Nasri, M., & Zouari, N. (2017). Opuntia ficus-indica cladodes as a functional ingredient: Bioactive compounds profile and their effect on antioxidant quality of bread. *Lipids in Health and Disease, 16*(1). https://doi.org/10.1186/s12944-016-0397-y

Porras Flórez, D., Albesiano, S., & Arrieta Violet, L. (2018). El género Opuntia (Opuntioideae–Cactaceae) en el departamento de Santander, Colombia. *Biota Colombiana, 18*(2). https://doi.org/10.21068/c2017.v18n02a07

Ramírez-Rodríguez, Y., Martínez-Huélamo, M., Pedraza-Chaverri, J., Ramírez, V., Martínez-Tagüeña, N., & Trujillo, J. (2020). Ethnobotanical, nutritional and medicinal properties of Mexican drylands Cactaceae Fruits: Recent findings and research opportunities. *Food Chemistry, 312*. https://doi.org/10.1016/j.foodchem.2019.126073

Reis, C.M.G., Gazarini, L.C., Fonseca, T.F., & Ribeiro, M.M. (2018). Above-ground biomass estimation of opuntia ficus-indica (L.) mill. for forage crop in a mediterranean environment by using non-destructive methods. *Experimental Agriculture, 54*(2). https://doi.org/10.1017/S0014479716000211

Reyes-Agüero, J.A., Aguirre-Rivera, J.R., & Hernández, H.M. (2005). Notas sistemáticas y una descripción detallada de Opuntia ficus-indica (L.) mill. (Cactaceae). *Agrociencia, 39*(4).

Reyes-Agüero, J.A., Aguirre, J.R., & Valiente-Banuet, A. (2006). Reproductive biology of Opuntia: A review. *Journal of Arid Environments, 64*(4). https://doi.org/10.1016/j.jaridenv.2005.06.018

Scalisi, A., Morandi, B., Inglese, P., & Lo Bianco, R. (2016). Cladode growth dynamics in Opuntia ficus-indica under drought. *Environmental and Experimental Botany, 122*. https://doi.org/10.1016/j.envexpbot.2015.10.003

Shackleton, S., Kirby, D., & Gambiza, J. (2011). Invasive plants - Friends or foes? contribution of prickly pear (Opuntia ficus-indica) to livelihoods in Makana Municipality, Eastern Cape, South Africa. *Development Southern Africa, 28*(2). https://doi.org/10.1080/0376835X.2011.570065

Silva, M.A., Albuquerque, T.G., Pereira, P., Ramalho, R., Vicente, F., Oliveira, M.B.P.P., & Costa, H.S. (2021). Opuntia ficus-indica (L.) mill.: A multi-benefit potential to be exploited. *Molecules, 26*(4). https://doi.org/10.3390/molecules26040951

Parthenium incanum

Allred Kelly, W. & Jercinovic Eugene, M. (2020). *Flora Neomexicana III: An Illustrated Identification Manual, Part 1* (2nd ed.). New Mexico: Allred, Kelly W. Las Cruces. https://floraneomexicana.org/flora-neomexicana-series/

Ebadi, M. (2006). *Pharmacodynamic Basis of Herbal Medicine*, Chapter 25 (2nd ed., p. 285). Boca Raton: CRC Press.

Gonzalez, F. M. (1998). *Plantas Medicinales del Noreste de Mexico*. Monterrey, NL, México: Editorial Grupo Vitro.

Gonzalez, L. C. R. (1988). *Estudio Preliminar del Uso y Aprovechamiento de especies Vegetales en los Municipios de R. Arizpe y Parras, Coahuila, Mexico*. Tesis de licenciatura Universidad Autónoma de Nuevo León, Monterrey, NL.

Kane, C.W. (2006). *Herbal Medicine of the American Southwest: A Guide to the Identification, Collection, Preparation and Use of Medicinal and Edible Plants of the Southwestern United States* (pp. 163–164). Tucson: Lincoln Town Press.

Kumar, S. Mishra, A., & Pandey, A.K. (2013). Antioxidant mediated protective effect of *Parthenium hysterophorus* against oxidative damage using *in vitro* models. *BMC complementary and alternative medicin, 13*(120), 1–9.

McVaugh, R. (1984). *Compositae. Flora Novo-Galiciana. A Descriptive Account of the Vascular Plants of Western Mexico* (Vol. 12). Ann Arbor: The University of Michigan Press.

Shah, W.A. (2014). Phytochemical review profile of sesquiterpene lactone parthenin. *International Journal Research in Pharmacy and Chemistry, 4*(2), 217–221.
Siddhardha, B., Ramakrishna, G., & Basaveswara Rao, M.V. (2012). In vitro antibacterial efficacy of a sesquiterpene lactone, parthenin from *Parthenium hysterophorus* L (Compositae) against enteric bacterial pathogens. *International Journal of Pharmaceutical, Chemical and Biological Sciences*, *2*(3), 206–209.
Villalobos, C. (2007). Estimating aboveground biomass of *Mariola* (*Parthenium incanum*) from plant dimensions. In Sosebee, R.E., Wester, D.B., Britton, C.M., McArthur, E.D., Kitchen, S.G. (eds.), Proceedings: Shrubland Dynamics—Fire and Water, Proceedings RMRS-P-47 (132–135). Lubbock, TX. Fort Collins, CO: U.S. Department of Agriculture, Forest Service, Rocky Mountain Research Station.
Villaseñor R., J.L. y F.J. Espinosa G. (1998). Catálogo de malezas de Mexico. Mexico, D.F: Universidad Nacional Autónoma de Mexico. Consejo Nacional Consultivo Fitosanitario. Fondo de Cultura Económica.

Pinus cembroides

Bailey, D.K., & Hawksworth, F.G. (1992). Change in status of *Pinus cembroides* Subsp. *orizabensis* (Pinaceae) from central Mexico. *Novon, 2*, 306–307.
Farjon, A., & Styles, B. (1997). *Pinus (Pinaceae). Flora Neotropica Monograph 75* (291 p) Nueva York: The New York Botanical Garden.
Flores Flores, J.D., & Diaz Esquivel, E. (1989). Factores asociados con la variacion anual en la produccion de cones y semillas en Pinus cembroides Zucc. en Saltillo, Coahuila. In J.D. Flores F. et al. (Comps.), *Memorias del III Simposio Nacional sobre Pinos Piñoneros* (pp. 136–144). Saltillo, Coah.: UAAAN-CIFAP-SARH-INIFAP-SARH.
Fonseca J.R.M. (2003). De piñas y piñones. *Ciencias (Universidad Autónoma de Mexico), 69*, 64–65.
Hernández, M.M., Islas, G.J., Guerra de la, C.V. (2011). Márgenes de comercialización del piñón (*Pinus cembroides* subsp. *orizabensis*) en Tlaxcala, Mexico. *Revista Mexicana de Ciencias Agrícolas., 2*, 265–279.
López Mata, L. (2001). Proteins, amino acids and fatty acids composition of nuts from the mexican endemic rarity, *Pinus maximartinezii*, and its conservation implications. *Interciencia, 26*(12). Recuperado de http://www.redalyc.org/ resumen.oa?id=33906305
Malusa, J. (1992). Phylogeny and biogeography of the pinyon pine (Pinussubsect. cembroides). *Systematic Botany, 17*(1), 42–66.
Menniger, E. A. (1977). *Edible Nuts of the World*. Stuart, FL: Horticultural Books (pp 153–157).
Mohedano-Caballero, L., Cetina, A.V.M., Vera, C.G., & Ferrera, C.R. (1999). Micorrización y poda aérea en la calidad de planta de pino piñonero en invernadero. *Revista Chapingo Serie Ciencias Forestales y del Ambiente, 5*, 141–148.
Passini, M.F. (1982). *Les fôrets de Pinus cembroides au Mexique. Etudes Mesoamericaines 11–5* (373 p). Paris: Ed. Recherche sur les Civilisations.
Passini, M. F. (1983). Un exemple de forêt tropicale sèche du Mexique: la forêt de Pinus cembroides Zucc. *Bulletin de la Société Botanique de France. Lettres Botaniques, 130*(1), 69–80.
Passini, M.F. (1996). Les pins mexicains de la sous-section cembroides Engelm.: distribution, cycle et phenologie, pollen. In Guillaume, J.L. et al. (ed.), *Phytogeographie tropicale, Realites et perspectives* (pp. 243–249). Paris: ORSTOM.
Romero Manzanares, A., García Moya, E., & Passini, M. F. (1996). Pinus cembroides sl y Pinus johannis del Altiplano Mexicano: una síntesis. *Acta botanica gallica, 143*(7), 681–693.
Rosengarten F (1984). *The Book of Edible Nuts* (pp. 309–315). New York: Walker and Company.
Sagrero-Nieves, L. (1992). Fatty acid composition of mexican pine nut (*Pinus cembroides*) oil from three seed coat phenotypes. *Journal of the Science of Food and Agriculture, 59*(3), 413–414.
Sen, F., Ozer, K.B., & Aksoy, U. (2016). Physical and dietary properties of in-shell pine nuts (Pinus pinea L.) and kernels. *American Journal of Experimental Agriculture, 10*(6), 1–9.
Wolf, F. (1985). Algunas propiedades de la madera de P. cembroides Zucc. In J. E. L. Flores (Ed.), *Memorias del 1er Simposio Nacional Sobre Pinos Piñoneros* (pp 69–82). Monterrey, NL: Universidad Autónoma de Nuevo León.
Wolff, R.L., & Bayard, C.C. (1995). Fatty acid composition of some pine seed oils. *Journal of American Oil Chemist Society, 72*(9), 1043–1046.

Prosopis spp.

Ahmad, A., Khan, K. A., Ahmad, V. U., & Qazi, S. (1986). Antibacterial activity of juliflorine isolated from Prosopis juliflora. *Planta Medica, 52*(4), 285–288.

Ahmad, V.U., & Mahmood Z.G. (1979). Studies on the structure of juliflorine. *J Chem Soc Pak, 1*, 137–138.

Aqeel, A., Khursheed, A.K., Viqaruddin, A., & Sabiha, Q. (1989). Antimicrobial activity of julifloricine isolated from *Prosopis juliflora. Arzneimittelforschung, 39*, 652–655.

Cowan, M.M. (1999). Plant products as antimicrobial agents. *Clinical microbiology reviews, 12*, 564–582.

European and Mediterranean Plant Protection Organization. (2019). *Prosopis juliflora* (Sw.) DC. *Bulletin OEPP EPPO Bulletin, 49*(2), 290–297.

Harzallah-Shhiri, F., & Jannet, H.B. (2005). Flavonoid diversification in organs of two *Prosopis farcta* (Banks and sol) (Leguminosae, Mimosoidae) populations occurring in the northeast and the southeast of Tunisia. *Journal of Applied Sciences Research, 1*, 130–136.

Khursheed, A.K., Arshad, H.F., Viqaruddin, A., Sabiha, Q., Sheikh, A.R., & Tahir, S.H. (1986). *In vitro* studies of antidermatophytic activity of juliflorine and its screening as carcinogen in Salmonella/ microsome test system. *Arzneimittelforschung, 36*, 17–19.

Leakey, R.R.B., & Last, F.T. (1980). Biology and potential of *Prosopis* species in arid environments, with particular reference to *P. cineraria. Journal of Arid Environments, 3*(1), 9–24. https://doi.org/10.1016/S0140-1963(18)31672-0.

Ott-Longoni, R., Viswanathan, N., & Hesse, M. (1980). The structure of the alkaloids juliprosopine from *Prosopis* juliflora A. DC. *Helvetica Chimica Acta, 63*, 2119–2129.

Patel, R.S., Patel, H.R., & Kanjariya, K.V. (2018). Observation on leguminous plants with their taxonomy and medicinally uses of Ahmedabad zoo, Gujarat, India. *International Journal of Scientific Research in Science and Technology, 4*(5), 590–602. http://ijsrst.com/paper/2807.pdf

Sirmah, P.K. (2000). *Towards Valorisation of Prosopis Juliflora as an Alternative to the Declining Wood Resources in Kenya* [PhD Dissertation. Nancy, France: Universite Henri Poincare].

Sirmah, P., Dumarçay, S., Masson, E., & Gerardin, P. (2009). Unusual amount of (-)-mesquitol from the heartwood of *Prosopis juliflora. Natural product research, 23*, 183–189.

Tapia, A., Egly Feresin, G., Bustos, D., Astudillo, L., Theoduloz, C., Schmeda-Hirschmann, G. (2000). Biologically active alkaloids and a free radical scavenger from Prosopis species. *Journal of ethnopharmacology, 71*, 241–246.

Quercus spp.

Aldrich, P.R., & Cavender-Bares, J. (2011). Quercus. In: Kole, C. (ed.), *Wild Crop Relatives: Genomic and Breeding Resources, Forest Trees* (pp. 89–129). Berlin, Germany: Springer-Verlag.

Custodio, L., Patarra, J., Alberício, F., Neng, d.R.N., Nogueira, M.F., Romano, A. (2015). Phenolic composition, antioxidant potential and in vitro inhibitory activity of leaves and acorns of Quercus suber on key enzymes relevant for hyperglycemia and Alzheimer's disease. *Industrial Crops and Products, 64*, 45–51.

Fried, M.A., & Sulla-Menashe, D. (2019). *MCD12Q1 MODIS/Terra + Aqua Land cover Type Yearly L3 Global 500 m SIN Grid V006. NASA EOSDIS Land Processes DAAC*. https://doi.org/10.5067/MODIS/MCD12 Q1.006

Galicia, L., Potvin, C., & Messier, C. (2015). Maintaining the high diversity of pine and oak species in Mexican temperate forests: A new management approach combining functional zoning and ecosystem adaptability. *Canadian Journal of Forest Research, 45*(10), 1358–1368. https://doi.org/10.1139/cjfr-2014-0561

García-Gómez, E., Pérez-Badia, R., Pereira, J., and Puri, R.K. (2017). The consumption of acorns (from *quercus* spp.) in the central west of the Iberian Peninsula in the 20th century. *Economic Botany, 71*(3), 256–268.

Gul, F., Khan, K.M., Adhikari, A., Zafar, S., Akram, M., Khan, H., & Saeed, M. (2017). Antimicrobial and antioxidant activities of a new metabolite from Quercus incana. *Natural Product Research, 31*(16), 1901–1909.

Kremer, A., & Hipp, A.L. (2019). Oaks: An evolutionary success story. *New Phytologist, 31*, 630–400.

Nixon, K.C. (2010). Global and geotropically distribution and diversity of oak (quercus) and oak forests. *Ecology and Conservation of Montane Oak Forests, 3*, 13–21.

Scareli-Santos, C., Sánchez-Mondragón, M.L., González-Rodríguez, A., & Oyama, K. (2013). Foliar micromorphology of Mexican oaks (Quercus: Fagacea). *Acta botánica mexicana, 104*, 31–52.

Taib, M., Rezzak, Y., Bouyazza, L., & Lyoussi, B. (2020). Medicinal uses, phytochemistry, and pharmocological activities of Quercus species. *Evidence-Based Complementary and Alternative Medicine*, 1920683. https://doi.org/10.1155/2020/1920683

Uddin, G., & Rauf, A. (2012). Phytochemical screeming, antimicrobial and antioxidant activities of aereal parts of Quercus robur L. *Middle-East Journal of Medicinal Plants Reseach*, *1*(1), 1–4.

Vihna, A.F., Costa, S.G., Barreira, C.M., Pacheco, R., & Oliveira, B.P. (2016). Chemical and antioxidant profiles of acorn tissues from *Quercus* spp.: Potential as new industria raw materials. *Industrial Crops and Products*, *94*, 143–151.

Selaginella spp.

Adnan, M., Siddiqui, A.J., Arshad, J., Hamadou, W.S., Awadelkareem, A.M., Sachidanandan, M., & Patel, M. (2021). Evidence-based medicinal potential and possible role of selaginella in the prevention of modern chronic diseases: Ethnopharmacological and ethnobotanical perspective. *Records of Natural Products*, *15*(5), 355.

Abdala, S., Martín-Herrera, D., Benjumea, D., & Pérez-Paz, P. (2008). Diuretic activity of Smilax canariensis, an endemic Canary Island species. *Journal of Ethnopharmacology*, *119*, 12–16.

Aguilar, M.I., Benítez, W.V., Colín, A., Bye, R., Ríos-Gómez, R., Calzada, F. (2015). Evaluation of the diuretic activity in two Mexican medicinal species: *Selaginella nothohybrida* and *S. lepidophylla* and its effects with ciclooxigenases inhibitors, *Journal of Ethnopharmacology*, *163*(2), 167–172.

Agra, M.F., Silva, K.N., Basílio, I.J.L.D., França de Freitas, P., Barbosa-Filho, J.M. (2008). Survey of medicinal plants used in the region of Northeast of Brazil. *Revista brasileira de farmacognosia*, *18*, 472–508.

Aguilar, M.I., Romero, M.G., Chávez, M.I., King–Díaz, B., Lotina-Hennsen, B. (2008). *Journal of Agricultural and Food Chemistry*, *56*, 6994–7000.

Aguilar, M.I., Mejía, I.A., Menchaca, C., Vázquez I., Navarrete, A., Chávez, M.I., Reyes-García, A., & Ríos-Gómez, R. (2013). Determination of biflavonoids in four Mexican medicinal species of selaginella by HPLC. *Journal of AOAC International*, *96*, 712–716.

Almeida, J.R.G., Sá, P.G.S., Macedo, L.A.R., Filho, J.A., Oliveira, V.R., Filho, J.M.B. (2013). Phytochemistry of the genus Selaginella (Selaginellaceae), *Journal of Medicinal Plants Research*, *7*, 1858–1868.

Argueta, A., Cano, L.M., & Rodarte E. (1994). *Atlas de las Plantas de la Medicina Tradicional Mexicana*. Mexico: Instituto Nacional Indigenista.

Hirai, R.Y., & Prado, J. (2000). Selaginellaceae Willk. no Estado de São Paulo, Brasil. *Brazilian Journal of Botany*, *23*, 313–339.

Judd, W.S., Campbell, C.S., Kellog, E.A., Stevens, P.F. (1999). Plant systematics: A phylogenetic approach. Sunderland: Sinauer Associates. https://www.nhbs.com/plant-systematics-book-5

Korall, P., & Kenrick, P. (2002). Phylogenetic relationships in Selaginellaceae based on rbcL sequences. *American Journal of Botany*, *89*, 506–517.

Korall, P., Kenrick, P., & Therrien, J.P. (1999). Phylogeny of Selaginellaceae: Evaluation of generic/subgeneric relationships based on rbcL gene sequences. *International Journal of Plant Science*, *160*, 585–594.

Márquez, A., Lara, F., Esquivel, B., Mata, R. (1999). *Plantas Medicinales de Mexico II. Composición, Usos y Actividad Biológica*. Mexico: UNAM.

Martínez, M. (1989). *Las plantas Medicinales de Mexico* (6th ed.). Mexico: Ediciones Botas.

Mickel, J. T., & Valdespino, I. A. (1992). Five new species of pteridophytes from Oaxaca, Mexico. *Brittonia*, *44*(3), 312–321. Springer.

Mukhopadhyay, R. (2001). A review of the work on the genus *Selaginella* P. Beauv, *Indian Fern Journal*, *18*, 1–44.

Quasim, M.A., Roy, S.K., Kamil, M., Llyas, M., (1985). Phenolic constituents of Selaginellaceae. *Indian Journal of Chemistry*, *24 B*, 220.

Queiroz, D.P., Carollo, C.A., Kadri, M.C., Rizk, Y.S., Araujo, V.C., Monteiro, P.E., Rodrigues, P.O., Oshiro, E.T., Matos Mde, F., Arruda, C.C. (2016). In vivo antileishmanial activity and chemical profile of polar extract from *Selaginella sellowii*. *Memórias do Instituto Oswaldo Cruz*, *111*(3), 147–154.

Richards, A.B., Krakowa, S., Dexter, L.B., Schmid, H., Wolterbeek, A.P.M., (2001). Trehalose: A review of properties, history of use and human tolerant and results of multiple safety studies. *Food and Chemical Toxicology*, *40*, 871–898.

Sá, P.G.S., Nunes, X.P., Lima, J.T., Siqueira-Filho, J.A., Fontana, A.P., Siqueira, J.S.,. Quintans- Júnior, L.J., Damasceno, P.K.F., Branco, C.R.C., Branco, A., & Almeida, J.R.G.S. (2012). Antinociceptive effect of ethanolic extract of Selaginella convoluta in mice. *BMC Complementary and Alternative Medicine, 12*, 187.

Singh, S., Singh, R. (2015). A review on endemic Indian resurrecting Herb *Selaginella bryopteris* (L.) Bak 'Sanjeevani', *International Journal of Pharmaceutical Sciences and Research, 6*, 50–56.

Tryon, R.M., & Tryon A.F. (1982). *Ferns and Allied Plants*. New York: Springer-Verlag.

Vázquez-Ramírez, M. D. L. Á., Meléndez-Camargo, M. E., & Arreguín Sánchez, M. D. L. L. (2005). Estudio etnobotánico de selaginella lepidophylla (hook. Et grev.) spring (selaginellaceae-pteridophyta) en San José Xicohténcatlmunicipio de Huamantla, *Tlaxcala, México. Polibotánica, 19*, 105–115.

Zhao, P., Chen, K., Zhang, G., Deng, G., & Li, J. (2017). Pharmacological basis for use of *Selaginella moellendorffii* in gouty arthritis: Antihyperuricemic, anti-inflammatory, and xanthine oxidase inhibition. *Evidence-Based Complementary and Alternative Medicine, 2017*, 2103254.

Simmondsia chinensis

Abdel-Wahhab, M.A., Joubert, O., El-Nekeety, A.A., Sharaf, H.A., Abu-Salem, F.M., & Rihn, B.H. (2016). Dietary incorporation of jojoba extract eliminates oxidative damage in livers of rats fed fumonisin-contaminated diet. *Hepatoma Research, 2*, 78–86. https://doi.org/10.4103/2394-5079.168078

Agarwal, S., Arya, D., & Khan, S. (2018). Comparative fatty acid and trace elemental analysis identified the best raw material of jojoba (*Simmondsia chinensis*) for commercial applications. *Annals of Agricultural Sciences, 63*(1), 37–45. https://doi.org/10.1016/j.aoas.2018.04.003

Al-Widyan, M.I., & Al-Muhtaseb, M.A. (2010). Experimental investigation of jojoba as a renewable energy source. *Energy Conversion and Management, 51*(8), 1702–1707. https://doi.org/10.1016/j.enconman.2009.11.043

Benzioni, A. (2010). Jojoba Domestication and Commercialization in Israel. *Horticultural Reviews*, 233–266. Retrieved February 18, 2022. https://www.researchgate.net/profile/Aliza-Benzioni/publication/229694602_Jojoba_Domestication_and_Commercialization_in_Israel/links/55118b990cf270fd7e3007c6/Jojoba-Domestication-and-Commercialization-in-Israel

Di Berardino, L., Di Berardino, F., Castelli, A., & Della Torre, F. (2006). A case of contact dermatitis from jojoba. *Contact Dermatitis, 55*(1), 57–58. https://doi.org/10.1111/j.0105-1873.2006.0847e.x

El-Mallah, M.H., & El-Shami, S.M. (2009). Investigation of liquid wax components of Egyptian jojoba seeds. *Journal of Oleo Science, 58*(11), 543–548. https://doi.org/10.5650/jos.58.543

Hassanein, A.M., Galal, E., Soltan, D., Abed-Elsaboor, K., Saad, G., M. Gaboor, G., & S. El Mogy, N. (2012). Germination of jojoba (Simmondsia chinensis L.) seeds under the influence of several conditions. *Journal of Environmental Studies, 9*, 29–35. https://doi.org/10.21608/jesj.2012.191446

Kara, Y. (2017). Phenolic contents and antioxidant activity of jojoba (*Simmondsia chinensis* (Link). *Schindler. International Journal of Secondary Metabolite, 4*(2), 142–147. https://doi.org/10.21448/ijsm.309538

Maldonado, C.M., Bermúdez, E.M., & Armendáriz, J.L.D. (1995). Caracterización ecológica de las poblaciones naturales de jojoba (*Simmonclsia chinensis* (Link) Schneider) en el estado de Sonora. *Revista Mexicana de Ciencias Forestales, 20*(78), 3–22. Retrieved February 18, 2022. http://cienciasforestales.inifap.gob.mx/index.php/forestales/article/download/1006/2312

Mokhtari, C., Malek, F., Manseri, A., Caillol, S., & Negrell, C. (2019). Reactive jojoba and castor oils-based cyclic carbonates for biobased polyhydroxyurethanes. *European Polymer Journal, 113*, 18–28. https://doi.org/10.1016/j.eurpolymj.2019.01.039

Reddy, P.M., & Chikara, J. (2010). Biotechnology advances in jojoba (*Simmondsia chinensis*). In Ramawat G.K. (ed.), *Desert Plants: Biology and Biotechnology* (pp. 407–421). Heidelberg, Germany: Springer. https://doi.org/10.1007/978-3-642-02550-1_19

Tietel, Z., Kahremany, S., Cohen, G., & Ogen-Shtern, N. (2021). Medicinal properties of jojoba (*Simmondsia chinensis*). *Israel Journal of Plant Sciences, 68*(1–2), 38–47. https://doi.org/10.1163/22238980-bja10023

Wagdy, S.M., & Taha, F.S. (2012). Primary assessment of the biological activity of jojoba hull extracts. *Life Science Journal, 9*(2), 244–253. Retrieved February 18, 2022. http://www.lifesciencesite.com/lsj/life0902/039_8503life0902_244_253.pdf

Taxodium mucronatun

Correa-Díaz, A., Cómez-Guerrero, A., Villanueva-Díaz, J., Castruita-Esparza, L.U., Martínez-Trinidad, T., & Cervantes-Martínez, R. (2014). Análisis dendroclimático de ahuehuete (Taxodium mucronatum Ten.) en el centro de Mexico. *Agrociencia*, *48*, 537–551.

Cortés-Arroyo, A.R., Domínguez-Ramírez, A.M., Gómez-Hernández, M., Medina López, J.R., Hurtado y de la Peña, M., López-Muñoz, F.J. (2011). Antispasmodic and bronchodilator activities of Taxodium mucronatum Ten leaf extract (en línea). *African Journal of Biotechnology*, *10*(1), 54–64. Consultado 22 jun. 2020. Disponible en https://www.researchgate.net/publication/266884463_Antispasmodic_and_bronchodilator_activities_of_Taxodium_mucronatum_Ten_leaf_extrac

Cortés Barrera, E. N., Villanueva Díaz, J., Estrada Ávalos, J., Pola, C. N. D. P., Cruz, V. G. D. L., & Vázquez Cuecuecha, O. (2010). Utilización de Taxodium mucronatum Ten para determinar la variación estacional de la precipitación en Guanajuato. *Revista mexicana de ciencias forestales*, *1*(1), 113–121. http://www.scielo .org.mx/pdf/remcf/ v1n1/v1n1a 13.pdf. Accessed 10 July 2019.

Enríquez-Peña, E.G., & Suzán-Azpiri, H. (2011). Estructura poblacional de Taxodium mucronatum en condiciones contrastantes de perturbación en el estado de Querétaro, Mexico (en línea). *Revista Mexicana de Biodiversidad*, *82*(1), 153–167.

Luján-Hidalgo, M.C., Gutiérrez-Miceli, F.A., Ventura-Canseco, L.M.C., Dendooven, L., Mendoza-López, M.R., Cruz-Sánchez, S., García-Barradas, O., Abud-Archila, M. (2012). Composición química y actividad antimicrobiana de los aceites esenciales de hojas de Bursera graveolens y Taxodium mucronatum de Chiapas, Mexico (en línea). *Gayana Botanica*, *69*, 7–14. Consultado 22 jun. 2020. Disponible en http://www2.udec.cl/~gvalencia/pdf/GB2012_69_ne_Lujan-Hidalgo_etal.pdf

Martínez-Sifuentes, A.R., Villanueva-Díaz, J., Crisantos de la Rosa, E., & Stahle, D.W. (2021). Modelado actual y futuro de la idoneidad de hábitat el ahuehuete (Taxodium mucronatum Ten.): Una propuesta para conservación en Mexico. *Botanical Sciences*, *99*(4), 752–770. Epub 18 de octubre de 2021.https://doi.org/10.17129/botsci.2772

Osorio-Osorio, J.A., Astudillo-Sánchez, C.C., Villanueva-Díaz, J., Soria-Díaz, L., & Vargas-Tristan, L.V. (2020). Reconstrucción histórica de la precipitación en la Reserva de la Biosfera El Cielo, Mexico, mediante anillos de crecimiento en Taxodium mucronatum (Cupressaceae). *Revista de Biología Tropical*, *68*(3), 818–832.

Stahle, D.W., Edmondson, J.R., Howard, I.M., Robbins, C.R., Griffin, R.D., Carl, A., & Torberson, M.C.A (2019). Longevity, climate sensitivity, and conservation status of wetland trees at Black River, North Carolina. *Environmental Research Communications*, *1*, 041002. https://doi.org/10.1088/2515-7620/abOc4a

Suzán-Azpiri, H., Enríquez-Peña, G., & Malda-Barrera, G. (2007). Population structure of the Mexican baldcypress (Taxodium mucronatum Ten.) in Queretaro, Mexico (en línea). *Forest Ecology and Management*, *242*, 243–249. Consultado 22 jun. 2020. Disponible en https://www.academia.edu/8433817/Population_structure_of_the_Mexican_baldcypress_Taxodium_mucronatum_Ten_in_Queretaro_Mexico

Villanueva-Díaz, J., Cerano-Paredes, J., Estrada-Ávalos, J., Constante-García, V., & Stahle, D.W. (2014). Reconstrucción de escurrimiento histórico de la cuenca alta del Río Nazas, Durango. *Revista Chapingo: Serie Zonas Áridas*, *13*, 21–25. https://doi.org/10.5154/r.rchsza.2012.06.036

Villanueva-Díaz, J., Stahle, D.W., Therrell, M.D., Beramendi-Orosco, L., Estrada-Avalos, J., Martinez-Sifuentes, A.R., ... & Cerano-Paredes, J. (2020). The climatic response of baldcypress (Taxodium mucronatum Ten.) in San Luis Potosi, *Mexico Trees*, *34*, 623–635. https://doi.org/10.1007/s00468-019-01944-0

Tecoma stans

Abere, T. A., & Enoghama, C. O. (2015). Pharmacognostic standardization and insecticidal activity of the leaves of Tecoma stans Juss (Bignoniaceae). *Journal of Science and Practice of Pharmacy*, *2*(1), 39–45.

Alonso-Castro, A.J., Zapata-Bustos, R., Romo-Yañez, J., Camarillo-Ledesma, P., Gómez-Sánchez, M., & Salazar-Olivo, L.A. (2010). The antidiabetic plants *Tecoma stans* (L.) Juss. ex Kunth (Bignoniaceae) and Teucrium cubenseJacq (Lamiaceae) induce the incorporation of glucose in insulin-sensitive and insulin-resistant murine and human adipocytes. *Journal of Ethnopharmacology*, *127*(1), 1–6. https://doi.org/10.1016/j.jep.2009.09.060

Anburaj, G., Marimuthu, M., Rajasudha, V., & Manikandan, R. (2016). Phytochemical screening and GC-MS analysis of ethanolic extract of *Tecoma stans* (Family: Bignoniaceae) Yellow Bell Flowers. *Journal of Pharmacognosy and Phytochemistry, 5*(6), 172.

Bhavan Kumar, A. (2017). *Scientific Validation of Antidiabetic Activity of Ethanolic Extract of Tecoma Stans (L) Juss. Leaf* (Doctoral dissertation, Karpagam College of Pharmacy, Coimbatore).

Calderón, G., & Rzedowski, J. (2001). *FloraFanerogámica del Valle de Mexico*. Valle de, Mexico: *Instituto de Ecología*.

Castillo, L., & Rossini, C. (2010). Bignoniaceae metabolites as semiochemicals. *Molecules, 15*, 7090–7015.

Comisión Nacional para el Conocimiento y uso de la Biodiversidad (CONABIO). (2018). Tecoma stans. http://www.conabio.gob.mx/conocimiento/info_especies/arboles/doctos/12-bigno8m.PDF (Consulta: 24-Enero-2022).

Dewangan, N., Satpathy, S., Shrivastava, A.K., & Shrivastava, R. (2017). In vitro evaluation of antimicrobial activity of Tecoma stansand Vitex negundo. *Indian Journal of Scientific Research, 13*(2), 248–253.

Gallardo Alba, C. (2019). Familia Bignoniaceae Tecoma stans. El parque de Málaga: Guía botánica. Consultado 14-mayo-2019 en: https://parquedemalaga.ddns.net/tecoma-stans

Hashem, F. A. (2008). Free radical scavenging activity of the flavonoids isolated from Tecoma radicans. *Journal of Herbs, Spices & Medicinal Plants, 13*(2), 1–10.

Laboratorio de Sistemática de Plantas Vasculares. (2017). *Bignoniaceae Juss. Departamento de Ecología y Ciencias Ambientales. Facultad de Ciencias*. Montevideo, Uruguay: Universidad de la República. http://thecompositaehut.com/ (Consulta: 16-septiembre-2018).

Martínez, E., & Ramos, C.H. (2012). *Bignoniaceae Juss. Fascículo 104. Flora del Valle de Tehuacán-Cuicatlán* (58 pp). Mexico: Instituto de Biología. Universidad Nacional Autónoma de Mexico.

Olivas, S.M. (1999). *Plantas medicinales del estado de Chihuahua* (Vol. I, 131 pp). Chihuahua, Mexico: Universidad Autónoma de Ciudad Juárez.

Prasanna, V.L., Lakshman, K., Hegde, M.M. and Bhat, V. (2013). Antinociceptive and anti-inflammatory activity of *Tecoma* stansleaf extracts. *Indian Journal of Research in Pharmacy and Biotechnology, 1*(2), 156–160.

Ramírez-Ortíz, M. E., Rodríguez-Carmona, O. Y., Hernández-Rodríguez, O. S., Chel-Guerrero, L. A., & Aguilar-Méndez, M. Á. (2017). Estudio de la actividad hipoglucemiante y antioxidante de tronadora, wereque y raíz de nopal. *OmniaScience Monographs*.

Salazar, R., & Soihet, C. (2001). *Tecoma stans (L.) C. Juss. ex Kunth* (No. CATIE ST MT-48). CATIE, Turrialba (Costa Rica). Programa de Investigación. Proyecto de Semillas Forestales Danida Forest Seed Centre, Humlebaek (Dinamarca).

Salem, M.Z.M., Gohar, Y.M., Camacho, L.M., El-Shanhorey, N.A. & Salem, A.Z.M. (2013). Antioxidant and antibacterial activities of leaves and branches extracts of Tecoma stans (L.) Juss. ex Kunth against nine species of pathogenic bacteria. In Mokche O, E Paul, & O Berry (eds.), *Nuevo catálogo de la flora de Venezuela* (pp. 270–275). Caracas-Venezuela: Fundación Instituto Botánico de Venezuela. "Dr. Tobias Lasser" (2008).

Salem, M. Z., Gohar, Y. M., Camacho, L. M., El-Shanhorey, N. A., & Salem, A. Z. M. (2013). Antioxidant and antibacterial activities of leaves and branches extracts of Tecoma stans (L.) Juss. ex Kunth against nine species of pathogenic bacteria. *African Journal of Microbiology Research, 7*(5), 418–426.

Sánchez de Lorenzo-Cáceres, J.M. (2015). Tecoma stans Juss. ex Kunth. Ayuntamiento de Murcia. España. https://www.murcia.es/medio-ambiente/parquesyjardines/material/Arbol_mes_2015/2015_05%20Tecoma%20stans.pdf (Consulta: 03-enero-2019)

Secretaría de Desarrollo y Medio Ambiente (SEDUMA). (2012). Xk'anlol. *Flora. Ficha técnica. Consutlado* 03- febrerp-2019 en: http://www.seduma.yucatan.gob.mx/flora/fichas-tecnicas/xkanlol.pdf

Zamora-Crescencio, P., Domínguez-Carrasco, M. D. R., Villegas, P., Gutiérrez-Báez, C., Manzanero-Acevedo, L. A., Ortega-Haas, J. J., ... Puch-Chávez, R. (2011). Composición florística y estructura de la vegetación secundaria en el norte del estado de Campeche, México. *Boletín de la Sociedad Botánica de México*, (89), 27–35.

Turnera diffusa

Aguilar, A., Argueta, A., Cano, L. coordinators, (1994). *Flora Medicinal Indígena de Mexico* (Vols. 1–3). Mexico: Instituto Nacional Indigenista.

Alcaraz-Meléndez, L., Delgado-Rodríguez, J., Real-Cosío, S. (2004). Analysis of essential oils from wild and micropropagated plants of damiana (*Turnera diffusa*). *Fitoterapia, 75*(7), 696–701. https:// doi.org/10.1016/j.fitote.2004.09.001

Alcaraz-Melendez, L., Real-Cosio, S., Suchý, V., & Švajdlenka, E. (2007). Differences in essential oil production and leaf structure in pheno-types of damiana (turnera diffusa willd. *Journal of Plant Biology, 50*, 378–382.

Alvarado-Cárdenas, O. (2006). Turneraceae. *Flora del valle de Tehuacán-Cuicatlán*. Mexico: Instituto de Biología. Universidad Nacional Autónoma de Mexico *43*, 1–7.

Arbo, M.M., & Espert, S.M. (2009). Morphology, phylogeny and biogeography of *Turnera* L. (Turneraceae). *Taxon, 58*, 457–467. http://doi.org/10.1002/tax.582011

Arletti, R., Benelli, A., Cavazzuti, E., Scarpetta, G., & Bertolini. (1999). A. Stimulating property of Turnera diffusa and Pfaffia paniculata extracts on the sexual-behavior of male rats. *Psychopharmacology, 143*(1), 15–19. https://doi.org/10.1007/s002130050913

Argueta, V.A. (1994). *Atlas de las Plantas de la Medicina Tradicional Mexicana* (Vol. 1). Mexico: Instituto Nacional Indigenista.

Domínguez, X.A., & Hinojosa, M. (1976). Isolation of 5-Hydroxy-7,3′4 - Trimethoxy -flavone from *Turnera diffusa*. *Planta Médica, 43*, 175–178.

Estrada-Castillón, E., Soto-Mata, B.E., Garza-López, M., Villarreal-Quintanila, J.Á., Jiménez-Pérez, J., Pando-Moreno, M., Sánchez-Salas, J., Scott-Morales, L., & Cotera-Correa, M. (2012). Medicinal plants in the southern region of the State of Nuevo León, Mexico. *Journal of Ethnobiology and Ethnomedicine, 8*, 45–57.

Estrada-Castillón, E., Garza-López, M., Villarreal-Quintanilla, J.A., Salinas-Rodríguez, M.M., Soto-Mata, B.E., González-Rodríguez, H., González-Uribe, D., Cantú-Silva, I., Carrillo-Parra, A., Cantú-Ayala, C. (2014). Ethnobotany in Rayones, Nuevo León, Mexico. *Journal of Ethnobiology and Ethnomedicine, 10*(1), 62. https://doi.org/10.1186/1746-4269-10-62

Estrada-Castillón, E., Villarreal-Quintanilla, J.Á., Encina-Domínguez, J.A., Jurado-Ybarra, E., Cuéllar-Rodríguez, L.G., Garza-Zambrano, P., Arévalo-Sierra, J.R., Cantú-Ayala, C.M., Himmelsbach, W., Salinas-Rodríguez, M.M., & Gutiérrez-Santillán, T.V. (2021). Ethnobotanical biocultural diversity by rural communities in the Cuatrociénegas Valley, Coahuila; Mexico. *Journal of Ethnobiology and Ethnomedicine, 17*, 21. https://doi.org/10.1186/s13002-021-00445-0

Garza-Juárez, A., Salazar-Cavazos de la Luz, M.,Salazar-Aranda, R., Pérez-Meseguer, J., & Waksman de Torres, N. (2011). Correlation between chromatographic finger- print and antioxidant activity of *Turnera diffusa* (Damiana). *PlantaMedica, 77*, 958–963.

González-Stuart, A.E. (2010). Use of medicinal plants in Monterrey, Mexico. *Notulae Scientia Biologicae, 2*, 7–11.

Hoffman, D. (1991). *The New Holistic Herbal* (3rd ed., pp. 146–147). Rockport, MA: Element Books Limited.

Martínez, M. (1969). *Las Plantas Medicinales de Mexico* (5th ed.). Mexico: Botas.

Osuna-Leal, E., Meza-Sánchez, R. (2000). Producción de plantas y establecimiento y manejo de plantaciones de Damiana Turnera diffusa Willd. *La Paz: INIFAP*. ISSN 1405-597X

Otsuka, R.D., GhilardiLago, J.H., Rossi, L., Fernández Galduróz, J.C., & Rodrigues, E. (2010). Psycho active plants described in a Brazilian literary work and their chemical compounds. *Central Nervous System Agents in Medicinal Chemistry, 10*, 218–237.

Pérez R.M., Ocegueda A., Munoz J.L., Avita J.G., Morrow W.W. (1984). A study of the hypoglycaemic effect of some Mexican plants. *Journal of Ethnopharmacology, 12*, 253–262.

Salazar, R., Pozos, M.E., Cordero, P., Pérez, J., Salinas, M.C., & Waksman, N. (2008). Determination of the antioxidant activity of plants from Northeast Mexico. *Pharmaceutical Biology, 46*, 166–170.

Spencer, K.C., & Seigler, D.S. (1981). Tetraphyllin B from *Turnera diffusa*. *Planta Médica, 43*(10), 175–178. https://doi.org/10.1055/s-2007-971495

Szewczyk, K., & Zidorn, C. (2014). Ethnobotany, phytochemistry, and bioactivity of the genus Turnera (Passifloraceae) with a focus on damiana-*Turnera diffusa*. *Journal of Ethnopharmacology, 152*(3), 424–443.

Thulin, M., Razafimandimbison, S.G., Chafe, P., Heidari, N., Kool, A., & Shore, J.S. (2012). Phylogeny of the Turneraceae clade (Passifloraceae sl): Trans-Atlantic disjunctions and two new genera in Africa. *Taxon, 61*(2), 308–323.

Villaseñor, J.L. (2016). Checklist of the native vascular plants of Mexico. *Revista Mexicana de Biodiversidad, 87*, 559–902. http://dx.doi.org/10.1016/j.rmb.2016.06.017
Zhao, J., Pawar, R.S., Ali, Z., Khan, I.A. (2007). Phytochemical investigation of *Turnera diffusa*. *Journal of Natural Products, 70*(2), 289–292. https://doi.org/10.1021/np060253r
Zhao, J., Dasmahapatra, A.K., Khan, S.I., & Khan, I.A. (2008). Anti-aromatase activity of the constituents from damiana (Turnera diffusa). *Journal of Ethnopharmacology, 120*, 387–393.

Yucca filifera

Arce-Montoya, M., Rodríguez-Álvarez, M., Hernández-González, J.A., & Robert, M.L. (2006). Micropropagation and field performance of *Yucca valida*. *Plant Cell Reports, 25*(8), 777–783. https://doi.org/10.1007/s00299-006-0144-3
Briones, O., & Villarreal, J.A.V. (2001). Vegetación y flora de un ecotono entre las provincias del altilpano y de la planicie costera del noreste de Mexico. *Acta Botanica Mexicana, 55*, 39–67. Retrieved February 20, 2022. https://dialnet.unirioja.es/descarga/articulo/2917719.pdf
Cambrón-Sandoval, V.H., Malda Barrera, G., Suzán Azpiri, H., & Díaz Salim, J.F. (2013). Comportamiento germinativo de semillas de *Yucca filifera* Chabaud con diferentes periodos de almacenamiento. *Cactáceas y Suculentas Mexicanas, 58*(3). Retrieved February 20, 2022. https://biblat.unam.mx/hevila/Cactaceasysuculentasmexicanas/2013/vol58/no3/2.pdf
Castillo, F., Aguilar, C.N., Hernández, D., Gallegos, G., & Rodríguez, R. (2012). Antifungal properties of bioactive compounds from plants. In Aguilar, C. (ed.), *Fungicides for Plant and Animal Diseases* (pp. 81–106). INTECH Open Access Publisher Press. Retrieved February 20, 2022. https://www.academia.edu/download/49504782/Antifungal_Properties_of_Bioactive_Compo20161010-14274-ke0x1u.pdf
Castillo-Reyes, F., Hernández-Castillo, F.D., Gallegos-Morales, G., Flores-Olivas, A., Rodríguez-Herrera, R., & Aguilar, C.N. (2015). Efectividad *in vitro* de *Bacillus* y polifenoles de plantas nativas de Mexico sobre *Rhizoctonia-Solani*. *Revista Mexicana de Ciencias Agrícolas, 6*(3), 549–562. Retrieved February 20, 2022. http://www.scielo.org.mx/pdf/remexca/v6n3/v6n3a9.pdf
Flores, G.J.A. (2001). Plantas nativas usadas como alimentos, condimentos y bebidas de las comunidades vegetacionales desérticas o semidesérticas en Nuevo León, Mexico. *Revista Salud Pública y Nutrición, 2*(1). Retrieved February 20, 2022. https://www.medigraphic.com/pdfs/revsalpubnut/spn-2001/spn011e.pdf
García-Berfon, L., Armijos-Riofrio, C., Aguilar-Ramírez, S., López-Cordova, C., Ramírez-Robles, J., Calva-Luzon, M., & Pogo-Tacuri, E. (2021). Estudio de especies no leñosas de la provincia de Loja (Ecuador) como potenciales materias primas para la fabricación de papel artesanal. *Ingeniería, Investigación y Tecnología, 22*(2), 1–13. https://doi.org/10.22201/fi.25940732e.2021.22.2.011
Granados-Sánchez, D., & López-Ríos, G.F. (1998). Yucca "izote" del desierto". *Revista Chapingo Serie Ciencias Forestales y del Ambiente, 4*(1), 179–192.
Granados-Sánchez, D., Sánchez-González, A., Granados Victorino, R.L., & Borja de la Rosa, A. (2011). Ecología de la vegetación del Desierto Chihuahuense. *Revista Chapingo serie Ciencias Forestales y del Ambiente, 17*(SPE), 111–130. ISSN e: 2007-4018; print: 2007-3828.
Knox, G.W. (2010). *Agave* and *Yucca*: Tough plants for tough times. *EDIS, 2010*(2). Retrieved February 20, 2022. https://journals.flvc.org/edis/article/download/118367/116299
Mulík, S., & Ozuna, C. (2020). Mexican edible flowers: Cultural background, traditional culinary uses, and potential health benefits. *International Journal of Gastronomy and Food Science, 21*, 100235. https://doi.org/10.1016/j.ijgfs.2020.100235
Oumama, K., Chliyeh, M., Chahdi, A.O., Touati, J., Touhami, A.O., Benkirane, R., & Douira, A. (2014). *Yucca filifera*, a new host of *Diplodia mutila* in Morocco. *International Journal of Recent Biotechnology, 2*(4), 13–17. Retrieved February 20, 2022. https://www.researchgate.net/profile/Mohamed-Chliyeh/publication/319266060_Yucca_filifera_a_new_host_of_Diplodia_mutila_in_Morocco/links/599f026fa6fdcc500355e16c/Yucca-filifera-a-new-host-of-Diplodia-mutila-in-Morocco.pdf
Pacheco-Pantoja, S.E., Cruz-Castillo, I.Y., Castro-Palafox, J., Ramos-Vargas, M.A., Mulík, S., Hernández-Carrión, M., & Ozuna, C. (2021). Consumo de flores comestibles mexicanas en el estado de Guanajuato: Un estudio preliminar. *Jóvenes en la Ciencia, 10*. Retrieved February 20, 2022. https://www.jovenesenlaciencia.ugto.mx/index.php/jovenesenlaciencia/article/download/3425/2922

Pellmyr, O. (2003). Yuccas, yucca moths, and coevolution: A review. *Annals of the Missouri. Botanical Garden*, 35–55. https://doi.org/10.2307/3298524

Renteria, L., & Cantú, C. (2003). El efecto de *Tegeticula yuccasella* Riley (Lepidoptera: Prodoxidae) sobre la fenología reproductiva de Yucca filifera Chabaud (Agavaceae) en Linares, NL, Mexico. *Acta Zoológica Mexicana*, *89*, 85–92. Retrieved February 20, 2022. http://www.scielo.org.mx/pdf/azm/n89/n89a7.pdf

Sánchez-Trinidad, L. (2017). *Las flores en la cocina Veracruzana*. Dirección general de culturas populares, indígenas y urbanas. ISBN: 978-607-745-787-9. Retrieved February 20, 2022. https://www.culturaspopulareseindigenas.gob.mx/pdf/2020/recetarios/Las%20flores%20en%20la%20cocina%20veracruzana.pdf

Sotelo, A., López-García, S., & Basurto-Peña, F. (2007). Content of nutrient and antinutrient in edible flowers of wild plants in Mexico. *Plant Foods for Human Nutrition*, *62*(3), 133–138. https://doi.org/10.1007/s11130-007-0053-9

Yucca carnerosana

Estrada, C.E., Brianda, E. Soto, M., Miriam Garza, L., José, Á. Villarreal, Q., Javier Jiménez, P., & Marisela Pando, M. (2013). Reseña del libro: Plantas útiles en el centro-sur del estado de Nuevo León. *Botanical Sciences*, *91*(4), 537–538. Retrieved February 20, 2022. http://www.scielo.org.mx/pdf/bs/v91n4/v91n4a13.pdf

Granados-Sánchez, D., & López-Ríos, G.F. (1998). Yucca "izote" del desierto". *Revista Chapingo Serie Ciencias Forestales y del Ambiente*, *4*(1), 179–192.

Jasso de Rodriguez, D., García, R.R., Castillo, F.H., González, C.A., Galindo, A.S., Quintanilla, J.V., & Zuccolotto, L.M. (2011). In vitro antifungal activity of extracts of Mexican Chihuahuan Desert plants against postharvest fruit fungi. *Industrial Crops and Products*, *34*(1), 960–966. https://doi.org/10.1016/j.indcrop.2011.03.001

López-Ramírez, Y., Cabañas-García, E., Areche, C., Trejo-Tapia, G., Pérez-Molphe-Balch, E., & Gómez-Aguirre, Y.A. (2021). Callus induction and phytochemical profiling of *Yucca carnerosana* (Trel.) McKelvey obtained from in vitro cultures. *Revista Mexicana de Ingeniería Química*, *20*(2), 823–837. https://doi.org/10.24275/rmiq/Bio2209

Maksimov, A.P., Kovalev, M.S., Trikoz, N.N., & Gil, A.T. (2019). The reproduction of *Yucca carnerosana* (Trel.) McKelvey in Crimea. *IOP Conference Series: Earth and Environmental Science*, *226*, 012029. IOP Publishing. https://doi.org/10.1088/1755-1315/226/1/012029

Medellín-Castillo, N.A., Hernández-Ramírez, M.G., Salazar-Rábago, J.J., Labrada-Delgado, G.J., & Aragón-Piña, A. (2017). Bioadsorción de Plomo (II) presente en solución acuosa sobre residuos de fibras naturales procedentes de la industria ixtlera (Agave lechuguilla Torr. y Yucca carnerosana (Trel.) McKelvey). *Revista internacional de contaminación ambiental*, *33*(2), 269–280. https://doi.org/10.20937/RICA.2017.33.02.08

Mellado, M., García, J.E., Arévalo, J.R., Dueñez, J., & Rodríguez, A. (2009). Effects of replacement of alfalfa by inflorescences of *Yucca carnerosana* in the diet on performance of growing goats. *Livestock Science*, *123*(1), 38–43. https://doi.org/10.1016/j.livsci.2008.10.004

Pellmyr, O. (2003). Yuccas, yucca moths, and coevolution: A review. *Annals of the Missouri Botanical Garden*, 35–55. https://doi.org/10.2307/3298524

Sheldon, S. (1980). Ethnobotany of *Agave lecheguilla* and *Yucca carnerosana* in Mexico's Zona Ixtlera. *Economic Botany*, *34*(4), 376–390. https://doi.org/10.1007/BF02858314

Sil-Berra, L.M., López, M.A., Márquez-Medero, M.A., & Cervantes-Cruz, J.M. (2022). De Mexico para el mundo… los murciélagos endémicos. *Therya Ixmana*, *1*(1), 29–31. https://doi.org/10.12933/therya_ixmana-22-186

Trenti-Very, L.C., González-Jácome, A., Landín-López, A.L., Mariaca Méndez, R., Jiménez-Ferrer, G., & Nahed-Toral, J. (2021). Caprinocultura, ambiente y economía campesina: Un análisis de los sistemas familiares ganaderos en el semidesierto potosino. *Revista de El Colegio de San Luis*, *22*, 12. https://doi.org/10.21696/rcsl112220211313

Chapter 3

Mexican Desert

Health and Biotechnological Properties Potential of Some Cacti Species (Cactaceae)

Joyce Trujillo, Sandra Pérez-Miranda, Alfredo Ramírez-Hernández, Alethia Muñiz-Ramírez, Abraham Heriberto Garcia-Campoy, and Yadira Ramírez-Rodríguez

CONTENTS

DOI: 10.1201/9781003251255-3

3.1 INTRODUCTION

Drylands (arid, semiarid, and dry sub-humid environments) cover one-third of the Earth's land surface and are the livelihood for more than 2 billion people. Mexico is a hotspot of biodiversity characterised by the variety of ecosystems in which drylands are the predominant environments because they cover nearly 60% of the country's territory (SEMARNAT) 2019). Mexican drylands host a wide variety of flora in which Cactaceae are dominant, followed by Agavaceae and Leguminosae, among others. Cacti in Mexican deserts became forests that dominate the landscape matrix.

Cacti are important for the ecosystem as well as humans. For instance, cacti exist in various shapes and sizes, from arboreal, shrubby, globular, columnar, crawlers, tall, or dwarf. Their flowers are large, and the spines and branches grow out of areoles. They are often used as ornamental plants, but many have been cultivated to obtain fodder vegetables and fruits (Secretaria de Medio Ambiente y Recursos Naturales (SEMARNAT) 2016).

Cultivated cactus, such as *Opuntia* spp. and *Hylocereus* spp. are highly technologically produced and industrialised. However, there are other cacti that are underutilised, which are found throughout the Mexican drylands, and they could be sustainably utilised in several industries, principally with the application for human goods.

This chapter describes the health and biotechnological properties of some Cactaceae. Currently, in Mexico, the *Opuntia ficus indica* is the most heavily commercialised (Torres-Ponce et al. 2015). However, there are other cacti of cultural and socio-economic importance with potential applications in health, food, cosmetic, agricultural, and bioenergy industries. This chapter represents a bibliographic review of the use, functional effects, and their application in biotechnological.

3.2 DIVERSITY AND CONSERVATION STATUS OF MEXICAN CACTI

Members of the Cactaceae Juss (1,789) family are one of the most charismatic and fascinating organisms because they occur in a wide range of shapes (e.g., cylindrical, globular, or flat (cladode) stems) and sizes (see Figure 3.1); moreover, these plants present anatomical and physiological characteristics that have adapted to live in dryland environments where humidity is scarce (Hernández-Hernández et al. 2014). Accordingly, cacti have modified their tissues to accumulate a large amount of water in them (Pavón et al. 2016) and have reduced or are practically absent of leaves and/or modified in spines to reduce evapotranspiration (Aliscioni et al. 2021), carrying out photosynthesis on the surface of their stems. Photosynthesis follows a peculiar metabolic pathway called as crassulacean acid metabolism (CAM), which saves water as gas exchange occurs at night when the ambient temperature is lower (Miller et al. 2021).

Accordingly, Cactaceae is a highly diversified family of xerophytes that are dominant throughout drylands of the Americas, which is its center of origin and diversification. Because the distribution range of cacti is from Canada to Patagonia, there are species also distributed across tropical and subtropical ecosystems (Hernández-Hernández et al. 2014). Cacti comprise approximately 1,400–1,800 described species in the world (Anderson 2001; Hunt 2006; Guerrero et al. 2019). Mexico is the country with the greatest diversity, with 52 genera and 850 species (Guerrero et al. 2019; González Medrano 2012). Cacti are plants of various shapes and sizes, succulent stems, and wide edible fruits. These plants are used as ornamental crops, in human food, and as fodder for animals, thus, guaranteeing food/nutritional security, economic development, and the livelihood of the local population (Ramírez-Rodríguez et al. 2020).

The state of San Luis Potosí has the most incredible diversity, with a record of 151 species. It is followed by the states of Coahuila with 126 species and Nuevo León and Oaxaca with 118 species each. San Luis Potosí is also the state with the most extraordinary richness of genera (33), followed by Oaxaca (32) and Tamaulipas (31) (see Figure 3.2).

According to Goettsch et al. (2015), changes in land use, the introduction of exotic species, and uncontrolled harvesting of these plants for use as food, raw material, and other purposes are some threats (see Figure 3.3); moreover, many species of cacti are at risk because of the remarkable endemism of their populations (Jiménez-Sierra 2011).

Consequently, nearly 31% of cacti are globally threatened (Goettsch et al. 2015) because of several factors, and some of them are listed by the International Union for the Conservation of

Figure 3.1 Morphology of some cacti species: a) *Lophophora williamsii* buried stem species, b) *Ferocactus pilosus*, c) *Echinocactus platyacanthus* simple succulent stem species, d) *Pelecyphora aselliformis* buried stem species, e) *Coryphanta maíz-tablasensis gregarious* species, f) *Myrtillocactus geometrizans* species with branched stem. Photographs by Joel Flores Rivas, PhD.

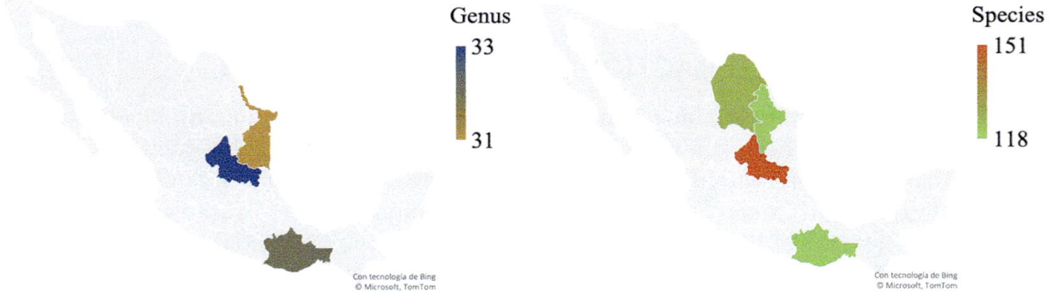

Figure 3.2 Mexican states with a high number of genus and cacti species. Source: Guzmán et al. (2003).

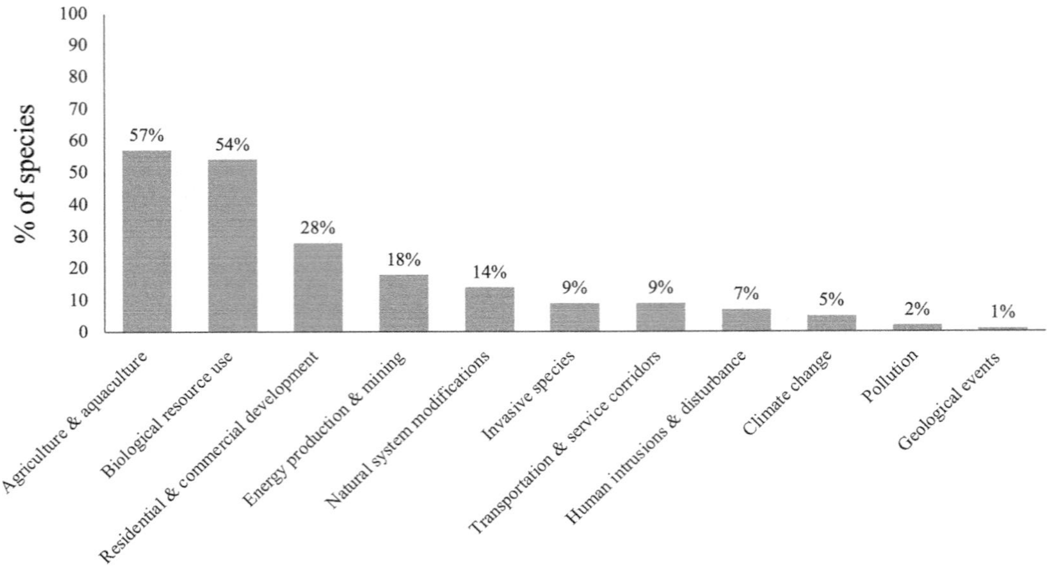

Figure 3.3 Percentage of threat process affecting cactus species (modified from Goettsch et al. 2015).

Nature (IUCN) under various threat categories (see Figure 3.4), which points to the need for conservation efforts.

In Mexico, 850 species have been described, of which an estimated 84% are endemic. However, the higher the endemism, along with the narrow distribution range of some species, and the more isolated its populations are from each other, each of them made up of a smaller number of individuals, the greater the risk of extinction is. All these factors increase the probability of loss of genetic diversity and, therefore, the probability that populations can adjust to biological and environmental changes in their environment. In addition, the slow growth of cacti due to CAM metabolism allows them to live successfully in environments with water scarcity, which means that their populations recover very slowly from population disturbances caused naturally or because of human activity. Notably, in Mexico the 'Norma Oficial Mexicana NOM-059-SEMARNAT-2010' offers the legal framework for the protection, use, and trends of native wild flora (SEMARNAT 2010).

3.3 ECOLOGICAL INTERACTIONS WITH MEXICAN CACTI

Despite investigations into the biology of many cacti species, the reproductive biology of only 2% of cacti species has been studied (Mandujano et al. 2010). It is likely that cactus diversified because of the interaction with novel pollinators or the expansion of the Sonoran Desert (further details in Hernández-Hernández et al. 2014) or the high prevalence of hybridisation and polyploidisation (Majure et al. 2012). Several adaptations can be found in cacti flowers; for instance, the *Nopalea* and *Consolea* clades contain species with flowers in which expanded floral nectaries for hummingbird pollination (Majure et al. 2012). Generally, columnar cacti with flowers are already differentiated for bats or hummingbirds (Mandujano et al. 2010; Tremlett et al. 2020).

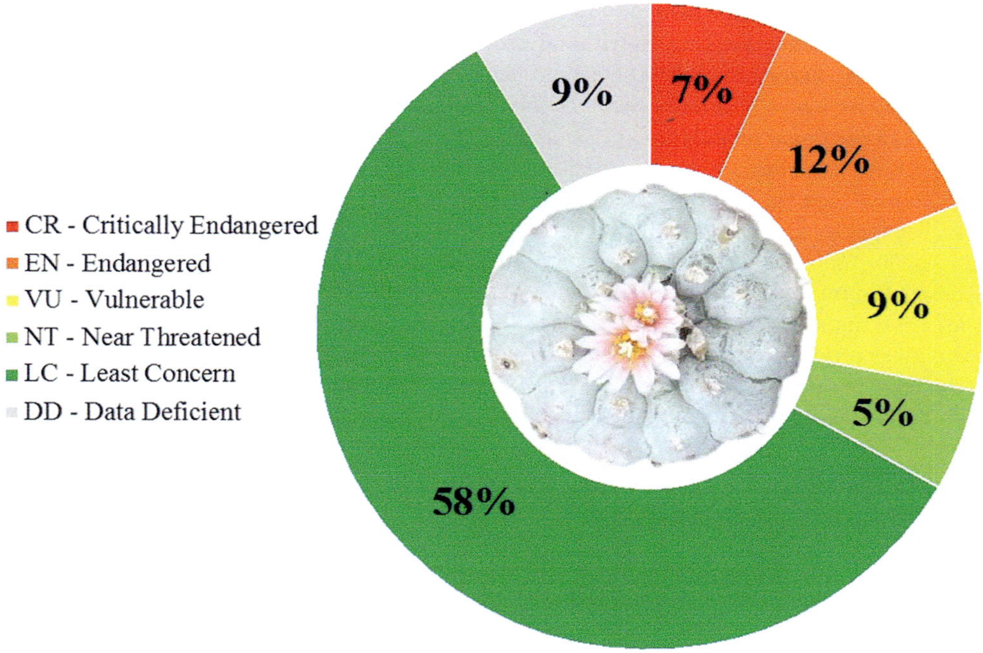

Figure 3.4 Percentage of cacti species under some red list categories. Photo in the center of the graph corresponds to *Lophophora williamsii*, an endemic species from Mexico under Vulnerable (VU) category, according to the International Union for Conservation of Nature (IUCN). Source: The IUCN Red List of Threatened Species (https://www.iucnredlist.org/). Photograph by Joel Flores Rivas, PhD.

Moreover, according to Mandujano et al. (2010), it can be found in mixed pollination systems (e.g., birds, bees), whereas other cacti species exhibit a more specialised system, for instance, *Cleistocactus baumannii*, which is predominately hummingbird pollinated.

Notwithstanding the economic, cultural, and medicinal importance of *Opuntia* spp., it is a complex group because of interspecific hybridisation, polyploidy, and morphological variability (Majure et al. 2012). Asexual propagation in *Opuntia* cacti occurs naturally by stem or cladode detachment (Rebman and Pinkava 2001). Nonetheless, sexual reproduction remains necessary for generating viable genetic pools in wild populations as well as to produce 'tunas' (Reyes-Agüero, Aguirre R., and Valiente-Banuet 2006). Sexual reproduction depends mainly on insects (Mandujano et al. 2010) and has probably been crucial for the diversification and success of *Opuntia* in colonising the American continent (Anderson 2001; Reyes-Agüero, Aguirre, and Valiente-Banuet 2006). Bees have been widely recognised as the main pollinators of *Opuntia* species, for which mellitophilous syndrome dominates (Mandujano et al. 2010). Visitation by insects depends on the flower's color and scent (which is usually hermaphroditic), taste and quality of the nectar, pollen production, the morphology of the visitors, and the spatio-temporal abundance of both flowers and flower visitors. However, Tenorio-Escandón et al. (2022) revealed the need for further research to understand the insect-pollination syndrome and to understand the competition between insect species for conservation purposes. Further research must pay attention to pollination's ecological and economic value because it is crucial for fruit production and crop productivity (Tenorio-Escandón et al. 2022; Tremlett et al. 2021).

The entomophagy that has been practiced for thousands of years has also been described. This activity, as well as the economic, nutritional properties of the use, preservation, and marketing of insects that are associated with the asparagaceae genus (*Agave salmiana* Otto ex Salm-Dyck), such as red agave worm (*Comadia redtenbacheri*), white maguey worm (*Aegiale hesperiaris*), and escamoles (*Liometopum apiculatum*) (Molina-Vega et al. 2021; Espinosa-García et al. 2018; Rostro et al. 2012), have alimentary options with nutritional benefits for humans. These insects are widely consumed in Mexico, the highest-ranking country worldwide in terms of insect diversity; although some insect species are considered pests of cacti. These insects are consumed principally in their larval stage for their exquisite, exotic flavor, high protein and lipids, easy digestion, and high content of essential minerals. Their production is seasonal and knowledge of their life cycle is scarce, so production under greenhouse conditions is unfeasible to date. However, studies are ongoing to generate mass reproduction strategies under sustainable greenhouse conditions that would promote the incorporation of these insects into functional foods and enact legislation that regulates their production, consumption, and health benefits (Molina-Vega et al. 2021).

3.4 CACTI AND MICROBIOME

To think of drylands and deserts is to assume the presence of cactus plants, ecosystem services such as oxygen production, carbon dioxide consumption, and transformation of heavy metals (Guillen-Cruz et al. 2021; Grodsky and Hernandez 2020; Sarria Carabalí et al. 2019); but despite the cactus plants' systems, today we know that microsystems help survival in this environment are significant (Davis et al. 2019). This niche contains microbial communities which turn out to be and have potential biotechnological uses where their principal associations are mutualists (Taketani et al. 2017).

As stated above, cactus plant microbial communities are responsible for many ecological processes and substantially impact agriculture, the pharmaceutical industry, and biotechnology (Adeleke and Babalola 2021; Salazar et al. 2020).

Microorganisms directly affect the environment, and with the availability of metagenomics technology today, we can know what inhabits cacti's interior and surroundings.

Some of the main microorganisms reported in metagenomic analysis has allowed the characterisation of the communities that inhabit cacti, for which the core microorganisms usually belong to the Actinobacteria (25%–42%), Proteobacteria (15%–33%), Bacteroidetes (4%–30%), Acidobacteria (5%–12%), Chloroflexi (2%–5%), Bacillota (1%–11%), Verrucomicrobia (1%–4%) phyla for bacteria (De la Torre-Hernández et al. 2020) and the Basidiomycota (3%–10%), Ascomycota (1%–10%) phyla for fungi (Zygomycota, Glomeromycota, and Chytridiomycota data not reported). However, their relative abundances can vary depending on soil type and use; Table 3.1 compares the abundance of operatinal taxonomic units from seed, phyllosphere, rhizosphere, and dryland soils according to the inhabiting phyla and their zones.

The genetic machinery available in metagenomes is the basis for the specific isolation of microorganisms that can help us solve problems (Rehman, Ijaz, and Mazhar 2019; Fonseca García et al. 2018). Some potential uses of microorganisms associated with cacti are plant growth promoters, antibiotics, antivirals, biocompounds, and dyes that may have a technological or biotechnological use (De Medeiros Azevedo et al. 2021; Verma et al. 2020).

In this regard, in agriculture and forestry in developed countries, the molecules produced by microorganisms are known as plant growth-promoting rhizobacteria (PGPR), e.g., amino acids, siderophores, indoleacetic acid, phosphate solubilisers, fungal growth inhibitors, and

TABLE 3.1 RANGE OF ABUNDANCES OF (A) BACTERIAL AND (B) FUNGAL MEASURABLE OPERATIONAL TAXONOMIC UNITS IN SEED, PHYLOSPHERE, RIZOSPHERE, AND DRYLANDS

Bacteria					
	Range of Abundances (%)				
Phyla	**Seed**	**Phylosphere**	**Rizosphere**	**Dryland**	**References**
Actinobacteria	< 1	7–52	25–42	38–49	Flores-Nuñez (2020, 2018) Larceda- Junior G. (2019) Lüneberg (2018) Wang (2017) Le (2016) Araya (2020) Neilson (2017) Mascot-Gómez (2021)
Protobacteria	5–33	11–52	15–33	10–31	
Bacteriodetes	NR	2–12	4–30	1.0–6.5	
Acidobacteria	NR	< 3	5-12	4.0–11	
Chloroflexi	NR	< 2	2–5	8–9	
Verrucomicrobia	NR	3–8	1–4	1–3	
Bacillota	2–15	5–12	1–11	1–8	
Fungi					
Phyla	**Range of Abundances (%)**				**References**
	Seed	Phylosphere	Rizosphera	Dryland	
Ascomycota	NR	1–50	1 10	15–34	Sommerman L, (2018) Ambardar (2016) Sun (2018) Gómez-Silva (2019) Tian (2017) Murgía M. (2019)
Zygomycota	NR	2–5	NR	1–2	
Basidiomycota	NR	NR	3 -10	1.5–25	
Glomeromycota	NR	NR	NR	<1	
Chytridiomycota	NR	NR	NR	<1	

NR: not reported

organic acids, with biofertiliser properties or pest control agents, which promotes crops and their productivity (Govindasamy et al. 2022; Flores-Núñez et al. 2020; Bezerra et al. 2017).

Mexico has begun analyising composition, structure, and molecule characteristics that can be used as PGPR from rhizospheric bacterial communities associated with wild and cultivated *Echinocactus platyacanthus* and *Neobuxbaumia polylopha*; the predominant phyla is Actinobacteria and Proteobacteria (De la Torre-Hernández et al. 2020). Using strains such as *Enterobacter sakazakii M2Pfe*, *Azobacter vinelandii M2Per*, and *Pseudomonas putida M5TSA* allowed the growth of *Mammillaria fraileana* to improve its development, favouring the accumulation of nitrogen and improving photosynthesis (Lopez et al. 2012). Furthermore, volatile compounds of fungi belonging to classes *Sordariomycetes*, *Eurotiomycetes*, and *Dothideomycetes* demonstrated their functionality as PGPR and the development of *Arabidopsis thaliana*, *Agave tequilana* and *A. Salmiana*, and the identified molecules were mainly terpenes, alcohols, and aliphatic compounds (Camarena-Pozos et al. 2021).

Making cactus extracts and testing their antimicrobial or antiviral effect against crop pathogens is something that Rasoulpour (2018) did and showed significant activity against cucumber mosaic virus. Also, this group showed the Opuntin B presence, an antiviral protein isolated from prickly pear (*O. ficus indica (L.) Miller*) extract, concluding that it is responsible for such effects. Nevertheless, the molecule's origin is confusing, given that it could come from the microbiome-metabolism plant and not from the extract (Rasoulpour et al. 2018).

The plant–microorganism interaction has led to studies at the biotechnological level related to the food, pharmaceutical, and energy combustion industries.

3.5 ETHNOBIOLOGY OF MEXICAN CACTUS

Humans have been in touch with nature since ancient times. Thanks to the contemplation of their surroundings, humans have obtained knowledge to establish methods for biodiversity management and suitable exploitation of the resources to attend to their own local needs. Cacti represent a great alternative because they provide fruits, pulp, seeds, peel, and roots for human food during the dry season and as animal fodder, medicine, or biotechnological compounds used as biopolymers or biofuels, among others. Accordingly, several benefits have been obtained from cacti (Anderson 2001).

Opuntia spp. are probably the most studied and used genus to obtain both goods and health benefits for humans, and that could be partly true because this is the richest genus within the Cactaceae family with nearly 200 described species (Anderson 2001). In Mexico, the genus has a marked historical and cultural importance because of the production of 'nopal' (the Mexican denomination for the edible young cladodes and developed by cacti species belonging to the genera *Opuntia* and *Nopalea*). The English name for the edible fruit of the cacti of these genera is called 'prickly pear' and 'tuna' is the Mexican name.

Indigenous communities have consumed fresh or fermented cactus fruits. Fermented beverages are considered nutritional sources and functional beverages with the presence of multiple compounds of biological importance (Robledo-Márquez et al. 2021) because of their bioactive product content, such as polyphenols, phytosterols, betalains, carotenoids, antioxidants, vitamins, and minerals, among others (Ramírez-Rodríguez et al. 2020; Bergantin et al. 2017). For example, 'colonche' (as well known as 'coloche' and '*nochoctli*') is a beverage produced by fermenting the fruits of different cactus species in drylands, mainly prepared with fruits of *Opuntia* spp. and several columnar cacti species such as *Pachycereus weberi*, *Escontria chiotilla*, *Stenocereus* spp., and *Polaskia* spp. (further details in (Ojeda-Linares et al. 2020)).

According to Fray Bernardino de Sahagún (1577), Mexican farmers developed a complex pattern of 'nopal' (*Opuntia* spp.) cultivation that provided essential sustenance to the valuable dye-producing insect (Sahagún 1577). Dye production became an activity of great economic importance in New Spain, particularly, in places where mining could not be carried out (Coll-Hurtado 1998). Some *Opuntia* species were introduced to other continents after the Spanish conquest because of their traditional uses as medicinal plants, fruits, vegetables, dyes, food ingredients, and forage, among others. Therefore, cacti species became a valuable genetic resource with promising potential applications in the cosmetic, pharmaceutical, and food industries as well as bioenergy production (Ciriminna et al. 2019; Le Houerou 2000).

In this regard, it has been estimated that the exploitation of *Opuntia* plants could generate jobs for approximately 2,000 families in rural areas (INEGI 2007). *Opuntia* species represent one of the most important crops because they cover about 30% of the country's land area, and they are mainly distributed throughout arid and semiarid regions (Gallegos-Vázquez, Méndez Gallegos, and Mondragón Jacobo 2013). Mexico is the primary producer of 'prickly pears', supporting 43% of annual world production, estimated at 1,060,000 tonnes (t) in an area of 100,000 hectares (ha) (Potgieter and D'Aquino 2018). The area planted with 'nopal' in Mexico in 2019 was 45,746 ha, mainly in the Mexican high plateau (central Mexico).

Other ethnobiologically important cacti are *Echinocactus Platyacanthus*, *Ferocactus* sp., *Hylocereus* sp., *Mammillaria* sp., *Myrtilocactus geomettizans*, and *Stenocereus sp.* because their stems, root, seeds, flowers, peel, and fruit are widely eaten and used in several purposes, e.g., animal forage, ornamentation, and a source of water and food. They are used for brines, alcoholic and nonalcoholic beverages, jelly, yogurt, jam, ice cream, popsicles, dried fruits (similar to raisins), tamales, tortillas, preserves, natural colorants, and encapsulating agents. Also, some are used for medicinal benefits for conditions including diabetes, menstrual pain, headaches, earache, chest pains, heart stimulant, painkiller, wound disinfectant, diarrhea, dysentery, inflammation, rheumatism, insect bites, snake bites, hemorrhages, and hemorrhoids. These are ailments are treated as infusions, gum, hot compresses, or as boiled, roasted, burned, or scorched plant parts (e.g., roots, peel, or stem) (Ramírez-Rodríguez et al. 2020).

3.6 PHYTOCHEMISTRY OF SOME CACTACEAE WITH ECONOMIC IMPORTANCE: BIOLOGICAL ACTIVITIES

3.6.1 *Hylocereus* spp.

Known as 'dragon fruit' or 'pitaya', it is native to southern Mexico (Mercado-Silva 2018). There are several genera of pitaya of which *H. undatus* and *H. polyrhizus* are the most widely cultivated. Research has reported that the fruit and seed of *H. polyrhizus* show mainly antioxidant activity (Luo et al. 2014) in addition to anti-inflammatory and antimicrobial activity (Tenore et al. 2012).

Mainly steroids and triterpenoids have been identified in the exocarp of *H. polyrhizus and H. Undatus*; they also contain octadecane, oleic acid, squalene, stigmasterol, and campesterol (Luo et al. 2014). Betalains are compounds identified in these species, found in a more significant proportion in the exocarp than in the pulp, concluding that the antioxidant potential in pitayas is due to the presence of betalains (Suh et al. 2014). *H. polyrizhus* pulp and exocarp have been reported to have anxiolytic effects and could, therefore, be used as an auxiliary in anxiety treatment. Some isolated compounds from this species are shown in Figure 3.5 (Lira et al. 2020).

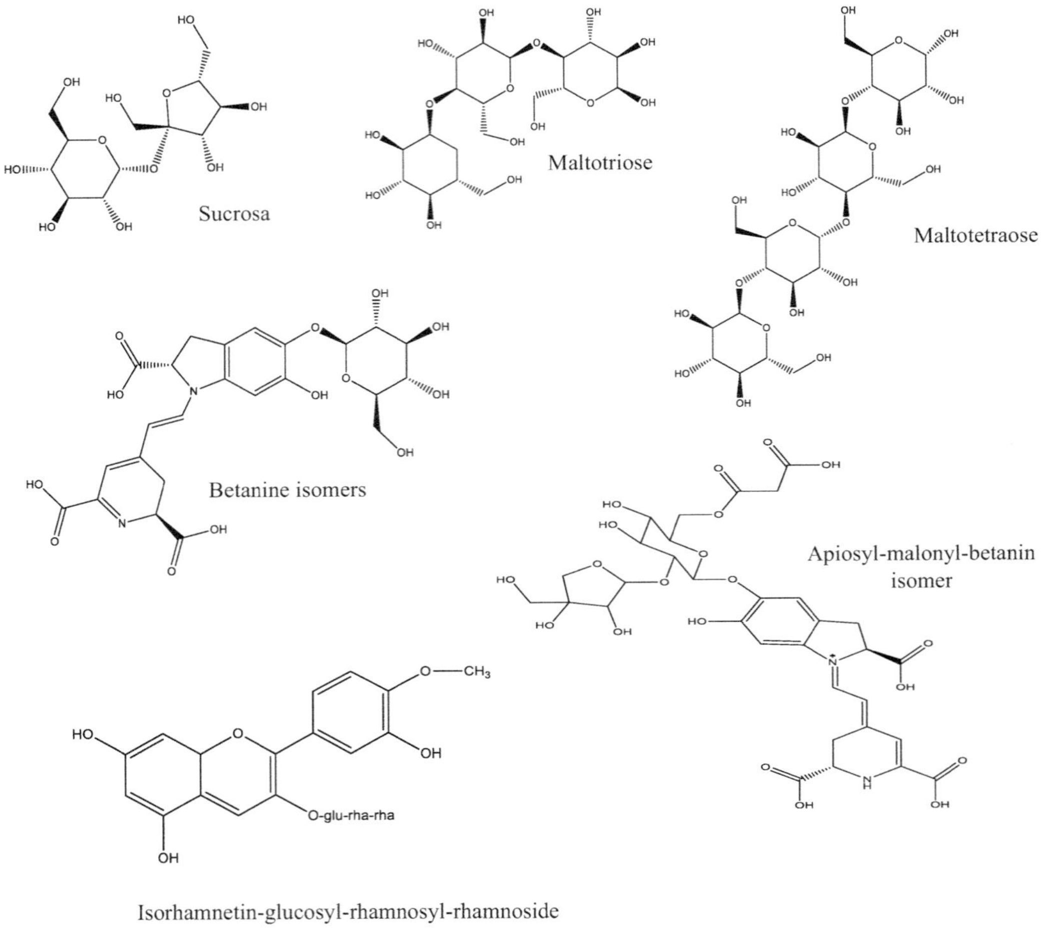

Figure 3.5 Isolated compounds of *Hylocereus polyrhizus*.

3.6.2 Opuntia Genus

Opuntia is one of the most commonly used plants in traditional medicine to treat cardiovascular diseases, gastric ulcers, and diabetes, among others (El-Mostafa et al. 2014). Various secondary metabolites have been identified from the different species of *Opuntia*, such as derivatives of pyrone and flavonoids such as rutin, kaempferol, orientin, isorhamnentin, vitexin, and myricetin, among others (Astello-García et al. 2015).

Aqueous extracts of *Opuntia* spp. have been reported to protect against liver damage caused by diclofenac, reducing oxidative stress (Villa-Jaimes et al. 2022). The polyphenols present in the cacti have been attributed to different therapeutic effects, e.g., the gallic acid found in their flowers shows antioxidant activity, which is responsible for reducing DNA damage (Khan et al. 2000). Gallic acid has also been attributed to cytotoxic activity against lung, leukemia, and prostate cancer tumor cells (You and Park 2010). The polyphenol family of molecules is widely distributed among different plants. Interest in these compounds has grown because of their antioxidant activity related to health benefits such as reducing inflammation and helping in the treatment of cancer and neurodegenerative diseases (Laughton et al. 1991).

Some therapeutic benefits of the cactus have been scientifically supported, showing that its metabolites help treat metabolic syndrome (including obesity and type 2 diabetes), bacterial and virological infections, rheumatism, different types of cancer, and cerebral ischemia (Ahmad et al. 1996).

3.6.2.1 *Opuntia ficus indica*

O. ficus indica is characterised by the presence of flavonoids and betalains, which show anti-inflammatory, hypoglycemic, and antioxidant activity (Gómez-Maqueo et al. 2019).

3.6.2.2 *O. ficus indica* Fruit

The fruits are usually 7 cm long and have a wide range of yellow, green, purple, white, and reddish colors. Flavonoids, phenolic acids (caffeic acid, ferulic acid, syringic acid) and betalains have been identified in the fruits (Aruwa et al. 2018). Betalain isolated from the pulp of *O. ficus indica* fights degenerative disorders that affect endothelial function, such as stroke, atherosclerosis, ischemia of the lower limbs, and atherothrombosis (Jang et al. 2008). It also has hepatoprotective, antidiabetic, anti-inflammatory, and neuroprotective properties (Serra et al. 2013). Extracts of ethyl acetate from the fruit suppresses human cervical cancer cell line proliferation (HeLa cells) (Hahm et al. 2015), and purple-red, purple, and green fruit have been reported to diminish liver, colon, and prostate cancer cells (Chavez-Santoscoy, Gutierrez-Uribe, and Serna-Saldívar 2009).

The fruit of *O. ficus indica* has various beneficial health effects, such as antioxidant, hypoglycemic and lipid-lowering (Osorio-Esquivel et al. 2011), antiulcerogenic, antiproliferative (Galati et al. 2003), antioxidant (Sreekanth et al. 2007), hepatoprotective (Kuti 2004; Galati et al. 2005), anticancer (Zou et al. 2005), and neuroprotective activity. This has been attributed to the different secondary metabolites it offers, such as antioxidant compounds (betacyanin, phenols, betaxanthin, and flavonoids), ascorbic acid, betalamic acid, amino acids, vitamin E, fiber, and carotenoids (see Figure 3.6) (El-Mostafa et al. 2014; Osorio-Esquivel et al. 2011). The prickly pear has antioxidant properties, serving as antibacterial and antifungal agents. It has been reported that the phenolic compounds present in the plant are what give it its antioxidant activity (Bargougui et al. 2019).

3.6.2.3 *O. ficus indica* Flowers

It has been shown that *O. ficus indica* flower infusions exhibit diuretic activity *in vitro* and *in vivo* assays related to the potassium content in the flower (Galati et al. 2002). The presence of flavonols in the flowers has been associated with anti-inflammatory (Benayad et al. 2014), antiatherosclerotic (Fuhrman et al. 2005) and anticancer properties (Piao et al. 2006). Studies carried out on the methanolic extract of *O. ficus indica* flowers showed that it contains flavonol glycosides (De Leo et al. 2010) (see Figure 3.7)

3.6.2.4 *O. ficus indica* Cladodes

The cladodes of *O. ficus indica* have a high content of nicotiflorin, which reduces the incidence of cerebral infarction (Li et al. 2006). In mice with multi-infarction dementia, nicotiflorin helps preserve spatial memory as well as protecting against oxidative stress and failure in energy metabolism (Huang et al. 2007).

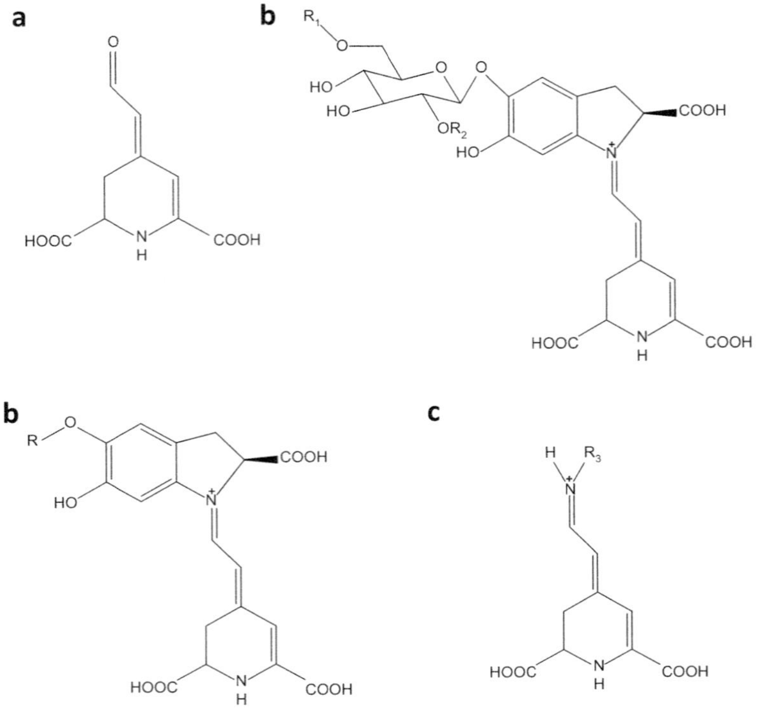

Figure 3.6 Generalstructure of betalamic acid (a), betacyanins (b) and (c) and betaxanthins (d).

It has also been reported that cladodes can lower cholesterol levels and have antiulcer, anti-inflammatory, and healing activity (Galati et al. 2003). Research has reported that isorhamnetin carbohydrates (see Figure 3.8) isolated from cladodes of this species exhibit cytotoxic activity in human colorectal adenocarcinoma cells (Caco-2 cells and HT-29 cells) (Antunes-Ricardo et al. 2017).

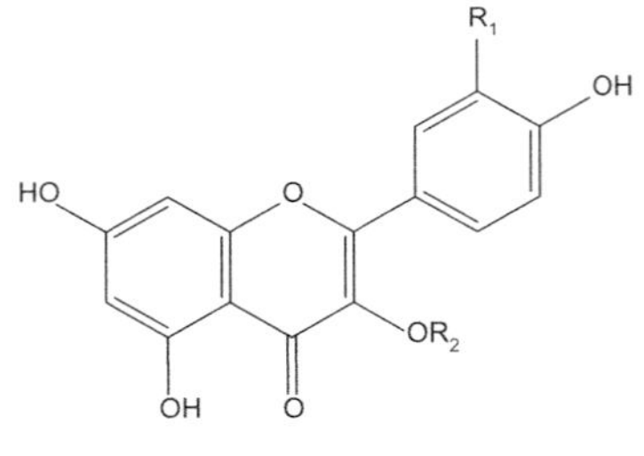

1	R_1=OH	R_2=Rha(1-6)Glc
2	R_1=H	R_2=Rha(1-6)Glc
3	R_1=OH	R_2=Glc
4	R_1=OCH_3	R_2=Rha(1-6)Glc
5	R_1=OCH_3	R_2=Gal
6	R_1=OCH_3	R_2=Glc
7	R_1=H	R_2=Ara

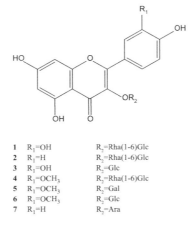

Figure 3.7 Flavonol glycosides isolated from the flowers of *Opuntia ficus indica*

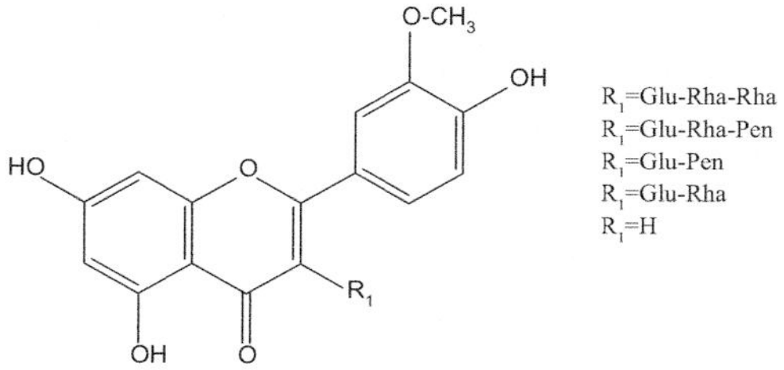

Figure 3.8 Chemical structures of isorhamnetin glycosides found in *Opuntia ficus indica* extract. Pen = pentose; Rha = rhamnose; Glc = glucose.

Non-alcoholic fatty liver disease is a pathology related to oxidative stress, inflammation, and cell death. When feeding obese rats with *O. ficus indica* for seven weeks, the rats decreased the concentration of liver triglycerides, observing a higher concentration of adiponectin and an increase in the genes involved in lipid peroxidation. In addition, treated animals showed a low concentration of postprandial serum insulin (Morán-Ramos et al. 2012).

Campylobacter is the most common cause of bacterial gastroenteritis. *O. ficus indica* extracts have been found to have antibacterial activity, inhibiting the growth of *Campylobacter jejuni* and *Campylobacter coli* (Castillo et al. 2011). The antimicrobial activity of polar extracts of *O. ficus indica* against *Vibrio cholerae* has been evaluated, showing that the methanol extract is the most efficient in preventing the growth of this microorganism, as it causes the rupture of its membrane (Sánchez et al. 2010).

3.6.2.5 *O. ficus indica* Exocarp

The exocarp, or peel, of the fruit of *O. ficus indica* contains high amounts of isorhamnetin (3'-methoxy-3,4', 5,7-tetrahydroxyflavone), which has anticancer activity; this compound has also presented a cardioprotective effect (Kim et al. 2011). Antioxidant metabolites such as gallic acid, oleide, epicatechin, and pyrogalol, among others, have also been isolated (Aruwa et al. 2018).

3.6.3 M. geometrizans

Also known as the 'blue candle' or 'blueberry' cactus, this species is native to Mexico's arid and semiarid regions, from which phenolic acids, flavonoids, betacyanins, and betaxanthins have been identified. Its fruit, which is known as the 'garambullo', contains various secondary metabolites such as vanillin, ellagic, caffeic, and gallic acid as well as quercetin and epicatechin, among others (Herrera-Hernández et al. 2011; Ramírez-Rodríguez et al. 2020). Its pulp has shown antioxidant activity due to its ability to absorb oxygen radicals, in addition to anti-inflammatory activity. Pulp and exocarp extracts inhibit the α-glucosidase enzyme by 85%, which could help reduce the digestion of starch and sucrose, thereby contributing to a better control of hyperglycemia (Montiel-Sánchez et al. 2021).

3.7 CURRENT APPLICATION IN THE DEVELOPMENT OF FOOD INDUSTRY AND BIOTECHNOLOGY FROM CACTACEAE

Studies have shown that Cactaceae members are good sources of nutrients and bioactive compounds. In addition, they are a resource that has a high agro-biotechnological potential, both as a food crop and as a base element for derivative products used in the food (human and animal), pharmacology, medicine, and agricultural industries (de Araújo et al. 2021; Ramírez-Rodríguez et al. 2020; Torres-Ponce et al. 2015).

3.7.1 Pharmaceutical Applications

The increased incidence of chronic diseases such as diabetes has represented a global and essential problem in recent years. That is why it is necessary to develop and explore alternative treatments to help reduce serum glucose levels. Among these treatments are medicinal plants with hypoglycemic effects. 'Nopal' (*Opuntia* spp.) is the most common plant used to control glucose because it has a high content of soluble fiber and pectin, affecting glucose uptake in the intestine it is, therefore, considered hypoglycemic. In animal studies, it has been reported to decrease postprandial glucose synergistic with insulin (Yeh et al. 2003). An ethnobotanical study that interviewed patients with diabetes and herbalists from Mexico confirmed that *Opuntia* spp. is traditionally used to treat non-insulin dependent diabetes (Andrade-Cetto and Wiedenfeld 2011). The medicinal parts are the tender cladodes, from which the spines are withdrawn, washed and cut, and eventually liquified with water to make a drink that is ingested before breakfast. The result is a decrease in postprandial glucose levels (Torres-Ponce et al. 2015).

Opuntia spp. is also known for its antioxidant capacity because of the presence of phytochemicals such as carotenoids, flavonoids, and other phenolic compounds as well as vitamin C and E (Torres-Ponce et al. 2015). The combination of dietary fiber associated with the phytochemicals described in the 'nopal' allows it to be used as a dietary supplement and food ingredient. In recent years, the commercialisation of dried 'nopal' fiber as an auxiliary in digestive disorders started. The dried pulp of 'nopal' is a fibrous material whose medicinal function is based, like any other natural fiber, on favoring the digestive process, reducing the risk of gastrointestinal problems, and helping in treatments against obesity (Bensadón et al. 2010).

Regarding the interaction of cacti with microorganisms, interestingly, the potential for obtaining antibiotics from cacti endophyte has also been described, as is the case of flavonoids from *Euphorbia caducifolia* (Singariya et al. 2018), metabolites in methanolic extracts from *Pachycereus marginatus* (Ramírez-Villalobos et al. 2021), and *Lophocereus marginatus* and *Metarhizium anisopliae* with cytotoxic activity against murine and human carcinogenic cells (Ramírez-Villalobos et al. 2021).

3.7.2 Food Applications

3.7.2.1 Supplement

The development of diverse foods with low caloric value and high dietary fiber content has occupied a preponderant place in the food industry in recent years because of the consumer's growing interest in healthy and nutritious diets. Pre-cooked or quick-to-prepare foods are attractive for the time they save; if such meals are nutritious, the attractiveness is even greater (Sáenz et al. 2002).

Studies have indicated that cladode flour from 'nopal' (*Opuntia* spp.) has a high fiber content, so it can be a good alternative source of fiber to enrich diets deficient in this compound (Maki-Díaz et al. 2015; Escobar Rodríguez 2017; Fernandes de Araújo et al. 2021). Sáenz et al. (2002) analysed different powder formulations to incorporate a dessert (flan) of 16%, 18%, and 20% of 'nopal' flour as a source of dietary fiber. It was observed that the flan with 16% of 'nopal' flour reached the best sensorial characteristics. More significantly, percentages of 'nopal' flour negatively affected the sensorial characteristics, mainly flavor, color, and texture. The analysis showed that the powder presented 5.7% of moisture, low water activity, as well as total recount of microorganisms. The content of protein was high (27.2%), the caloric contribution was low (40 Kcal/portion), and the total dietary fiber was high (9.8%) compared to a flan without 'nopal' flour. The contribution of dietary fiber in the selected formulation was more incredible than that of other similar commercial products, which, together with low energy content, made it a food that benefits human health (Sáenz et al. 2002).

While the 'tortilla de maíz' (*Zea mays L.*), a food consumed daily in a Mexican diet, has also been one of the targets to enrich with 'nopal' fiber. This food is the basis of a large part of the food culture because of its high nutritional value; however, its high caloric content has led to the search for alternatives that reduce this content; one of them is 'nopal' fiber (Ceja-León et al. 2021). A study carried out a bromatological analysis of the 'tortilla de maíz' 1:1 with 'nopal' flour (by cladodes dehydration). The results obtained in bromatological analysis indicated that the addition of 'nopal' flour decreased the caloric content and increased minerals (such as calcium) and fiber crude contents, although fat and protein content did not show an increase concerning control. Nevertheless, given the high-water activity of the 'nopal', the shelf life of the tortilla considerably decreases, causing significant deterioration in a few days (Hernández Hernández 2003).

Due to the high demand for healthy products and the excellent qualities of nutrients in 'nopal', among which its calcium and fiber content and its easy production, this has been used to enrich other products such as energy bars (Escobar Rodríguez 2017), pancakes (Bautista-Justo et al. 2010; Marino 2018), ice creams, yogurt, sweets, and beverages among other products (Fernandes de Araújo et al. 2021), which suggests that, because of its low caloric intake, it can be consumed without negative consequences to health (Bautista-Justo et al. 2010).

3.7.2.2 Natural Additive

The current demand for the consumption of minimally processed natural foods has launched a search for new natural alternatives for its conservation through reducing microbial attack and oxidation prevention. Mucilage is a compound in the cladodes, skin, and pulp of some species of the Cactaceae family. Polysaccharide with a high molecular weight, with food industry studies about changes in viscosity, elasticity, and water retention. Besides this, the polysaccharide has a high gelling and emulsifying power (Martínez et al. 2016). The rheofluidising or pseudoplastic behavior and viscosity of the mucilage in aqueous or saline suspensions, depending on polysaccharide concentration, is like that of some gums, such as xanthan, locust bean, guar, and others, which are used as thickening agents and to modify the rheology of foods (López-Palacios et al. 2016).

Buendía et al. (2020), studied adding 'nopal' mucilage to a cream-type corn soup and evaluating its thickening and sensory acceptance. Mucilage was extracted from cladodes of the *Atlixco, Milpa Alta, Toluca, Tobarito,* and *Tuna Blanca* variants of *Opuntia* spp. The evaluated concentrations were 0.7% and 1.0% and with starch as the control (commercial starch). The results did not show significant differences between treatments on their pH (6.09 and 6.32) and their viscosity (1.623 and 6.043 Pa s) with respect to the controls. With regard to their chromaticity (17.47° and 18.49°), it was similar to controls and tended to have higher luminosity (46.15° and 47.72°) and lower tone (78.36° and 79.52°). The sensorial soup viscosity acceptability was like controls, except with mucilage of *Atlixco* and *Toluca* variants (17% and 7% lower), and the overall sensorial acceptability of the soups with mucilage of the five variants was up to 19% lower compared to controls. The authors concluded that the 'nopal' mucilage (*Opuntia* spp.) of the *Atlixco, Milpa Alta, Toluca, Tobarito,* and *Tuna Blanca* variants is a candidate to add to cream-type corn soup because, in general, it equals physical characteristics, such as pH, color, and viscosity of commercial starch thickened soup. Because of its functional characteristics, mucilage could be considered more suitable as a thickener than starch (Reyes-Buendía et al. 2020).

In another study, the effect of adding celery juice and 'nopal' mucilage to Hass avocado puree for preservation was evaluated. Avocados with a maturity of three days were added to celery juice and 'nopal' mucilage at 2.5% and 5% p/p, respectively. Puree samples were stored at −18 °C for 50 days and were evaluated for puree samples, assessing the oxidation, pH, and color. The best treatment corresponded to avocado puree with 'nopal' mucilage at 5% p/p, which significantly modified the natural color of the avocado; colorimetric measurements showed an increase in luminosity and a decrease in chromaticity and the pH of the preparation during storage, and the titratable acidity was not significantly affected by the treatment, a result that indicates that lipid oxidation was low. In this study, the uses of natural vegetables for the conservation of avocados such as celery and 'nopal' mucilage, because of their properties, showed good conservation results in combination with freezing, which increased the life of Hass avocados for 50 days, considering that this fruit undergoes strong enzymatic oxidation in a matter of minutes (Martínez et al. 2016).

3.7.2.3 Alcoholic Beverages

Notably, the Cactaceae family has been described since pre-Hispanic times as a source of raw material for fermentation of traditional Mexican beverages. The characterisation of the genomic data from the microorganisms associated with the production of Mexican fermented beverages and the development of high a quality product, or bioproducts with biotechnological applications in the food or the pharmaceutical industries, which will have an economic impact on all producer communities and on the transmission of knowledge in our culture (Robledo-Márquez et al. 2021).

For example, the microbiota native to cactus and their bioactive metabolites produced during the fermentation of the *Agave* spp. or *Opuntia streptacantha* sap (alcoholic beverages 'pulque' and 'colonche,' respectively). Such as agavins that promote iron and calcium absorption, alleviate chest and gastrointestinal pain, and induce menstruation and milk production. In addition, betalains have been associated with chain fatty acids production and leptin generation (Oleszek and Oleszek 2020), besides containing other bioactive compounds. These compounds are produced by several microorganisms such as *Proteobacteria*, *Firmicutes*, *Acidobacteria*, *Actinobacteria*, *Cyanobacteria*, *Fusobacteria*, and *Nitrospira* (Robledo-Márquez et al. 2021).

3.7.2.4 Advanced Material: Biopolymers or Edible Films

Depending on the type of product, the Food and Agriculture Organization of the United Nations has estimated that postharvest losses of fresh fruits and vegetables represent between 5% and 25% of production in developed countries and between 20% and 50% in developing countries (Bello-Lara et al. 2016). Various techniques have been implemented to help solve the problem of postharvest losses. An alternative that has gained acceptance in the fruit and vegetable market is the application of biopolymers or edible films (Fernández Valdés et al. 2015).

These edible films can retard moisture migration and the loss of volatile compounds, reduce the respiration rate, and delay changes in textural properties. Also, they are excellent barriers to fats and oils and have a highly selective gas permeability ratio CO_2/O_2 compared to conventional synthetic films (Del-Valle et al. 2005). Their composition, based on carbohydrates, proteins, and lipids, makes them edible and easy to apply directly to the surface of food products.

The 'nopal' (*O. ficus indica*), whose young cladodes ('nopalitos') are rich in polysaccharides of the structural type. These macromolecules are monosaccharide units, which their physicochemical characteristics have grouped into mucilages, pectins, hemicelluloses, and celluloses (Bello-Lara et al. 2016). 'Nopal' mucilage is considered a hydrocolloid, matrix-forming substance, a characteristic necessary to obtain a coating (Peña-Valdivia et al. 2012). Pectins are hydrocolloids with a complex composition and an essential function as primary cell wall constituents. These substances function as a moisturising agent and cementing material of the cellulose networks. In addition, one of the most attractive characteristics of pectin is that it is a non-toxic, biocompatible, and biodegradable biopolymer (Bello-Lara et al. 2016).

Rambutan (*Nephelium lappaceum L.*) is a tree that offers an exotic fruit, native to Malaysia and Indonesia, that is mainly marketed fresh because of its attractive color and shape. However, its good appearance is preserved for a short time, mainly due to dehydration and darkening of the epicarp. Brindis-Trujillo et al. (2020) evaluated the effect of coating this fruit with 'nopal' mucilage (*O. ficus indica*). The factors were the type of container (polyethylene bag and polystyrene container), coating (with and without coating), and time (0, 3, 6, 10, and 12 days). The coating consisted of mucilage obtained from developing cladodes (15–21 cm) and applied by immersion. All treatments were stored at 5 °C. Total soluble solids, firmness, and color were evaluated at each storage period. Likewise, 40 untrained judges (47% men and 53% women) evaluated sensory acceptability, consumption intention, and acceptance/rejection. The coating and the polyethylene container at 5 °C, had a significant effect on the color and acceptability of the product. The intention to consume was greater, and they were maintained for 10 days compared to the 6 days of the fruits without coating (Brindis-Trujillo et al. 2020).

The preservation of the parameters of color, firmness, and shelf life because of the coating with edible films of 'nopal' mucilage or pectin has also been reported in strawberries (Del-Valle et al. 2005), Hass avocados (Bello-Lara et al. 2016), blueberries (Ginez Povez and Godoy Hernández 2018), guavas (Fernández Paredes 2019), and 'nopal' (González-González 2011).

3.7.2.5 Animal Nutrition Application

The demand for food that can be used in both human and animal diets has increased recently. Factors such as drought and climate change can affect the production of cereals and increase their prices on the international market, thus, hampering access to this raw material (Fernandes de Araújo et al. 2021). Given this, the use of nutritious species such as cactus can be an alternative for agribusiness, as these crops are multifunctional, resistant to drought conditions, and grow in semiarid and arid regions (Abidi et al. 2009; Morales n.d.).

Within this context, the addition of *O. ficus indica* in Tifton 85 silage impacted the morphometric measurements, carcass compactness index, loin eye area, and the weight of casting of commercial cuts. They also observed a reduction in meat color parameters and pH and changes in fatty acid composition in the muscle of lambs (do Nascimento Souza et al. 2020). Another study found the ingestion of *O. ficus indica* reduced the fat concentration in the goat carcass; however, it was responsible for a higher concentration of linoleic acid and a higher amount of polyunsaturated fatty acids in the meat (Mahouachi et al. 2012).

Regarding milk composition, Catunda et al. (2016) evaluated the influence of supplying five cacti species from the Brazilian semiarid northeast region on the physical-chemical sensory characteristics and the profile of fatty acids of Saanen goat milk. The goats were supplemented with 473–501 g/kg of a Cactaceae mix (*Pilosocereus gounellei, Cereus jamacaru, Cereus squamosus, Nopalea cochenillifera*, or *Opuntia stricta*). The supplementation of the mix did not influence the milk´s sensory characteristics and lipid profile, but there were changes in the levels of protein, lactose, non-fat solids, and cryoscopy point (Catunda et al. 2016).

3.7.3 Water Treatment

Chemical coagulants, widely used in water treatment to remove colloids and fine particles, despite being very efficient, can negatively impact the environment. Thus, some studies have shown the potential of natural coagulating agents derived from plants to replace chemical flocculating agents (Fernandes de Araújo et al. 2021). The Cactaceae mucilage is a natural polymer that is soluble in water, which forms a gel that can attract particles and/or metals, forming biofilms (known as flocs). Through infrared spectroscopy, it was confirmed that the 'nopal' mucilage contains functional groups, such as hydroxyl and carboxyl groups, which allow the biopolymer to be cross-linked, generating more resistance and being more helpful in the removal of heavy metals dissolved in water (Ovando Franco 2012). Bouaouine et al. (2018) described that the 'nopal' mucilage (*Opuntia* spp.) treatment in the clarification of wastewater reduces water turbidity by more than 90% (Bouaouine et al. 2019). Another study revealed that mucilage treatment reduced the turbidity of wastewater by about 70%, which was compared with aluminum sulfate, a metal coagulant, which reduced turbidity by approximately 83%. However, although the effect was greater with the aluminum sulfate, mucilage is an alternative to traditional coagulants and an environmentally friendly option for treating turbid waters (Lisintuña et al. 2020).

Currently, there are no studies on the use of mucilage from other cacti, specifically that use the fruit peels or stems of cacti and with the properties discussed in this section.

3.7.4 Other Applications

Showing that biotechnological developments associated with cacti are relevant, and it is necessary to promote the application of these biotechnologies sustainably in different industries such as cosmetics, food, and pharmaceuticals. In a review, the authors used the terms Cactaceae, cactus, palm fruit, *O. ficus indica*, among other words, to check the number of patent applications granted, observing a total of 15,184 inventions between 2001 and 2020 patented in Questel Intellectual Property Portal on the internet. The technological domain is represented mainly by basic materials in chemistry, biotechnology, medical technology, environmental technology, food chemistry, fine organic chemistry, other particular machines, and pharmaceuticals (Fernandes de Araújo et al. 2021).

Some of the applications related to these species involve the development of new food products (Akanni et al. 2014), production of polymeric material for application in the manufacture of plastic parts (Villada et al. 2009) and natural dyes (Carmona Rodríguez 2020), development of conservation of architectural finishes, objects, and mural works (Torres-Soria et al. N.d.), and development of energy production (Morales n.d.), organic fertiliser (Medina Díaz and Borrero Ortiz 2017), and cosmetics and skin moisturisers (Nazareno et al. 2013), among others.

A very peculiar case is Cochineal (*Dactylopius coccus Costa*). It is considered one of the main sources of natural coloring (carminic acid) worldwide, which has a tremendous economic impact. Nevertheless, insect behavior and coloring quality obtained until now from hosts of different geographic origins caused differential effects on survival, reproduction, and carminic acid contents. So, there is a field of opportunities to carry out biotechnological studies and designs in the future.

There is another trend for the benefit of the plant–microorganism interaction, and they are bioenergy crops, in which they seek to generate crops for the creation of biomass for biofuels (Howe et al. 2022). However, there is still a lack of depth in elucidating plant microbiomes. For example, *Pectobacterium cacticida*, a soil-derived eubacterium, is capable of degrading sugars during cladode fermentation and efficient ethanol production (Blair et al. 2021). In the case of biobatteries, the microbiome system of the plant is also involved; the efficiency demonstrated for the microbial fuel of the *Opuntia albicarpa* plant was 285.2 J (Apollon et al. 2020).

3.8 CONCLUSION AND PERSPECTIVES

In this chapter, we have reported that some Cactaceae or their portions offer a wide variety of uses that may contribute to the sustainable development of the drylands' inhabitants. As we have noted, these cacti have high fiber, protein, and mineral content and a low caloric intake, which makes them an excellent food alternative for various at-risk populations. Their medicinal properties can be attributed to their bioactive product content: polyphenols, phytosterols, betalains, antioxidants, vitamins, and minerals, among others.

Nutritional and pharmacological research needs to be done to identify the specific molecules with biological activity that will improve human well-being. Recent evidence has shown that consumption of cacti portions (e.g., fruit, cladodes, 'nopal' juice, 'nopal' flour) confer antioxidant, lipid-lowering, and hypoglycemic activity in experimental studies. Thus, their introduction into a daily diet could be an alternative for reducing the progression or incidence of metabolic diseases.

As indicated in this chapter, Cactaceae opens a new scope for industrial applications, improving exploitation and sustainable production with multiple applications in the food, pharmaceutical, chemical, and bioenergy industries. It will benefit the development of new edible products with high nutritional value, fiber supplements, isolation and development of new drugs, and production of natural colorants, among others, which will be advantageous for the people who live in the Mexican drylands. Therefore, collaborations are needed in which researchers from multiple disciplines work together with the members from local communities (ethnobotanical knowledge holders must provide permission and receive benefits) and from different sectors to generate projects for sustainable production of these cacti in the food, pharmaceutical, and biotechnology industries.

ACKNOWLEDGMENTS

The authors would like to thank the Instituto Potosino de Investigación Científica y Tecnológica, AC for the use of its facilities and Consejo Nacional de Ciencia y Tecnología (CONACyT) for its financial support (FORDECYT-PRONACES 101732). Joel Flores Rivas for the use photography's. Joyce Trujillo, Alfredo Ramírez-Hernández, Alethia Muñiz-Ramírez and Sandra Pérez-Miranda are supported as research fellows in the CONACyT Cathedra's program (project numbers 615 and 887). Yadira Ramírez-Rodríguez is supported by grants from CONACyT (637582). CONACyT had no role in the design, analysis, or writing of this chapter. #CienciaBajoProtesta

REFERENCES

Abidi, Sourour, Hichem Ben Salem, Valentina Vasta, and Alessandro Priolo. 2009. "Supplementation with Barley or Spineless Cactus (Opuntia Ficus Indica f. Inermis) Cladodes on Digestion, Growth and Intramuscular Fatty Acid Composition in Sheep and Goats Receiving Oaten Hay." *Small Ruminant Research* 87 (1–3): 9–16. https://doi.org/10.1016/j.smallrumres.2009.09.004.

Adeleke, Bartholomew Saanu, and Olubukola Oluranti Babalola. 2021. "The Endosphere Microbial Communities, a Great Promise in Agriculture." *International Microbiology* 24 (1): 1–17. https://doi.org/10.1007/s10123-020-00140-2.

Ahmad, Afshan, Judith Davies, Sharon Randall, and Gordon Skinner. 1996. "Antiviral Properties of Extract of Opuntia Streptacantha." *Antiviral Research* 30 (2–3): 75–85. https://doi.org/10.1016/0166-3542(95)00839-X.

Akanni, Gabriel, Victor Ntuli, and James C Preez. 2014. "Review Article Cactus Pear Biomass , a Potential Lignocellulose Raw Material for Single Cell Protein Production (SCP): A Review." *International Journal of Current Microbiology and Applied Sciences* 3 (7): 171–197.

Aliscioni, Nayla Luján, Natalia E Delbón, and Diego E Gurvich. 2021. "Spine Function in Cactaceae, a Review." *Journal of the Professional Association for Cactus Development* 23: 1–16.

Ambardar, Sheetal, Heikham Russiachand Singh, Malali Gowda, and Jyoti Vakhlu. 2016. "Comparative Metagenomics Reveal Phylum Level Temporal and Spatial Changes in Mycobiome of Belowground Parts of Crocus Sativus." *PLOS ONE* 11 (9): e0163300. https://doi.org/10.1371/JOURNAL.PONE.0163300.

Anderson, Edward F. 2001. *The Cactus Family*. Timber Press, 2001.

Andrade-Cetto, Adolfo, and Helmut Wiedenfeld. 2011. "Anti-Hyperglycemic Effect of Opuntia Streptacantha Lem." *Journal of Ethnopharmacology* 133 (2): 940–43. https://doi.org/10.1016/j.jep.2010.11.022.

Antunes-Ricardo, Marilena, César Rodríguez-Rodríguez, Janet Gutiérrez-Uribe, Eduardo Cepeda-Cañedo, and Sergio Serna-Saldívar. 2017. "Bioaccessibility, Intestinal Permeability and Plasma Stability of Isorhamnetin Glycosides from Opuntia Ficus-Indica (L.)." *International Journal of Molecular Sciences* 18 (8): 1816. https://doi.org/10.3390/ijms18081816.

Apollon, Wilgince, Sathish-Kumar Kamaraj, Héctor Silos-Espino, Catarino Perales-Segovia, Luis L. Valera-Montero, Víctor A. Maldonado-Ruelas, Marco A. Vázquez-Gutiérrez, Raúl A. Ortiz-Medina, Silvia Flores-Benítez, and Juan F. Gómez-Leyva. 2020. "Impact of Opuntia Species Plant Bio-Battery in a Semi-Arid Environment: Demonstration of Their Applications." *Applied Energy* 279 (December): 115788. https://doi.org/10.1016/j.apenergy.2020.115788.

Araya, Juan Pablo, Máximo González, Massimiliano Cardinale, Sylvia Schnell, and Alexandra Stoll. 2020. "Microbiome Dynamics Associated With the Atacama Flowering Desert." *Frontiers in Microbiology* 10 (January). https://doi.org/10.3389/FMICB.2019.03160.

Aruwa, Christiana Eleojo, Stephen O. Amoo, and Tukayi Kudanga. 2018. "Opuntia (Cactaceae) Plant Compounds, Biological Activities and Prospects: A Comprehensive Review." *Food Research International* 112 (October): 328–44. https://doi.org/10.1016/j.foodres.2018.06.047.

Astello-García, Marizel G., Ilse Cervantes, Vimal Nair, María del Socorro Santos-Díaz, Antonio Reyes-Agüero, Françoise Guéraud, Anne Negre-Salvayre, Michel Rossignol, Luis Cisneros-Zevallos, and Ana P. Barba de la Rosa. 2015. "Chemical Composition and Phenolic Compounds Profile of Cladodes from *Opuntia* Spp. Cultivars with Different Domestication Gradient." *Journal of Food Composition and Analysis* 43 (November): 119–30. https://doi.org/10.1016/j.jfca.2015.04.016.

Bargougui, Ahlem, Hend Maarof Tag, Mohamed Bouaziz, and Saida Triki. 2019. "Antimicrobial, Antioxidant, Total Phenols and Flavonoids Content of Four Cactus (Opuntiaficus-Indica) Cultivars." *Biomedical and Pharmacology Journal* 12 (3). https://doi.org/10.13005/bpj/1764.

Bautista-Justo, México, Pineda Torres, Rosa Inés, Da Mota, Victor Manuel, and José Eleazar. 2010. "Redalyc. El Nopal Fresco Como Fuente de Fibra y Calcio En Panqués." *Acta Universitaria* 20 (3). http://www.redalyc.org/articulo.oa?id=41615947002.

Bello-Lara, J Esteban, Rosendo Balois-Morales, M Teresa Sumaya-Martínez, Porfirio Juárez-López, Edgar I Jiménez-Ruíz, Leticia M Sánchez-Herrera, Graciela G López-Guzmán, and J Diego García-Paredes. 2016. "Biopolímeros de Mucílago, Pectina de Nopalitos y Quitosano, Como En Almacenamiento y Vida de anaquelde Frutos de Aguacate "Hass."" *Acta Agrícola y Pecuaria* 2 (2): 43–50.

Benayad, Zakia, Cristina Martinez-Villaluenga, Juana Frias, Carmen Gomez-Cordoves, and Nour Eddine Es-Safi. 2014. "Phenolic Composition, Antioxidant and Anti-Inflammatory Activities of Extracts from Moroccan Opuntia Ficus-Indica Flowers Obtained by Different Extraction Methods." *Industrial Crops and Products* 62 (December): 412–20. https://doi.org/10.1016/j.indcrop.2014.08.046.

Bensadón, Sara, Deisy Hervert-Hernández, Sonia G. Sáyago-Ayerdi, and Isabel Goñi. 2010. "By-Products of Opuntia Ficus-Indica as a Source of Antioxidant Dietary Fiber." *Plant Foods for Human Nutrition* 65 (3): 210–16. https://doi.org/10.1007/s11130-010-0176-2.

Bergantin, Caterina, Annalisa Maietti, Alberto Cavazzini, Luisa Pasti, Paola Tedeschi, Vincenzo Brandolini, and Nicola Marchetti. 2017. "Bioaccessibility and HPLC-MS/MS Chemical Characterization of Phenolic Antioxidants in Red Chicory (Cichorium Intybus)." *Journal of Functional Foods* 33 (June): 94–102. https://doi.org/10.1016/j.jff.2017.02.037.

Bezerra, Jadson Diogo Pereira, João Lucio de Azevedo, and Cristina Maria Souza-Motta. 2017. "Why Study Endophytic Fungal Community Associated with Cacti Species?" In *Diversity and Benefits of Microorganisms from the Tropics*, 21–35. Cham: Springer International Publishing. https://doi.org/10.1007/978-3-319-55804-2_2.

Blair, Brittany Braden., Won Cheol Yim, and John C. Cushman. 2021. "Characterization of a Microbial Consortium with Potential for Biological Degradation of Cactus Pear Biomass for Biofuel Production." *Heliyon* 7 (8): e07854. https://doi.org/10.1016/j.heliyon.2021.e07854.

Bouaouine, O., Bourven, I., Khalil, F. et al. 2018. Identification of functional groups of Opuntia ficus-indica involved in coagulation process after its active part extraction. *Environmental Science and Pollution Research* 25: 11111–19. https://doi.org/10.1007/s11356-018-1394-7

Bouaouine, Omar, Isabelle Bourven, Fouad Khalil, Philippe Bressollier, and Michel Baudu. 2019. "Identification and Role of Opuntia Ficus Indica Constituents in the Flocculation Mechanism of Colloidal Solutions." *Separation and Purification Technology* 209 (January): 892–99. https://doi.org/10.1016/j.seppur.2018.09.036.

Brindis-Trujillo, Raúl Alberto, Ma Rosa Salinas-Hernández, and Hortensia Brito-Vega. 2020. "Efecto Del Recubrimiento Con Mucílago de Nopal (Opuntia Ficus-Indica) En La Conservación de Rambután (Nephelium Lappaceum L.) Mínimamente Procesado." *Revista Iberoamericana de Tecnología Postcosecha* 21 (1). https://www.redalyc.org/articulo.oa?

Camarena-Pozos, David A., Victor Manuel Flores-Núñez, M. G. López, and Laila Pamela Partida-Martínez. 2021. "Fungal Volatiles Emitted by Members of the Microbiome of Desert Plants Are Diverse and Capable of Promoting Plant Growth." *Environmental Microbiology* 23 (4): 2215–29. https://doi.org/10.1111/1462-2920.15395.

Carmona Rodríguez, Juan Carlos. 2020. *Micropartículas de pulpa de tuna anaranjada Opuntia ficus indica con mucílago de nopal y su aplicación como colorante en alimentos*. Santiago, Chile: Universidad de Chile.

Castillo, Sandra L., Norma Heredia, Juan F. Contreras, and Santos García. 2011. "Extracts of Edible and Medicinal Plants in Inhibition of Growth, Adherence, and Cytotoxin Production of Campylobacter Jejuni and Campylobacter Coli." *Journal of Food Science* 76 (6): M421–26. https://doi.org/10.1111/j.1750-3841.2011.02229.x.

Catunda, Karen Luanna Marinho, Emerson Moreira de Aguiar, Pedro Etelvino de, Góes Neto, José Geraldo Medeiros da Silva, José Aparecido Moreira, Adriano Henrique do Nascimento Rangel, and Dorgival Morais de Lima Júnior. 2016. "Gross Composition, Fatty Acid Profile and Sensory Characteristics of Saanen Goat Milk Fed with Cacti Varieties." *Tropical Animal Health and Production* 48 (6): 1253–59. https://doi.org/10.1007/s11250-016-1085-7.

Ceja León, Edna Sofía, Ma. Guadalupe Sánchez Saavedra, Ricardo Pérez Cárdenas, Pedro Damian Loeza Lara, and Verónica Núñez Oregel. 2021. "Evaluación de Aditivos Químicos y Biológicos En La Vida de Anaquel de La Tortilla de Nopal (Opuntia Ficus-Indica)." *South Florida Journal of Development* 2 (4): 6133–41. https://doi.org/10.46932/sfjdv2n4-087.

Chavez-Santoscoy, Rocío Alejandra, Janet. A. Gutierrez-Uribe, and Sergio O. Serna-Saldívar. 2009. "Phenolic Composition, Antioxidant Capacity and In Vitro Cancer Cell Cytotoxicity of Nine Prickly Pear (*Opuntia* Spp.) Juices." *Plant Foods for Human Nutrition* 64 (2): 146–52. https://doi.org/10.1007/s11130-009-0117-0.

Ciriminna, Rosaria, Norberto Chavarría-Hernández, Adriana I. Rodríguez-Hernández, and Mario Pagliaro. 2019. "Toward Unfolding the Bioeconomy of Nopal (*Opuntia* Spp.)." *Biofuels, Bioproducts and Biorefining* 13 (6): 1417–27. https://doi.org/10.1002/bbb.2018.

Coll Hurtado, Atlántida. 1998. "Oaxaca: Geografía Histórica de La Grana Cochinilla." *Investigaciones Geográficas* 36: 71–82.

Davis, Sarah C, June Simpson, Katia del Carmen Gil-Vega, Nicholas A Niechayev, Evelien van Tongerlo, Natalia Hurtado Castano, Louisa V Dever, and Alberto Búrquez. 2019. "Undervalued Potential of Crassulacean Acid Metabolism for Current and Future Agricultural Production." *Journal of Experimental Botany* 70 (22): 6521–37. https://doi.org/10.1093/jxb/erz223.

De Leo, Marinella, Abreu M. De Bruzual, Agata Maria Pawlowska, Pier Luigi Cioni, and Alessandra Braca. 2010. "Profiling the Chemical Content of Opuntia Ficus-Indica Flowers by HPLC-PDA-ESI-MS and GC/EIMS Analyses." *Phytochemistry Letters* 3 (1). https://doi.org/10.1016/j.phytol.2009.11.004.

Del-Valle, Valeria, Pilar Hernández-Muñoz, Abél Guarda, and María José Galotto. 2005. "Development of a Cactus-Mucilage Edible Coating (Opuntia Ficus Indica) and Its Application to Extend Strawberry (Fragaria Ananassa) Shelf-Life." *Food Chemistry* 91 (4): 751–56. https://doi.org/10.1016/j.foodchem.2004.07.002.

De la Torre-Hernández, María Eugenia, Leilani I. Salinas-Virgen, J. Félix Aguirre-Garrido, Antonio J. Fernández-González, Francisco Martínez-Abarca, Daniel Montiel-Lugo, and Hugo C. Ramírez-Saad. 2020. "Composition, Structure, and PGPR Traits of the Rhizospheric Bacterial Communities Associated With Wild and Cultivated Echinocactus Platyacanthus and Neobuxbaumia Polylopha." *Frontiers in Microbiology* 11 (June). https://doi.org/10.3389/fmicb.2020.01424.

El-Mostafa, Karym, Youssef El Kharrassi, Asmaa Badreddine, Pierre Andreoletti, Joseph Vamecq, M' Hammed, Saïd El Kebbaj, et al. 2014. "Molecules Nopal Cactus (Opuntia Ficus-Indica) as a Source of Bioactive Compounds for Nutrition, Health and Disease." *Molecules* 19: 14879–901. https://doi.org/10.3390/molecules190914879.

Escobar, Rodríguez, and Luz María. 2017. *Optimización de Barra de Nopal de "Alto Contenido de Fibra."* Rodríguez: Universidad Autónoma de Barcelona.

Espinosa-García, N., C. Llanderal-Cázares, K. Miranda-Perkins, M. Vargas-Hernández, H. González-Hernández, and J. Romero-Nápoles. 2018. "Induced Infestation of Red Worm Comadia Redtenbacheri in Agave Salmiana." *Southwestern Entomologist* 43 (4). https://doi.org/10.3958/059.043.0418.

Fernandes de Araújo, Fábio, David de Paulo Farias, Iramaia Angélica Neri-Numa, and Glaucia Maria Pastore. 2021. "Underutilized Plants of the Cactaceae Family: Nutritional Aspects and Technological Applications." *Food Chemistry*. Elsevier Ltd. https://doi.org/10.1016/j.foodchem.2021.130196.

Fernández Paredes, Manuel Enrique. 2019. *Conservación de La Guayaba (Psidium Guajava l) Mediante La Aplicación de Un Comestible a Base de Mucílago de Nopal (Opuntia Ficus Indica) Con Aceite de Tomillo.* Ambato, Ecuador: UNIVERSIDAD TÉCNICA DE AMBATO.

Fernández Valdés, Daybelis, Silvia Bautista Baños, Dayvis Fernández Valdés, Arturo Ocampo Ramírez, Annia García Pereira, and Alejandro Falcón Rodríguez. 2015. "Películas y Recubrimientos Comestibles: Una Alternativa Favorable En La Conservación Poscosecha de Frutas y Hortalizas." *Revista Ciencias Técnicas Agropecuarias* 24 (3): 52–7.

Flores-Núñez, Víctor Manuel, Citlali Fonseca-García, Damaris Desgarennes, Emiley Eloe-Fadrosh, Tanja Woyke, and Laila Pamela Partida-Martínez. 2020. "Functional Signatures of the Epiphytic Prokaryotic Microbiome of Agaves and Cacti." *Frontiers in Microbiology* 10 (January). https://doi.org/10.3389/fmicb.2019.03044.

Fonseca García, Citlali, Damaris Desgarennes, Víctor Manuel Flores-Núñez, and Laila Pamela Partida-Martínez. 2018. "The Microbiome of Desert CAM Plants: Lessons From Amplicon Sequencing and Metagenomics." *Metagenomics: Perspectives, Methods, and Applications* January: 231–54. https://doi.org/10.1016/B978-0-08-102268-9.00012-4.

Fuhrman, Bianca, Nina Volkova, Raymond Coleman, and Michael Aviram. 2005. "Grape Powder Polyphenols Attenuate Atherosclerosis Development in Apolipoprotein E Deficient (E0) Mice and Reduce Macrophage Atherogenicity." *The Journal of Nutrition* 135 (4): 722–28. https://doi.org/10.1093/jn/135.4.722.

Galati, Enza Maria, Maria Rita Mondello, Eugenia Rita Lauriano, Maria Fernanda Taviano, M. Galluzzo, and Natalizia Miceli. 2005. "Opuntia Ficus Indica (L.) Mill. Fruit Juice Protects Liver from Carbon Tetrachloride-Induced Injury." *Phytotherapy Research* 19 (9): 796–800. https://doi.org/10.1002/ptr.1741.

Galati, Enza Maria, Maria Rita Mondello, Daniele Giuffrida, Giacomo Dugo, Natalizia Miceli, Simona Pergolizzi, and Maria Fernanda Taviano. 2003. "Chemical Characterization and Biological Effects of Sicilian *Opuntia Ficus Indica* (L.) Mill. Fruit Juice: Antioxidant and Antiulcerogenic Activity." *Journal of Agricultural and Food Chemistry* 51 (17): 4903–8. https://doi.org/10.1021/jf030123d.

Galati, Enza Maria, Maria Marcella Tripodo, Ada Trovato, Natalizia Miceli, and Maria Teresa Monforte. 2002. "Biological Effect of Opuntia Ficus Indica (L.) Mill. (Cactaceae) Waste Matter." *Journal of Ethnopharmacology* 79 (1): 17–21. https://doi.org/10.1016/S0378-8741(01)00337-3.

Gallegos-Vázquez, Clemente, Santiago de Jesús Méndez Gallegos, and Candelario Mondragón Jacobo. 2013. *Producción Sustentable de Tuna En San Luis Potosí*. 1st ed. San Luis Potosí, México: Colegio de Postgraduados–Fundación Produce San Luis Potosí.

Ginez Povez, Patricia, and Melissa Sue Godoy Hernández. 2018. *Formulación de un recubrimiento comestibles preservante de arándano empleando mucílago extraído de la penca de tuna (Opuntia Ficus-Indica)*. Perú: Universidad Nacional Del Callao.

Goettsch, Bárbara, Craig Hilton-Taylor, Gabriela Cruz-Piñón, James P. Duffy, Anne Frances, Héctor M. Hernández, Richard Inger, et al. 2015. "High Proportion of Cactus Species Threatened with Extinction." *Nature Plants* 1 (10): 15142. https://doi.org/10.1038/nplants.2015.142.

Gómez-Silva, Benito, Claudia Vilo-Muñoz, Alexandra Galetović, Qunfeng Dong, Hugo G. Castelán-Sánchez, Yordanis Pérez-Llano, María Del Rayo Sánchez-Carbente, et al. 2019. "Metagenomics of Atacama Lithobiontic Extremophile Life Unveils Highlights on Fungal Communities, Biogeochemical Cycles and Carbohydrate-Active Enzymes." *Microorganisms* 7 (12): 2064–77. https://doi.org/10.3390/MICROORGANISMS7120619.

Gómez-Maqueo, Andrea, Tomás García-Cayuela, Rebeca Fernández-López, Jorge Welti-Chanes, and M Pilar Cano. 2019. "Inhibitory Potential of Prickly Pears and Their Isolated Bioactives against Digestive Enzymes Linked to Type 2 Diabetes and Inflammatory Response." *Journal of the Science of Food and Agriculture* 99 (14): 6380–91. https://doi.org/10.1002/jsfa.9917.

González-González, Leandro Rodrigo. 2011. "Desarrollo y Evaluación de Una Película Comestible Obtenida Del Mucílago de Nopal (Opuntia Ficus-Indica) Utilizada Para Reducir La Tasa de Respiración de Nopal Verdura." *Investigación Universitaria Multidisciplinaria*, 2011, 1–8.

González Medrano, Francisco. 2012. *Las Zonas Áridas y Semiáridas de México y Su Vegetación. INE-SEMARNAT.* México: INE-SEMARNAT.

Govindasamy, Venkadasamy, Priya George, S. V. Ramesh, P. Sureshkumar, Jagadish Rane, and P. S. Minhas. 2022. "Characterization of Root-Endophytic Actinobacteria from Cactus (Opuntia Ficus-Indica) for Plant Growth Promoting Traits." *Archives of Microbiology* 204: 150. https://doi.org/10.1007/s00203-021-02671-2.

Grodsky, Steven M, and Rebecca R. Hernandez. 2020. "Reduced Ecosystem Services of Desert Plants from Ground-Mounted Solar Energy Development." *Nature Sustainability* 3 (12): 1036–43. https://doi.org/10.1038/s41893-020-0574-x.

Guerrero, Pablo C, Lucas C Majure, Amelia Cornejo-Romero, and Tania Hernández-Hernández. 2019. "Phylogenetic Relationships and Evolutionary Trends in the Cactus Family." *Journal of Heredity* 110 (1): 4–21. https://doi.org/10.1093/jhered/esy064.

Guillen-Cruz, G., A.L. Rodríguez-Sánchez, F. Fernández-Luqueño, and D. Flores-Rentería. 2021. "Influence of Vegetation Type on the Ecosystem Services Provided by Urban Green Areas in an Arid Zone of Northern Mexico." *Urban Forestry & Urban Greening* 62 (July): 127135. https://doi.org/10.1016/j.ufug.2021.127135.

Guzmán, U., S. Arias, and P. Dávila. 2003. *Catálogo de Cactáceas Mexicanas*. México, DF: Universidad Nacional Autónoma de México and Comisión Nacional para el Conocimiento y Uso de la Biodiversidad.

Hahm, Sahng-Wook, Jieun Park, Se-Yeong Oh, Chul-Won Lee, Kun-Young Park, Hyunggee Kim, and Yong-Suk Son. 2015. "Anticancer Properties of Extracts from *Opuntia Humifusa* Against Human Cervical Carcinoma Cells." *Journal of Medicinal Food* 18 (1): 31–44. https://doi.org/10.1089/jmf.2013.3096.

Hernández-Hernández, Tania, Joseph W. Brown, Boris O. Schlumpberger, Luis E. Eguiarte, and Susana Magallón. 2014. "Beyond Aridification: Multiple Explanations for the Elevated Diversification of Cacti in the New World Succulent Biome." *New Phytologist* 202 (4): 1382–97. https://doi.org/10.1111/nph.12752.

Hernández Hernández, Elvira. 2003. *Evaluación Del Efecto de La Adición de Harina de Nopal (Opuntia Ssp.) Natural y Libre de Clorofila En La Elaboración de Tortillas de Maíz*. Coahuila, México: Universidad Autónoma Agraria "Antonio Narro."

Herrera-Hernández, María G., Fidel Guevara-Lara, Rosalía Reynoso-Camacho, and Salvador H. Guzmán-Maldonado. 2011. "Effects of Maturity Stage and Storage on Cactus Berry (Myrtillocactus Geometrizans) Phenolics, Vitamin C, Betalains and Their Antioxidant Properties." *Food Chemistry* 129 (4): 1744–50. https://doi.org/10.1016/j.foodchem.2011.06.042.

Houerou, Henry N. Le. 2000. "Utilization of Fodder Trees and Shrubs in the Arid and Semiarid Zones of West Asia and North Africa." *Arid Soil Research and Rehabilitation* 14 (2): 101–35. https://doi.org/10.1080/089030600263058.

Howe, The Inter-BRC Microbiome Workshop Consortium, Adina, Gregory Bonito, Ming-Yi Chou, Melissa A. Cregger, Anna Fedders, John L. Field, Hector Garcia Martin, et al. 2022. "Frontiers and Opportunities in Bioenergy Crop Microbiome Research Networks." *Phytobiomes Journal* February. https://doi.org/10.1094/PBIOMES-05-21-0033-MR.

Huang, Jin-Ling, Shou-Ting Fu, Yuan-Ying Jiang, Yong-Bing Cao, Mei-Li Guo, Yan Wang, and Zheng Xu. 2007. "Protective Effects of Nicotiflorin on Reducing Memory Dysfunction, Energy Metabolism Failure and Oxidative Stress in Multi-Infarct Dementia Model Rats." *Pharmacology Biochemistry and Behavior* 86 (4): 741–48. https://doi.org/10.1016/j.pbb.2007.03.003.

Hunt, D. 2006. *The New Cactus Lexicon*. Edited by DH Books. Vol. 1 y 2. Milborne Port, UK: DH Books.

INEGI. 2007. "Características Principales Del Cultivo Del Nopal En El Distrito Federal Caso Milpa Alta." In *Censo Agropecuario*. Aguascalientes, México: Instituto Nacional de Estadística y Geografía.

Jang, Moon Hee, Xiang Lan Piao, Jong Moon Kim, Sung Won Kwon, and Jeong Hill Park. 2008. "Inhibition of Cholinesterase and Amyloid-β Aggregation by Resveratrol Oligomers FromVitis Amurensis." *Phytotherapy Research* 22 (4): 544–49. https://doi.org/10.1002/ptr.2406.

Jiménez-Sierra, C.L. 2011. "Las Cactáceas Mexicanas y Los Riesgos Que Enfrentan." In *Revista Digital Universitaria*. México: Universidad Nacional Autónoma de México, 2011.

Khan, Nelofer S, Aamir Ahmad, and S.M Hadi. 2000. "Anti-Oxidant, pro-Oxidant Properties of Tannic Acid and Its Binding to DNA." *Chemico-Biological Interactions* 125 (3): 177–89. https://doi.org/10.1016/S0009-2797(00)00143-5.

Kim, Jong-Eun, Dong-Eun Lee, Ki Won Lee, Joe Eun Son, Sang Kwon Seo, Jixia Li, Sung Keun Jung, et al. 2011. "Isorhamnetin Suppresses Skin Cancer through Direct Inhibition of MEK1 and PI3-K." *Cancer Prevention Research* 4 (4): 582–91. https://doi.org/10.1158/1940-6207.CAPR-11-0032.

Kuti, Joseph O. 2004. "Antioxidant Compounds from Four Opuntia Cactus Pear Fruit Varieties." *Food Chemistry* 85 (4): 527–33. https://doi.org/10.1016/S0308-8146(03)00184-5.

Lacerda-Júnior, Gileno V., Melline F. Noronha, Lucélia Cabral, Tiago P. Delforno, Sanderson Tarciso Pereira De Sousa, Paulo I. Fernandes-Júnior, Itamar S. Melo, and Valéria M. Oliveira. 2019. "Land Use and Seasonal Effects on the Soil Microbiome of a Brazilian Dry Forest." *Frontiers in Microbiology* 10 (APR): 648. https://doi.org/10.3389/FMICB.2019.00648/BIBTEX.

Laughton, Miranda J, Patricia J. Evans, Michele A. Moroney, J.Robin S. Hoult, and Barry Halliwell. 1991. "Inhibition of Mammalian 5-Lipoxygenase and Cyclo-Oxygenase by Flavonoids and Phenolic Dietary Additives." *Biochemical Pharmacology* 42 (9): 1673–81. https://doi.org/10.1016/0006-2952(91)90501-U.

Le, Phuong Thi, Thulani P. Makhalanyane, Leandro D. Guerrero, Surendra Vikram, Yves Van De Peer, and Don A. Cowan. 2016. "Comparative Metagenomic Analysis Reveals Mechanisms for Stress Response in Hypoliths from Extreme Hyperarid Deserts." *Genome Biology and Evolution* 8 (9): 2737–47. https://doi.org/10.1093/GBE/EVW189.

Li, Runping, Meili Guo, Ge Zhang, Xiongfei Xu, and Quan Li. 2006. "Nicotiflorin Reduces Cerebral Ischemic Damage and Upregulates Endothelial Nitric Oxide Synthase in Primarily Cultured Rat Cerebral Blood Vessel Endothelial Cells." *Journal of Ethnopharmacology* 107 (1): 143–50. https://doi.org/10.1016/j.jep.2006.04.024.

Lira, Sandra Machado, Ana Paula Dionísio, Marcelo Oliveira Holanda, Chayane Gomes Marques, Gisele Silvestre da Silva, Lia Coêlho Correa, Glauber Batista Moreira Santos, et al. 2020. "Metabolic Profile of Pitaya (Hylocereus Polyrhizus (F.A.C. Weber) Britton & Rose) by UPLC-QTOF-MSE and Assessment of Its Toxicity and Anxiolytic-like Effect in Adult Zebrafish." *Food Research International* 127 (January): 108701. https://doi.org/10.1016/j.foodres.2019.108701.

Lisintuña, Welington F, Edwin F Cerda, Mario A García, Simón Rodríguez, Barrio El Ejido, San Felipe, Ecuador Latacunga, and Mario A. García-Pérez. 2020. "Tratamiento de aguas residuaes de una industria láctea con mucílago de nopal (Opuntia ficus-indica [L.] MILL.)." *Ciencia y Tecnología de Alimentos* 30 (2): 52.

López-Palacios, Cristian, Cecilia Beatriz Peña-Valdivia, Adriana Inés Rodríguez-Hernández, and Juan Antonio Reyes-Agüero. 2016. "Rheological Flow Behavior of Structural Polysaccharides from Edible Tender Cladodes of Wild, Semidomesticated and Cultivated "Nopal" (Opuntia) of Mexican Highlands." *Plant Foods for Human Nutrition* 71 (4): 388–95. https://doi.org/10.1007/s11130-016-0573-2.

Lopez, Blanca Rafaela, Clara Tinoco-Ojanguren, Macario Bacilio, Alberto Mendoza, and Yoav Bashan. 2012. "Endophytic Bacteria of the Rock-Dwelling Cactus Mammillaria Fraileana Affect Plant Growth and Mobilization of Elements from Rocks." *Environmental and Experimental Botany* 81. https://doi.org/10.1016/j.envexpbot.2012.02.014.

Lüneberg, Kathia, Dominik Schneider, Christina Siebe, and Rolf Daniel. 2018. "Drylands Soil Bacterial Community Is Affected by Land Use Change and Different Irrigation Practices in the Mezquital Valley, Mexico." *Scientific Reports* 2018 8:1 8 (1): 1–15. https://doi.org/10.1038/s41598-018-19743-x.

Luo, Honglin, Guangyao Xiong, Kaijing Ren, Sudha R. Raman, Zhe Liu, Qiuping Li, Chunying Ma, Deying Li, and Yizao Wan. 2014. "Air DBD Plasma Treatment on Three-Dimensional Braided Carbon Fiber-Reinforced PEEK Composites for Enhancement of in Vitro Bioactivity." *Surface and Coatings Technology* 242 (March): 1–7. https://doi.org/10.1016/j.surfcoat.2013.12.069.

Mahouachi, Mokhtar., Naziha Atti, and H. Hajji. 2012. "Use of Spineless Cactus (Opuntia Ficus Indica f. Inermis) for Dairy Goats and Growing Kids: Impacts on Milk Production, Kid's Growth, and Meat Quality." *The Scientific World Journal* 2012. https://doi.org/10.1100/2012/321567.

Majure, Lucas C., Raul Puente, M. Patrick Griffith, Walter S. Judd, Pamela S. Soltis, and Douglas E. Soltis. 2012. "Phylogeny of *Opuntia* s.s. (Cactaceae): Clade Delineation, Geographic Origins, and Reticulate Evolution." *American Journal of Botany* 99 (5): 847–64. https://doi.org/10.3732/ajb.1100375.

Maki-Díaz, Griselda, Cecilia B Peña-Valdivia, Rodolfo García-Nava, M Lourdes Arévalo-Galarza, Guillermo Calderón-Zavala, and Socorro Anaya-Rosales. 2015. "Características físicas y químicas de nopal verdura (*Opuntia Ficus-Indica*) para exportación y consumo nacional." *Agrociencia* 49: 31–51.

Mandujano, Maria del Carmen, I.G. Carrillo-Angeles, C. Martínez-Peralta, and J. Golubov. 2010. *Reproductive Biology of Cactaceae*. Edited by K.G. Ramawat. Berlin, Germany: Desert Plants – Biology and Biotechnology.

Marino, Julieta. 2018. *GALLETAS AGREGADO DEN O P A L*. Madrid: Universidad Fasta.

Martínez, Veronica, José Tejero, Gaudalupe Luna, and Rosalia Cerecero. 2016. "Estudio de La Adición de Concentrado de Vegetal Como Conservador de Puré deaguacate Hass." *Revista de Simulación y Laboratorio* 3 (8): 22–28.

Mascot-Gómez, Ernesto, Joel Flores, and Nguyen E. López-Lozano. 2021. "The Seed-Associated Microbiome of Four Cactus Species from Southern Chihuahuan Desert." *Journal of Arid Environments* 190 (July): 104531. https://doi.org/10.1016/J.JARIDENV.2021.104531.

Medeiros Azevedo, Thamara De, Flávia, Figueira Aburjaile, José Ribamar, Costa Ferreira-Neto, Valesca Pandolfi, Ana, and Maria Benko-Iseppon. 2021. "The Endophytome (Plant-Associated Microbiome): Methodological Approaches, Biological Aspects, and Biotech Applications." *World Journal of Microbiology and Biotechnology* 37 (3): 206. https://doi.org/10.1007/s11274-021-03168-2.

Medina Díaz, Gabriel José, and Yirley Dayana Borrero Ortiz. 2017. *Factibilidad de abono orgánico a base de cactus de nopal*. Colombia: Universidad Santo Tomas, Bucaramanga.

Mercado-Silva, Edmundo M. 2018. "Pitaya— Hylocereus Undatus (Haw)." In *Exotic Fruits*, 339–49. New York: Elsevier. https://doi.org/10.1016/B978-0-12-803138-4.00045-9.

Miller, Greta, Samantha Hartzell, and Amilcare Porporato. 2021. "Ecohydrology of Epiphytes: Modelling Water Balance, CAM Photosynthesis, and Their Climate Impacts." *Ecohydrology* 14 (3). https://doi.org/10.1002/eco.2275.

Molina-Vega, Aracely, Edna María Hernández-Domínguez, Matilde Villa-García, and Jorge Álvarez-Cervantes. 2021. "Comadia Redtenbacheri (Lepidoptera: Cossidae) and Aegiale Hesperiaris (Lepidoptera: Hesperiidae), Two Important Edible Insects of Agave Salmiana (Asparagales: Asparagaceae): A Review." *International Journal of Tropical Insect Science*. https://doi.org/10.1007/s42690-020-00396-1.

Montiel-Sánchez, Mara, Tomás García-Cayuela, Andrea Gómez-Maqueo, Hugo S. García, and M. Pilar Cano. 2021. "In Vitro Gastrointestinal Stability, Bioaccessibility and Potential Biological Activities of Betalains and Phenolic Compounds in Cactus Berry Fruits (Myrtillocactus Geometrizans)." *Food Chemistry* 342 (April): 128087. https://doi.org/10.1016/j.foodchem.2020.128087.

Morán-Ramos, Sofía, Azalia Avila-Nava, Armando R. Tovar, José Pedraza-Chaverri, Patricia López-Romero, and Nimbe Torres. 2012. "Opuntia Ficus Indica (Nopal) Attenuates Hepatic Steatosis and Oxidative Stress in Obese Zucker (Fa/Fa) Rats." *The Journal of Nutrition* 142 (11): 1956–63. https://doi.org/10.3945/jn.112.165563.

Murgia, Manuela, Maura Fiamma, Aleksandra Barac, Massimo Deligios, Vittorio Mazzarello, Bianca Paglietti, Pietro Cappuccinelli, et al. 2019. "Biodiversity of Fungi in Hot Desert Sands." *MicrobiologyOpen* 8 (1). https://doi.org/10.1002/MBO3.595.

Nascimento Souza, Aelson Fernandes do, Gherman Garcia Leal de Araújo, Edson Mauro Santos, Paulo Sérgio de Azevedo, Juliana Silva de Oliveira, Alexandre Fernandes Perazzo, Ricardo Martins Araujo Pinho, and Anderson de Moura Zanine. 2020. "Carcass Traits and Meat Quality of Lambs Fed with Cactus (Opuntia Fícus-Indica Mill) Silage and Subjected to an Intermittent Water Supply." *PLoS ONE* 15 (4). https://doi.org/10.1371/journal.pone.0231191.

Nazareno, Mónica A., María Ochoa, and José Dubeaux Jr. 2013. "Actas de La Segunda Reunión Para El Aprovechamiento Integral de La Tuna y Otras Cactáceas y I Reunión Sudamericana CACTUSNET FAO-ICARDA." In *Proceedings of the Second Meeting for the Integral Use of Cactus Pear and Other Cacti and 1st South American Meeting of the FAO-ICARDA CACTUSNET*, edited by Moica A. Nazareno, María Judith Ochoa, and Dubeaux Jr. José, 13th ed. Santiago del Estero, Argentina: FAO-ICARDA.

Neilson, Julia W., Katy Califf, Cesar Cardona, Audrey Copeland, Will van Treuren, Karen L. Josephson, Rob Knight, et al. 2017. "Significant Impacts of Increasing Aridity on the Arid Soil Microbiome." *MSystems* 2 (3). https://doi.org/10.1128/MSYSTEMS.00195-16.

Ojeda-Linares, César I., Mariana Vallejo, Patricia Lappe-Oliveras, and Alejandro Casas. 2020. "Traditional Management of Microorganisms in Fermented Beverages from Cactus Fruits in Mexico: An Ethnobiological Approach." *Journal of Ethnobiology and Ethnomedicine* 16 (1): 1. https://doi.org/10.1186/s13002-019-0351-y.

Oleszek, Marta, and Wieslaw Oleszek. 2020. "Saponins in Food." In *Handbook of Dietary Phytochemicals*, 1–40. Singapore: Springer Singapore. https://doi.org/10.1007/978-981-13-1745-3_34-1.

Osorio-Esquivel, Obed, Alicia-Ortiz-Moreno, Valente B. Álvarez, Lidia Dorantes-Álvarez, and M. Mónica Giusti. 2011. "Phenolics, Betacyanins and Antioxidant Activity in Opuntia Joconostle Fruits." *Food Research International* 44 (7): 2160–68. https://doi.org/10.1016/j.foodres.2011.02.011.

Ovando Franco, Monserrat. 2012. *Modificación de Biopolímero Extraído de Nopal (Opuntia Ficus Indica) y Su Aplicación Para La Remoción de Metales En Agua*. San Luis Potosí, México: Instituto Potosino de Investigación Científica y Tecnológica A.C.

Pavón, Numa P., Christian O. Ayala, and Ana Paola Martínez-Falcón. 2016. "Water and Carbon Storage Capacity in *I Solatocereus Dumortieri* (Cactaceae) in an Intertropical Semiarid Zone in Mexico." *Plant Species Biology* 31 (3): 240–43. https://doi.org/10.1111/1442-1984.12102.

Peña-Valdivia, Cecilia Beatriz, Carlos Trejo, V. Baruch Arroyo-Peñ, Adriana Beatriz Sá Nchez Urdaneta, and Rosendo Balois Morales. 2012. "Diversity of Unavailable Polysaccharides and Dietary Fiber in Domesticated Nopalito and Cactus Pear Fruit (*Opuntia* Spp.)." *Chemistry & Biodiversity* 9(8): 1599–610.

Piao, Meihua, Daisuke Mori, Tosimi Satoh, Yasuo Sugita, and Osamu Tokunaga. 2006. "Inhibition of Endothelial Cell Proliferation, in Vitro Angiogenesis, and the down-Regulation of Cell Adhesion-Related Genes by Genistein Combined with a CDNA Microarray Analysis." *Endothelium: Journal of Endothelial Cell Research* 13 (4). https://doi.org/10.1080/10623320600903940.

Potgieter, J., and S. D'Aquino. 2018. "Fruit Production and Post-Harvest Management." In *Ecologia Del Cultivo, Manejo y Usos Del Nopal*, edited by P. Inglese, C. Mondragon, A. Nefzaoui, C. Saenz, M. Taguchi, H. Makkar, and M Louhaichi, 1:51–71. (ICARDA), Organización de las Naciones Unidas para la Alimentación y la Agricultura (FAO); Centro Internacional de Investigaciones Agrícolas en Zonas Áridas.

Ramírez-Rodríguez, Yadira, Miriam Martínez-Huélamo, José Pedraza-Chaverri, Victoria Ramírez, Natalia Martínez-Tagüeña, and Joyce Trujillo. 2020. "Ethnobotanical, Nutritional and Medicinal Properties of Mexican Drylands Cactaceae Fruits: Recent Findings and Research Opportunities." *Food Chemistry* 312 (May): 126073. https://doi.org/10.1016/j.foodchem.2019.126073.

Ramírez-Villalobos, Jesica M., César I. Romo-Sáenz, Karla S. Morán-Santibañez, Patricia Tamez-Guerra, Ramiro Quintanilla-Licea, Alonso A. Orozco-Flores, Ricardo Romero-Arguelles, Reyes Tamez-Guerra, Cristina Rodríguez-Padilla, and Ricardo Gomez-Flores. 2021. "In Vitro Tumor Cell Growth Inhibition Induced by Lophocereus Marginatus (Dc.) s. Arias and Terrazas Endophytic Fungi Extracts." *International Journal of Environmental Research and Public Health* 18 (18). https://doi.org/10.3390/ijerph18189917.

Rasoulpour, Rasoul, Alireza Afsharifar, and Keramat Izadpanah. 2018. "Antiviral Activity of Prickly Pear (Opuntia Ficus-Indica (L.) Miller) Extract: Opuntin B, a Second Antiviral Protein." *Crop Protection* 112 (October): 1–9. https://doi.org/10.1016/j.cropro.2018.04.017.

Rebman, Jon Paul, and Donald John Pinkava. 2001. "Opuntia Cacti of North America: An Overview." *The Florida Entomologist* 84 (4): 474. https://doi.org/10.2307/3496374.

Rehman, Abdul, Muhammad Ijaz, Komal Mazhar, Sami Ul-Allah, and Qasim Ali. 2019. "Metagenomic Approach in Relation to Microbe–Microbe and Plant–Microbiome Interactions." In *Microbiome in Plant Health and Disease*, 507–34. Singapore: Springer Singapore. https://doi.org/10.1007/978-981-13-8495-0_22.

Reyes-Agüero, Jose Antonio, Juan Rogelio Aguirre Rivera, and Alfonso Valiente-Banuet. 2006. "Reproductive Biology of Opuntia: A Review." *Journal of Arid Environments* 64 (4): 549–85. https://doi.org/10.1016/j.jaridenv.2005.06.018.

Reyes-Buendía, Claudia, José Joel E. Corrales- García, Cecilia B. Peña-Valdivia, Arturo Hernández Montes, and Ma. Carmen Ybarra-Moncada. 2020. "Sopa de Elote (Zea Mays) Tipo Crema Con Mucílago de Nopal (*Opuntia* Spp.) Como Espesante, Sus Características Físicas y Aceptación Sensorial." *TIP Revista Especializada En Ciencias Químico-Biológicas* 23 (October). https://doi.org/10.22201/fesz.23958723e.2020.0.257.

Robledo-Márquez, Karina, Victoria Ramírez, Aarón Fernando González-Córdova, Yadira Ramírez-Rodríguez, Luis García-Ortega, and Joyce Trujillo. 2021. "Research Opportunities: Traditional Fermented Beverages in Mexico. Cultural, Microbiological, Chemical, and Functional Aspects." *Food Research International* 147 (September): 110482. https://doi.org/10.1016/j.foodres.2021.110482.

Rostro, Beverly Ramos, Baciliza Quintero Salazar, Julieta Ramos-Elorduy, José Manuel Pino Moreno, Sergio C.Angeles Campos, Agueda García Pérez, and V. Daniela Barrera García. 2012. "Analisis Quimico y Nutricional de Tres Insectos Comestibles de Interés Comercial En La Zona Arqueologica Del Municipio de San Juan Teotihuacan y En Otumba, En El Estado de México." *Interciencia* 37 (12).

Sáenz, Carmen, Elena Sepúlveda, Elly Pak, and Ximena Vallejos. 2002. "Uso de Fibra Dietética de Nopal En La Formulación de Un Polvo Para Flan." *Archivos latinoamericanos de nutrición Organo Oficial de La Sociedad Latinoamericana de Nutrición*. Vol. 52.

Sahagún, Bernardino. 1577. "Historia General de Las Cosas de Nueva España." https://www.wdl.org/en/item/10096/view/1/1/.

Salazar, Juan Rodrigo, Marco Antonio Loza-Mejía, and Diego Soto-Cabrera. 2020. "Chemistry, Biological Activities and In Silico Bioprospection of Sterols and Triterpenes from Mexican Columnar Cactaceae." *Molecules* 25 (7): 1649. https://doi.org/10.3390/molecules25071649.

Sánchez, Eduardo, Santos García, and Norma Heredia. 2010. "Extracts of Edible and Medicinal Plants Damage Membranes of Vibrio Cholerae." *Applied and Environmental Microbiology* 76 (20). https://doi.org/10.1128/AEM.03052-09.

Sarria Carabalí, Margarita María, Felipe García-Oliva, Luis Enrique Cortés Páez, and Nguyen E. López Lozano. 2019. "The Response of Candy Barrel Cactus to Zinc Contamination Is Modulated by Its Rhizospheric Microbiota." *Rhizosphere* 12 (December): 100177. https://doi.org/10.1016/J.RHISPH.2019.100177.

Secretaria de Medio Ambiente y Recursos Naturales (SEMARNAT). 2010. *NOM-059-SEMARNAT-2010. Diario Oficial de La Federación*.

———. 2016. "Cactáceas, Riqueza Natural de México." Cactáceas, Riqueza Natural de México. 2016. https://www.gob.mx/semarnat/articulos/cactaceas-riqueza-natural-de-mexico.

———. 2019. "Suelos." El Medio Ambiente En México 2013–2014. 2019. http://apps1.semarnat.gob.mx/dgeia/informe_resumen14/03_suelos/3_3.%0Ahtml.

Serra, Ana Teresa, Joana Poejo, Ana A. Matias, Maria R. Bronze, and Catarina M.M. Duarte. 2013. "Evaluation of *Opuntia* spp. Derived Products as Antiproliferative Agents in Human Colon Cancer Cell Line (HT29)." *Food Research International* 54 (1). https://doi.org/10.1016/j.foodres.2013.08.043.

Singariya, Premlata, Mourya Krishan Kumar, and Padma Kumar. 2018. "Comparative Study of Antibacterial Properties of Flavonoids of Leaves from Different Cactus, Perennial Grasses and Medicinal Plant." *Asian Journal of Pharmaceutical Research and Development* 6 (3). https://doi.org/10.22270/ajprd.v6i3.372.

Sommermann, Loreen, Joerg Geistlinger, Daniel Wibberg, Annette Deubel, Jessica Zwanzig, Doreen Babin, Andreas Schlüter, and Ingo Schellenberg. 2018. "Fungal Community Profiles in Agricultural Soils of a Long-Term Field Trial under Different Tillage, Fertilization and Crop Rotation Conditions Analyzed by High-Throughput ITS-Amplicon Sequencing." *Plos One* 13 (4): e0195345. https://doi.org/10.1371/JOURNAL.PONE.0195345.

Sreekanth, Devalraju, Marasanapalli Kalle Arunasree, Karnati Rammohan Roy, T. Chandramohan Reddy, Gorla Venkateswara Reddy, and Pallu Reddanna. 2007. "Betanin a Betacyanin Pigment Purified from Fruits of Opuntia Ficus-Indica Induces Apoptosis in Human Chronic Myeloid Leukemia Cell Line-K562." *Phytomedicine* 14 (11). https://doi.org/10.1016/j.phymed.2007.03.017.

Suh, Dong Ho, Sunmin Lee, Do Yeon Heo, Young-Suk Kim, Somi Kim Cho, Sarah Lee, and Choong Hwan Lee. 2014. "Metabolite Profiling of Red and White Pitayas (Hylocereus Polyrhizus and Hylocereus Undatus) for Comparing Betalain Biosynthesis and Antioxidant Activity." *Journal of Agricultural and Food Chemistry*. https://doi.org/10.1021/jf5020704.

Sun, Qiqi, Rui Wang, Yaxian Hu, Lunguang Yao, and Shengli Guo. 2018. "Spatial Variations of Soil Respiration and Temperature Sensitivity along a Steep Slope of the Semiarid Loess Plateau." *PLOS ONE* 13 (4): e0195400. https://doi.org/10.1371/JOURNAL.PONE.0195400.

Taketani, Rodrigo Gouvêa, Vanessa Nessner Kavamura, and Suikinai Nobre dos Santos. 2017. "Diversity and Technological Aspects of Microorganisms from Semiarid Environments." In *Diversity and Benefits of Microorganisms from the Tropics*, 3–19. Cham: Springer International Publishing. https://doi.org/10.1007/978-3-319-55804-2_1.

Tenore, Gian Carlo, Ettore Novellino, and Adriana Basile. 2012. "Nutraceutical Potential and Antioxidant Benefits of Red Pitaya (Hylocereus Polyrhizus) Extracts." *Journal of Functional Foods*. https://doi.org/10.1016/j.jff.2011.09.003.

Tenorio-Escandón, Perla, Alfredo Ramírez-Hernández, Joel Flores, Jorge Juan-Vicedo, and Ana Paola Martínez-Falcón. 2022. "A Systematic Review on Opuntia (Cactaceae; Opuntioideae) Flower-Visiting Insects in the World with Emphasis on Mexico: Implications for Biodiversity Conservation." *Plants* 11 (1): 131. https://doi.org/10.3390/plants11010131.

Tian, Qin, Takeshi Taniguchi, Wei Yu Shi, Guoqing Li, Norikazu Yamanaka, and Sheng Du. 2017. "Land-Use Types and Soil Chemical Properties Influence Soil Microbial Communities in the Semiarid Loess Plateau Region in China." *Scientific Reports* 7 (1): 1–9. https://doi.org/10.1038/srep45289.

Torres-Ponce, Lizeth Reyna, Dayanira Morales-Corral, María de Lourdes Ballinas-Casarrubias, and Guadalupe Virginia Nevárez-Moorillón. 2015. "El Nopal: Planta Del Semidesierto Con Aplicaciones En, Alimentos y Nutrición Animal." *Revista Mexicana de Ciencias Agrícolas*, 6(5): 1129–1142, 2015.

Torres-Soria, Pablo, Sandra Cruz Flores, Norma Cristina, Peña Peláez, Sara Eugenia, Fernández Mendiola, Moisés Adrián, Rodríguez Ibarra, and Alfonso Cruz Becerril. n.d. "La Baba y El Mucílago de Nopal,Una Alternativa Natural La Conservación deacabados Arquitectónicos de Tierra." *Antropología*, Revista Interdisciplinaria del INAH, 99: 92–114.

Tremlett, Constance J., Kelvin S.-H. Peh, Veronica Zamora-Gutierrez, and Marije Schaafsma. 2021. "Value and Benefit Distribution of Pollination Services Provided by Bats in the Production of Cactus Fruits in Central Mexico." *Ecosystem Services* 47 (February): 101197. https://doi.org/10.1016/j.ecoser.2020.101197.

Tremlett, Constance J., Mandy Moore, Mark A. Chapman, Veronica Zamora-Gutierrez, and Kelvin S.-H. Peh. 2020. "Pollination by Bats Enhances Both Quality and Yield of a Major Cash Crop in Mexico." *Journal of Applied Ecology* 57 (3): 450–59. https://doi.org/10.1111/1365-2664.13545.

Verma, J. P, C Macdonald, V. K Gupta, and A. R Podile. 2020. *New and Future Developments in Microbial Biotechnology and Bioengineering: Phytomicrobiome for Sustainable Agriculture*. Edited by Elsevier. Amsterdam: Elsevier.

Villa-Jaimes, Gloria Stephanie, Fabio Alejandro Aguilar-Mora, Herson Antonio González-Ponce, Francisco Javier Avelar-González, Ma Consolación Martínez Saldaña, Manon Buist-Homan, and Han Moshage. 2022. "Biocomponents from Opuntia Robusta and Opuntia Streptacantha Fruits Protect against Diclofenac-Induced Acute Liver Damage in Vivo and in Vitro." *Journal of Functional Foods* 89 (October 2021). https://doi.org/10.1016/j.jff.2022.104960.

Villada, Hector S, Harol A Acosta, and Reinado J Velasco. 2009. "Materiales Naturales Para Empaques Biodegradables." *Mundo Alimentario* 2009, 27–31.

Wang, Jianming, Tianhan Zhang, Liping Li, Jingwen Li, Yiming Feng, and Qi Lu. 2017. "The Patterns and Drivers of Bacterial and Fungal β-Diversity in a Typical Dryland Ecosystem of Northwest China." *Frontiers in Microbiology* 8 (Nov): 2126. https://doi.org/10.3389/FMICB.2017.02126/BIBTEX.

Wayland Morales, Rodrigo. 2020. "Usos de Cáctus Para La Alimentación, Energía y Creación de Empleo." Santiago, Chile. www.elquiglobalenergy.com.

Yeh, Gloria Y, David M Eisenberg, Ted J Kaptchuk, and Russell S Phillips. 2003. "Systematic Review of Herbs and Dietary Supplements for Glycemic Control in Diabetes." *Diabetes Care* 26 (4): 1277–94. http://diabetesjournals.org/care/article-pdf/26/4/1277/659287/dc0403001277.pdf.

You, Bo Ra, and Woo Hyun Park. 2010. "Gallic Acid-Induced Lung Cancer Cell Death Is Related to Glutathione Depletion as Well as Reactive Oxygen Species Increase." *Toxicology in Vitro* 24 (5). https://doi.org/10.1016/j.tiv.2010.04.009.

Zou, Da-Ming, Molly Brewer, Francisco Garcia, Jean M Feugang, Jian Wang, Roungyu Zang, Huaguang Liu, and Changping Zou. 2005. "Cactus Pear: A Natural Product in Cancer Chemoprevention." https://doi.org/10.1186/1475-2891-4-25.

Chapter 4

Potential of Plants from the Arid Zone of Coahuila in Mexico for the Extraction of Essential Oils

Orlando Sebastian Solis-Quiroz, Adriana Carolina González-Machado, Jorge Alejandro Aguirre-Joya, David Ramiro Aguillón-Gutierrez, Agustina Ramírez-Moreno, and Cristian Torres-León

CONTENTS

4.1 INTRODUCTION

Plants are producers of a wide variety of biological compounds. Among these substances, some are a product of their metabolism, known as metabolites, and formed during the plant's growth cycle (Badui, 2006). The production of these compounds in the biochemical cycles of plants (glycolysis or Krebs cycle) is associated with genetics that determines the plant's type of metabolism. In addition, the climatic conditions to which the plant in question is exposed and the practices carried out on it (irrigation, fertilisation, etc.) play an equally important role in determining this profile (Badui, 2006).

A large number of substances participate in the formation of this complex mixture of components (Rojas et al., 2009). These substances are called essential oils, volatile oils, or essences. This range of compounds is synthesised through six main routes:

DOI: 10.1201/9781003251255-4

1) Transformation of carbohydrates with an increase in the concentration of soluble sugars and degradation of the cell wall.
2) Amino acid conversion.
3) Fatty acids are utilised to synthesise alcohols, esters, ketones, and acids.
4) Enzymatic oxidation of linoleic acid.
5) Conversion of L-phenylalanine to phenolic esters.
6) Synthesis of terpenes and carotenoid derivatives.

The process is focused on the synthesis of terpene and carotenoid derivatives. These compounds are derived from isoprene and constitute an important group of essences. Geranyl pyrophosphate is a byproduct of isoprene and participates in the metabolic pathway of some plants, giving rise to a wide range of terpenes by the action of oxidoreductase enzymes (Dergal, 2006).

Various methods of extracting these oils have been extensively studied to obtain high-performance processes that generate a good quality product. Conventional methods such as steam distillation, hydrodistillation, and extraction of oils in chemical solvents have been used. These methods contribute to the largest amount of essential oil production currently marketed. Although, emerging technologies that consist of conventional systems assisted by microwave or ultrasound heating techniques or techniques with high specificity and performance, such as enzyme extraction, usually have high implementation costs, making them complex options to use on an industrial scale.

At present, various communities of the Coahuilense semi-desert base their livelihood on the collection of plants; in the Sierra y Cañon de Jimulco Municipal Ecological Reserve, located in Torreón, Coahuila, studies have been carried out and have concluded that the commercialisation structure favors intermediaries and industrialists and leaves a very low-profit margin for collectors. This problem exists in various municipalities of Viesca, Coahuila, where the conditions of the semi-desert prevent regular activities of agriculture and livestock, so revaluation of the value-added to products such as essential oils from various plants, including *Lippia graveolens* Kunth (oregano), *Flourensia cernua* DC. (hojasén), *Allium sativum* L. (ajo) and *Larrea tridentata* (Sessé & Moc. ex DC.) Coville (Gobernadora), that are collected by the locals can increase rural communities' economic income (Orona Castillo et al., 2017; Torres León et al., 2023). In this context, the objective of this chapter is to present an overview of the extraction of essential oils from plants in arid areas of Coahuila.

4.2 METHODS OF OBTAINING ESSENTIAL OILS

A literature review revealed that various methods of obtaining essential oils have been studied. Some methods are conventionally used, such as hydrodistillation, steam distillation, and solvent extraction (Saldaña-Mendoza et al., 2022).

4.2.1 Steam Distillation and Hydrodistillation

Steam distillation is the most widely used on a large-scale; approximately 90% of the marketed essential oils are produced using this technology (Radwan et al., 2020). It is a modified distillation system in which steam is produced using a boiler, and later it is put in contact with the vegetable matter; the high temperature of the water vapor generates interactions with the low molecular weight molecules that make up the essential oils, which allows them to volatilise. Consequently, a cooling system with recirculation is used, which allows the recovery of essential oils and the hydrolate, or residual aromatic water, which is later separated by decantation because of the effect of the difference in densities (Estrada Jiron, 2015). It has been highlighted that two important parameters that determine the efficiency and performance of the process are the flow of steam supplied – that is, the heat transfer in the system – and the batch size of the vegetable matter (Radwan et al., 2020; Kant and Kumar, 2021). This type of technology has multiple advantages, such as the easy process instrumentation and the low cost of the extraction agent (steam).

On the other hand, in hydrodistillation processes, the raw material is directly combined with the extraction agent without this being a solvent and is subsequently brought to boiling temperature; the vapors produced by the boiling of the mixture contain the essential oils (Lahlou, 2004). This technique has been used for many years to obtain floral waters, so essential oils were discarded as they were considered a byproduct. It is worth mentioning that this technique, like steam distillation, can cause product degradation and odors due to high process temperatures (Perino-Issartier et al., 2013). Furthermore, it consumes large amounts of energy, raw materials, and time to obtain considerable yields.

4.2.2 Extraction by Chemical Solvents and Green Solvents

Solvent extraction is also a popular technique for obtaining essential oils. Plant cells tend to increase the permeability of their cell walls, thus reaching a state of swelling that allows the diffusion of the solvent inside the cell, releasing the active ingredients. This extraction process depends directly on the raw material and the solvent used (Aguirre et al., 2012).

Some commonly used solvents are mainly petrochemical solvents, among which n-hexane stands out, which has shown high efficiency and has been preferred for its high oil selectivity and low boiling point. However, prolonged exposure has been associated with health problems (Zhuang et al., 2018). Despite its relative simplicity, this process is not used on a large scale because there are problems with the storage and disposal of solvents on an industrial scale (Koubaa et al., 2016).

At present, efforts have been added in the formulation of green solvents, that is, solvents that meet improvements in availability, price, recyclability, lower toxicity, simplicity of synthesis, ease of storage, and renewability (Saldaña-Mendoza et al., 2022). Using these solvents reduces the energy required during the separation process, improving biocompatibility and the extraction and nutritional value of oils compared to traditional chemical solvents (Zhuang et al., 2018).

4.2.3 Emerging Essential Oil Extraction Technologies

Today, various emerging technologies for essential oil extraction include supercritical fluid extraction, ohmic-assisted steam distillation, microwave-assisted solvent extraction, ultrasonic-assisted solvent extraction, and enzymatic extraction. These innovative technologies improve the quality of the extracted product by keeping the volatile compounds' molecular structures intact. However, they suggest a greater technical complexity of the operation and a higher economic cost of implementation (Saldaña-Mendoza et al., 2022).

4.3 POTENTIALLY USABLE COAHUILENSE SEMI-DESERT PLANTS

There is a wide variety of raw materials that could be used to extract essential oils; many of them are present in the Coahuilense semi-desert; in this document, the medicinal properties of *Lippia graveolens* Kunth (oregano), *Flourensia cernua* DC. (hojasén), *Allium sativum* L. (ajo), *Larrea tridentata* (sessé & moc. ex DC.) Coville (gobernadora) and *Euphorbia antisyphilitica* (candelilla) will be described.

4.3.1 *Lippia graveolens* Kunth (Oregano)

L. graveolens is an aromatic plant that grows wild in semi-arid climates in the following states of Mexico: Querétaro, Guanajuato, Hidalgo, Oaxaca, Jalisco, San Luís Potosí, Zacatecas, Chihuahua, Sinaloa, Coahuila, and Durango. This plant has broad participation in the international and national markets with a production of more than 4,000 tons per year. Coahuila falls within the states that participate in a large proportion of the production that is exported annually.

L. graveolens, or Mexican oregano, is one of the highest quality raw materials in the aromatic plant market, with a high market value because of the concentration of thymol and carvacrol. In 2002, a liter of essential oil from this plant cost 170 dollars.

Currently, the primary source of use of the essential oil of oregano could be found in the antioxidant activity that it presents thanks to its content of polyphenols (secondary metabolites of plants and vegetables with physiological importance). The polyphenols are mainly thymol and carvacrol. Additionally, various studies have been published on the antimicrobial effect of different types of oregano on pathogenic bacteria that affect the food industry and the health sector because of thymol and carvacrol. Antimicrobial effect has been individually verified on microorganisms such as *Alternaria alternata*, *Escherichia coli*, *Staphylococcus aureus*, *Staphylococcus epidermidis*, *Enterococcus faecalis*, *Proteus vulgaris*, *Candida albicans*, and *Aspergillus niger* (Amadio et al., 2011).

Regarding the health sector, preparations using the essential oil have been proposed to combat fungal infections caused by *Candida albicans*. It effectively inhibits the pathogenic fungus that presents a recurring problem in humans and has begun to develop resistance to currently available pharmacological treatments (López-Rivera et al., 2018).

The effect of Mexican oregano extract was studied to evaluate its toxic activity (Soto-Dominguez et al., 2012), and the results obtained for the extracts at different concentrations proved to be non-toxic, which validates their safe use in the applications described above.

4.3.2 *Flourensia cernua* DC. (Hojasén)

F. cernua is a branched bush that can reach 2 meters in height. It has thin, resinous, light brown to gray branches approximately 17–25 mm long and 6.5–11.5 mm wide (Gutierrez-Ortega, 2004). The geographical location favors its growth in semi-arid areas and the deserts of Chihuahua and Sonora, covering different Mexican states such as Coahuila, Chihuahua, Durango, Nuevo León, San Luis Potosí, Sonora, and Zacatecas (Alvarez-Perez et al., 2020). This plant is used to treat common gastrointestinal discomforts such as stomach pain, indigestion, diarrhea, and dysentery; it is also used as a purgative and expectorant to treat rheumatism (Mata et al., 2003).

The Asteraceae family contains an oil rich in terpenes, giving favorable properties against pathogenic fungi with a high antioxidant capacity (Alvarez-Perez et al., 2020). Likewise, it has the ability to inhibit the growth of *S. aureus* and infectious agents that cause tuberculosis (Castro et al., 2012). Over time, the genus *Flourensia* has generated significant interest because their resins have a potential economic value, and critical studies have characterised their chemical compounds of interest to find possible uses (Delbon, 2014). Studies carried out by Salas-Méndez et al.

(2019) showed that *F. Cernua* can be used to prepare nanolaminate coatings used to prolong the life of tomatoes (Salas-Méndez et al., 2019).

4.3.3 *Allium sativum L.* (A jo)

A. sativum L. is a bulb belonging to the Amaryllidaceae family; this plant is used to prepare food and treat diseases. It has a root system made up of a bulbous root composed of 6 to 12 bulbils gathered at their base with a thin film to form the 'garlic head' (Juarez-Segovia et al., 2019).

A. sativum L. contains essential oil made up of aromatics such as diallyl disulfide, diallyl trisulfide, and other sulfur compounds such as allicin (Casella et al., 2013). Studies show that allicin is the most active compound in garlic (Rahman et al., 2012). Beneficial properties of antimicrobial and antimycotic capacity are attributed to garlic (Hernández-Suarez, 2005).

The German chemist Wertheim first obtained the essential oil (1844) after steam distillation of the volatile constituents. Because of his research, he proposed the name allyl for the hydrocarbon contained in the oil, and today this term is still used to describe the group $CH_2 = CH\text{-}CH_2$ (Casella et al., 2013).

A study conducted by Cerda Morales (2018) showed that dry garlic essential oil has greater antioxidant capacity when using the steam entrainment technique, as handling low temperatures prevents the denaturation of the oil. The essential oil of garlic can be used to preserve meat in the food industry (Nieto et al., 2012).

4.3.4 *Larrea tridentata* (Sessé & Moc. ex DC.) Coville (Gobernadora)

L. tridentata is a plant of the Zygophyllaceae family, widely distributed in the deserts of North America, especially in the drier places of northern Mexico. The plant's name comes from its ability to inhibit the growth of other plants around it to obtain more water, which is why it forms exclusive and extensive communities. It grows like a bush from 1 to 3 meters high with small leaves and, in its flowering (throughout the year, but more frequent between February and April), it gives a yellow flower with five petals. All its characteristics make the governor plant highly resistant to extreme conditions, which results in problems: Because it acquires invasive propagation and, once established in a territory, it is difficult to control its density.

Different and very diverse uses have been attributed to all plant parts. The flower bud is used as a condiment, and the branches, roots, and bark treat ailments such as kidney pain and bladder inflammation. The leaf has been used as an adhesive, as food for animal consumption because of its protein content, and for medicinal use to treat a wide variety of conditions such as urinary tract infections, to dissolve kidney stones, as a treatment for dermatitis, fungal infections, and as an antiseptic. Its resin is used to dye leather and produce paints and plastics in making fats, oils, and rubber.

Among its components, *L. tridentata* contains thymol and carvacrol in high concentrations that, as mentioned above, have diverse and essential functions; thymol is antibacterial, antifungal, anti-inflammatory, antioxidant, antirheumatic, and antiseptic. Carvacrol is antibacterial, antifungal, anti-inflammatory, antiseptic, antispasmodic, and expectorant. In this case, the plant has antimicrobial activity against many bacteria, fungi, and intestinal parasites, in addition to containing flavonoids (secondary plant metabolites) that act against viruses of great medical importance such as polio and herpes (Delgadillo Ruiz et al., 2017). The antifungal activity of the extracts of *L. tridentata* has been proven in the inhibition of *Fusarium oxysporum,* one of the main causes of crown rot in the tomato plant, which is grown and exported in Mexico, causing significant losses in production systems (Peñuelas-Rubio et al., 2017).

4.3.5 *Euphorbia antisyphilitica* (Candelilla)

E. antisyphilitica is a bush with cylindrical stems covered with wax, no leaves, an approximate diameter of 90 cm, and small roots. This plant is endemic to the semi-desert regions of Coahuila, Chihuahua, Durango, Hidalgo, Nuevo León, San Luis Potosí, Tamaulipas, Zacatecas, and Puebla. In the twentieth century, the extraction of candelilla wax was one of the main economic activities in the semi-desert of Chihuahua. Currently, Coahuila is the primary producer of candelilla wax. In the medicinal field, candelilla has been used as a laxative and to treat toothache.

E. antisyphilitica has good plasticising and essential oil retention properties because its chemical structure is composed of esters of fatty acids and long-chain fatty acids, such as hydroxylated esters, free acids, diesters, hydrocarbons, and free alcohols. Essential oils have disadvantages such as degradation by oxidation, by the action of light, and loss by volatilisation, so strategies are required to increase their shelf life and reduce degradation. To this problem, candelilla wax nanoparticles as an encapsulating matrix have been shown to increase the encapsulation capacity of essential oils to provide photo protection (Navarro Guajardo, 2019).

Additionally, little-used byproducts are obtained from the extraction of candelilla wax, a source of bioactive compounds; examples are the polyphenolic compounds ellagitannins and ellagic acid with antioxidant properties in the scavenging of free radicals. This suggests a new window for the use of candelilla with industrial and cosmetic relevance, as it has potential against skin damage induced by free radicals (Bautista-Hernández et al., 2021).

CONCLUSIONS

The essential oils of semi-desert plants represent an excellent potential for revaluating various species collected by the communities of the Coahuila municipalities of Parras, Matamoros, Torreón, and Viesca. The steam distillation plants are an economical and efficient option for producing essential oils. Using instruments that are easy to operate and maintain, the antifungal and antimicrobial properties and diverse applications of the essential oils of the species above have been demonstrated. The management of the commercialisation of products could bring a notable improvement in the income of collecting communities and directly impact their quality of life.

REFERENCES

Aguirre, V et al. (2012) "Obtaining and in vitro evaluation of the efficiency of extracts with active ingredients from eucalyptus (eucalyptus globulus), garlic (allium sativum) and chrysanthemum (chrysanthemum cinerariaefolium) as natural fungicides for the control of botrytis cinerea, phra", *Centro de Scientific Research (ESPE)*, 1(3), pp. 1–17. Available at: http://repositorio.espe.edu.ec/handle/21000/6840.

Alvarez-Perez, OB et al. (2020) "Valorization of Flourensia cernua DC as source of antioxidants and antifungal bioactives", *Industrial Crops and Products*. Elsevier, 152(April), p. 112422. doi: 10.1016/j.indcrop.2020.112422.

Amadio, C et al. (2011) "Essential oil of oregano: A potential food additive", *Magazine of the Faculty of Agricultural Sciences UNCUYO*, 43(1), pp. 237–245. doi: 10.1016/j.jare.2013.12.007.

Badui Dergal, S (2006) *Food Chemistry*. Pearson Education. 738 p.

Bautista-Hernández, I et al. (2021) "Antioxidant activity of polyphenolic compounds obtained from Euphorbia antisyphilitica by-products", *Heliyon*, 7(4). doi: 10.1016/j.heliyon.2021.e06734.

Casella, S et al. (2013) "The role of diallyl sulfides and dipropyl sulfides in the in vitro antimicrobial activity of the essential oil of garlic, Allium sativum L., and leek, Allium porrum L.", *Phytotherapy Research*, 27(3), pp. 380–383. doi: 10.1002/ptr.4725.

Castro, V et al. (2012) "In vitro antibacterial activity of the ethanolic and chloroform extracts of Flourensia polycephala Dillon (phauka) and preparation of a topical pharmaceutical form for its in vivo evaluation in dermal infections by Staphylococcus aureus", 8(8), pp. 1967–1969.

Cerda, EA (2018) "Comparative study of the quality of garlic extract obtained by steam dragging and hydrodistillation previously subjected to a dehydration process". Food Science and Technology Engineer. Autonomous Agrarian University Antonio Narro.

Delbón, NE (2014) "Biology and conservation of flourensia dc species. (asteraceae) endemic of central argentina".

Delgadillo Ruiz, L et al. (2017) "Chemical composition and in vitro antibacterial effect of extracts from Larrea tridentata, Origanum vulgare, Artemisa ludoviciana and Ruta graveolens", *Nova Scientia*, 9(19), p. 273. doi: 10.21640/ns.v9i19.1019.

Estrada Jirón, JB (2015) "Extraction of the essential oil from the flavelus of the sweet orange (Citrus sinensis L.) valencia variety, coming from agro-industrial waste, using the method of distillation by steam at pilot scale, for its application in la fo", *Emecanica.Ingenieria.Usac.Edu.Gt*, pp. 1–75. Available at: http://emecanica.ingenieria.usac.edu.gt/sitio/wp-content/subidas/6ARTÍCULO-III-INDESA-SIE.pdf.

Gutiérrez-Ortega, J A (2004) "Antifungal effect of flourensia microphylla, flourensia cernua and flourensia retionophylla on Alternaria solani, fysarium exysporum and rhizctonia". Parasitologist Agronomist Engineer. Autonomous Agrarian University Antonio Narro.

Hernández-Suárez, M (2005) "*Antibacterial and Antifungal Effectiveness of Quitosan and Larrea tridentata against Microorganisms that Affect Humans and Agricultural Products*". Biology degree. University Center of Biological and Agricultural Sciences. University of Guadalajara, pp. 12–30.

Juarez-Segovia, K et al. (2019) "Effect of raw garlic extracts (Allium sativum) on the in vitro development of Aspergillus parasiticus and Aspergillus niger", *Polibotánica*, 47, pp. 99–111. doi: 10.18387/polibotanica.47.8.

Kant, R and Kumar, A (2021) "Advancements in steam distillation system for oil extraction from peppermint leaves", *Materials Today: Proceedings*. Elsevier Ltd., 47(xxxx), pp. 5794–5799. doi: 10.1016/j.matpr.2021.04.123.

Koubaa, M et al. (2016) "Oilseed treatment by ultrasounds and microwaves to improve oil yield and quality: An overview", *Food Research International*. Elsevier Ltd, 85, p. 59–66. doi: 10.1016/j.foodres.2016.04.007.

López-Rivera, R et al. (2018) "Antifungal effect of emulsions based on essential oil of Mexican oregano (Lippia graveolens), against Candida albicans", *Revista Salud Jalisco*, 1, p. 4. Available at: http://www.medigraphic.com/pdfs/saljalisco/sj-2018/sj181g.pdf.

Mata, R et al. (2003) "Phytotoxic compounds from Flourensia cernua", *Phytochemistry*, 64(1), pp. 285–291. doi: 10.1016/S0031-9422(03)00217-6.

Mouhssen Lahlou (2004) "Methods to Study the Phytochemistry and Bioactivity of Essential Oils", *Phytotherapy Research*, 448(November 2003), pp. 435–448.

Navarro Guajardo, NB (2019) "Development of New Hybrid Biopolymer Systems as a Matrix for Encapsulation with Improved Hydrophobicity and Ultraviolet Absorption Properties". Master in Polymer Technology. Applied Chemistry Research Center.

Nieto, G, Skibsted, LH, Andersen, ML and Ros, G (2012) "Antioxidant and pro-oxidant activity of garlic essential oil by electron spin resonance", *Anales de Veterinaria de Murcia*, 28, pp. 23–33.

Orona Castillo Ignacio, SA and Jazmín, A, Espinoza Arellano José de Jesús, VC (2017) "Collection and commercialization of oregano (Lippia spp) in the Mexican semi-desert, a case study: sierra y cañon de jimulco municipal ecological reserve, Mexico", *Mexican Journal of Agribusiness*, 41(2017), pp. 684–695.

Peñuelas-Rubio, O et al. (2017) "Extracts of Larrea tridentata as an ecological strategy against Fusarium oxysporum radicis-lycopersici in tomato plants under greenhouse conditions", *Revista Mexicana de Fitopatología, Mexican Journal of Phytopathology*, 35(3), pp. 360–376. doi: 10.18781/r.mex.fit.1703-3.

Perino-Issartier, S et al. (2013) "A comparison of essential oils obtained from lavandin via different extraction processes: Ultrasound, microwave, turbohydrodistillation, steam and hydrodistillation", *Journal of Chromatography A*. Elsevier BV, 1305, pp. 41–47. doi: 10.1016/j.chroma.2013.07.024.

Radwan, MN et al. (2020) "A solar steam distillation system for extracting lavender volatile oil", *Energy Reports*. Elsevier Ltd, 6, p. 3080–3087. doi: 10.1016/j.egyr.2020.11.034.

Rahman, MM, Fazlic, V and y Saad, NW (2012) "Antioxidant properties of raw garlic (Allium sativum) extract", *International Food Research Journal*, 19(2), pp. 589–591.

Rojas, JPL, Perea, AV and Stashenko, EE (2009) "Obtaining of essential oils and pectins from by-products of citrus juice", *Vitae*, 16(1), pp. 110–115.

Salas-Méndez, E de J et al. (2019) "Application of edible nanolaminate coatings with antimicrobial extract of Flourensia cernua to extend the shelf-life of tomato (Solanum lycopersicum L.) fruit", *Postharvest Biology and Technology*. Elsevier, 150(June 2018), pp. 19–27. doi: 10.1016/j.postharvbio.2018.12.008.

Saldaña-Mendoza, SA et al. (2022) "Technological trends in the extraction of essential oils", *Environmental Quality Management*, (January), pp. 1–10. doi:10.1002/tqem.21882.

Soto-Dominguez, A et al. (2012) "The aqueous extract of Oregano (Lippia graveolens HBK) from Northern Mexico has antioxidant activity without showing a toxic effect in vitro and in vivo", *International Journal of Morphology*, 30(3), pp. 937–944. doi: 10.4067/S0717-95022012000300029.

Torres-Léon, C et al. (2023) "Medicinal plants used by rural communities in the arid zone of Viescaand Parras Coahuila in northeast Mexico", *Saudi Pharmaceutical Journal*, 31, pp. 21–28.

Zhuang, X et al. (2018) "The effect of alternative solvents to n-hexane on the green extraction of Litsea cubeba kernel oils as new oil sources", *Industrial Crops and Products*. Elsevier, 126(September), pp. 340–346. doi: 10.1016/j.indcrop.2018.10.004.

Chapter 5

Ethnopharmacology of Important Aromatic Medicinal Plants of the Caatinga, Northeastern Brazil

Sikiru Olaitan-Balogun, Mary Anne Medeiros-Bandeira,
Karla do Nascimento-Magalhães, and Igor Lima-Soares

CONTENTS

5.1 CAATINGA BIOME: THE "SILVER-WHITE FOREST" RESTRICTED TO BRAZIL

Brazil, with its continental dimensions (8,516,000 km^2), occupies almost half of South America and is the country with the greatest biodiversity in the world, with 15%–20% of the total number of species of living beings, making it the most relevant among the 17 megadiverse countries and

DOI: 10.1201/9781003251255-5

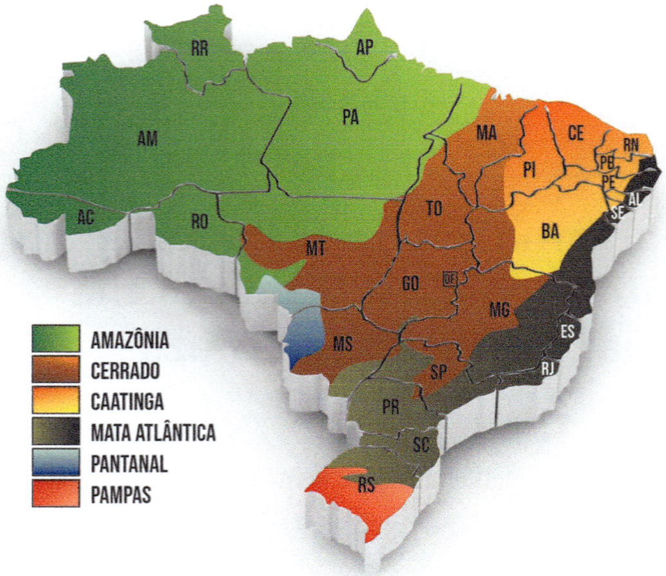

Figure 5.1 The Brazilian Biomes Map featuring the Amazon, Cerrado, Caatinga, Atlantic Forest, Pantanal and Pampas. States – AC – Acre, AL – Alagoas, AP – Amapá, AM – Amazonas, BA – Bahia, CE – Ceará, ES – Espírito Santo, GO – Goiás, MA – Maranhão, MT – Mato Grosso, MS – Mato Grosso do Sul, MG – Minas Gerais, PA – Pará, PB – Paraíba, PR – Paraná, PE – Pernambuco, PI – Piauí, RJ – Rio de Janeiro, RN – Rio Grande do Norte, RS – Rio Grande do Sul, RO – Rondônia, RR – Roraima, SC – Santa Catarina, SP – São Paulo, SE – Sergipe, TO – Tocantins. Source: Authors (2022)

the largest remainder of tropical ecosystems. There are more than 116,000 animal species and 46,000 plant species known in the country, with high ecological, genetic, scientific, economic, social, educational and cultural value spread over the six terrestrial biomes and three large marine ecosystems (Brasil, 2018; Ulloa et al., 2017). According to Forzza et al. (2012), Brazil has the highest global endemism rate (56%) of plant species.

Knowledge about the biodiversity of Brazil is still far from having a complete inventory of the animal and plant species it harbors. An example of this is the marked evolution in the numbers of new taxa described for groups of fauna every year (Varjabedian, 2010).

Brazil is not only rich in diversity of genetic resources but also in cultures (more than 200 indigenous peoples and several communities such as *quilombolas*, *caiçaras* and rubber tappers), different people who manage their environment, knowing it in detail and in the whole of its connections and interrelationships (Posey, 1983; Brasil, 2017).

Figure 5.1 represents the geographical distribution of the six Brazilian biomes: Amazon, Atlantic Forest, Pantanal, Cerrado, Pampa and Caatinga. These biomes are defined by a homogeneous area in relation to its vegetational, climatic, pedological and altimetric characteristics, arranged on a regional scale and influenced by the same formation processes (Coutinho, 2006; Nascimento & Ribeiro, 2017).

According to the online database of the Flora do Brasil website, 49,979 species are recognised as species of Brazilian flora (native, cultivated and naturalised), of which 4,993 are algae, 35,539 are angiosperms, 1,610 are bryophytes, 6,320 are fungi, 114 are gymnosperms and 1,403 are ferns and lycophytes.

In this context, the semi-arid biome of Brazil that is the focus of this chapter. The name *caatinga* is of Tupi-Guarani origin, and means "white forest", which characterises well

Figure 5.2 Caatinga biome, Paraíba State (PB), Brazil. Source: Photo by Jaime Dantas on Unsplash

the aspect of the vegetation in the dry season, when the leaves fall and only the white and shiny trunks of trees and bushes remain in the dry landscape (Figure 5.1) (Albuquerque & Bandeira, 1995).

The Caatinga (Figure 5.2) occupies about 11% of the country (844,453 km^2) and is the main ecosystem of the Northeast Region of Brazil. There are about 27 million inhabitants in the original Caatinga area, and 80% of its original ecosystems have already been altered (Silva et al., 2004; Brasil, 2002). This biome represents the fourth largest vegetation formation in the country, covering about 60% of the Northeastern territory, extending to a small part of the state of Minas Gerais (Sampaio & Rodal, 2002; Castelletti et al., 2005).

Although Caatinga has a high level of environmental heterogeneity and endemic species and genera with rare taxa, it has, in the past, been neglected by conservation policies in the face of Brazil's vast biodiversity, and its richness and importance have not been the target of policies for the study and conservation of biodiversity in the country (Brasil, 2008a, 2008b, 2011; Teixeira, 2018; Sampaio, 1995).

Although the diversity of plants and animals in arid and semi-arid environments is lower than in lush tropical forests, deserts have plants and animals adapted to their extreme conditions, making them environments with a high rate of endemic flora and fauna (Craveiro et al., 1994; Vieira & Martins, 1998; Leal, Silva & Tabarelli, 2003; Queiroz, Conceição & Giulietti, 2006; Anon, 1995).

There is a mistaken view that the Caatinga is synonymous with poverty in biodiversity, which certainly reflects how little is known about the region, as it is the Brazilian biome with the fewest inventories and underestimated biological diversity (Barros, 2004; Brasil, 1998). According to the online database of the Flora do Brasil website (JBRJ, 2022), there are currently 5,022 species of angiosperms cataloged that are distributed in 177 families and 1,249 genera in this biome, with 2,677 species endemic to the Caatinga.

The idea of unproductivity is always associated with the arid and semi-arid areas of the world, and because of this misconception, the conservation of the biome is relegated to the background (Castelletti et al., 2005; Albuquerque & Andrade, 2002a, 2002b). Caatinga is not a desert but a unique ecosystem, and, for this reason, it has animals and plants that only survive there. In fact, in this biome, there is a greater wealth of species than any other biome in the world with similar environmental conditions (Castelletti et al., 2005; Silva, 2003; Andrade-Lima, 1981; Brasil, 2011).

The harsh climate that dominates the region determines vegetation with a high frequency of xerophytic elements, especially cacti and bromeliads, which defines the physiognomy Steppe Savanna as the most characteristic of the Caatinga, with fascinating adaptations to semi-arid habitats (thorns, aculus, succulent leaves and stems) (Zappi, 2009; Giuletti et al., 2009).

Data from 2010 revealed the floristic survey of the entire Brazilian territory, in which the Caatinga biome presented a total of 4,322 species of seed plants, 744 of which are endemic to this biome, corresponding to 17.2% of the total taxa recorded (Forzza et al., 2010).

The most frequent families in the Caatinga are Leguminosae, Euphorbiaceae, Cactaceae, Asteraceae and Malpighiaceae (Andrade-Lima, 1981).

The unique and still little-known characteristics of the Caatinga, as well as the fragility of its most arid system, have not been reflected in its protection. In 2008, according to the Brazilian Environmental Office (Brasil, 2008b), in a collaborative map with The Nature Conservancy of Brazil (TNC), the Caatinga had 7.12% of its area protected by Conservation Units (UCs), with only 0.99% in Full Protection and 6.04% in Sustainable Use, representing a very small number of UCs between 2008 and 2014 (Brasil, 2008c; Hauff, 2010).

The Caatinga has been greatly modified by anthropic action, either by the advance of agriculture or road construction, reducing the populations of native species (Trombulak & Frissell, 2000; Castelletti et al., 2005; Giullietti et al., 2004). It is important to point out the intense desertification process that the northeastern soils have been suffering mainly due to deforestation (which reaches 46% of the biome area) by burning (Araújo & Souza, 2011). The scenario of desertification in Brazil, whether in the conceptual, scientific and combat scope of the problem, is still marked by challenges (Albuquerque et al., 2020).

The conservation of the Caatinga is essential for the maintenance of climatic patterns, availability of drinking water, fertile and productive soils and part of the planet's biodiversity (Brasil, 2011). For Kiill (2010), there are at least 19 threatened, vulnerable or endangered species in the Caatinga. Water stress is one of the most limiting factors for productivity and geographical distribution of plant species (Costa et al., 2010; Holanda et al., 2015).

Tabarelli and Vicente (2002), when evaluating the geographic distribution of plant collections and floristic and phytosociological studies developed in the Caatinga, estimated that 80% of the biome's area would be under-sampled, and for half of this area (40%) there were no records of collections. Among the determining factors for this situation are indiscriminate deforestation for the formation of new plantations, the trade in wood for improvements and charcoal production, successive burnings, overgrazing and inadequate soil use. All these practices have contributed to the compromising of its balance (Albuquerque, Lombardi-Neto & Srinivasan, 2001), reflecting in the alteration of 80% of its original ecosystems and the susceptibility of 62% of its territory to desertification processes (Brasil, 2011).

As in any dry forested region in the world today, the legacy of the northeastern semi-arid region is far from being properly documented, appreciated and conserved. The trajectory of degradation that has marked the region's history since the arrival of Europeans in Brazil remains unchanged, and the climate changes predicted for the region, particularly reduced precipitation, indicate a bleak future (Tabarelli et al., 2018). Although it is, indeed, much altered, especially in the lowlands, the Caatinga contains a great variety of vegetational types, with a high number of species and also remnants of still well-preserved vegetation, which include an expressive number of rare and endemic taxa (Giulietti et al., 2004).

The urgency to define a policy for the conservation of the Caatinga biodiversity becomes evident when considering that the biome is one of the most populated semi-arid regions in the world, thus making it a great challenge to promote the development of the region with environmental protection (Tabarelli et al., 2018). The study and conservation of the biological diversity of the Caatinga is one of the biggest challenges of Brazilian science, warns Leal, Silva and Tabarelli (2003).

5.2 ETHNOPHARMACOLOGY IN THE BRAZILIAN NORTHEAST AND THE IMPORTANT ROLE OF PROFESSOR FRANCISCO JOSÉ DE ABREU MATOS

In Brazil, the use of medicinal plants in the treatment of diseases is influenced by African, indigenous and European cultures (Martins et al., 2000). Indigenous peoples, traditional peoples and communities and family-based farmers are important holders of traditional knowledge, an integral part of Brazil's cultural heritage, and safeguard part of Brazil's genetic heritage (Brazil, 2017). Many factors contribute to the expressive use of plants for medicinal purposes in Brazil, notably the high cost of industrialised medicines, the difficult access of the population to medical and pharmaceutical assistance and the tendency to use products of natural origin (Badke et al., 2012). Research shows that about 91.9% of the country's population has already made use of some kind of medicinal plant, and 46% cultivate some medicinal species at home (Abifisa, 2022).

5.2.1 The Economic and Socio-cultural Diversity of the Caatinga

The Caatinga biome, besides possessing rich biodiversity, is home to great socio-diversity – cowboys, ox-drivers, *quilombolas*, indigenous people, among other groups – that have vast knowledge about local plant resources (Diegues, 2000; Gomes & Bandeira, 2012). The Brazilian semi-arid region is densely populated and has the lowest life expectancy, lowest income per capita and highest illiteracy rate in the country. The rural inhabitants of the Caatinga, called *sertanejo*, has developed a peculiar socio-cultural structure and has a strong relationship with the use of natural resources (Giulietti et al., 2009).

A large part of the population living in the Caatinga area is poor and depends on the resources of its biodiversity to survive, and many of these species are found in the forest fragments explored by the native population. Thus, the practice of using medicinal plants in self-care is deeply rooted, sometimes incorporating sympathies and prayers, in a mixture of beliefs and faith, inherited from the shamans and the Jesuits (Silva, 2003; Silva et al., 2004; Brasil, 2002; Gomes et al., 2008; Jha, 1995; Gera, Blsht & Rana, 2003).

5.2.2 Ethnobotanical Studies on the Caatinga Biome: A Brief Synopsis

Ethnobotanical studies in the Brazilian semi-arid region are still relatively scarce, which reflects the lack of interest of researchers in dry forests. The current forms of land use are extremely precarious and do not respect the complexity of these delicate ecosystems (Albuquerque & Hanazaki, 2006; Amorozo & Gely, 1988; Pereira-Júnior et al., 2014).

It is relevant to note that even though the state of Ceará has practically all of its territorial area within the Caatinga biome, it is lagging behind in terms of published articles. One of the greatest contributions occurred with the publication to the scientific community of the ethnopharmacopeia of Professor Francisco José Matos in 2019. This work brings the compilation of data from the reports of ethnobotanical expeditions conducted by his team to the Caatinga between the years of 1979–1982, including native species that are extinct or threatened with extinction (Magalhães et al., 2019). The number of species obtained in this study can be considered high when compared to other studies conducted in the Caatinga, with reports ranging from 22 to 119 species (Albuquerque & Oliveira, 2007, Albuquerque & Andrade, 2002a, Alcântara Júnior et al., 2005; Almeida et al., 2005, Cartaxo, Souza & Albuquerque, 2009; Morais et al., 2005, Silva; Andrade & Albuquerque, 2005).

5.2.3 A Brief Insight into the Ethnopharmacopeia of the Late Prof. Francisco José de Abreu Matos

The aromatic medicinal plants discussed in this chapter were selected based on the ethnopharmacopoeia of the late Prof. Francisco José de Abreu Matos. For the purpose of brevity, being concise and in line with the book theme, we limited our selection to only aromatic medicinal plants of the Caatinga.

The ethnopharmacopeia of Prof. Francisco José de Abreu Matos includes data on 272 plant species. It describes botanical nomenclature revised by the Angiosperm Phylogeny Group IV System (2016), popular names, origin, popular therapeutic uses, part used, preparation method, form of use, voucher number, citations, number of therapeutic properties associated with the species, number of body systems according to International Classification of Primary Care (WHO, 2012) and relative importance value. The term *ethnopharmacopoeia* by Professor Francisco José de Abreu Matos was chosen to honor the unique scientific collection that portrays the associated traditional knowledge of the northeastern Brazilian people about medicinal plants from the Caatinga (especially from the state of Ceará) in the 1980s.

For Matos (1999), in the Northeast Region, the use of medicinal plants and homemade preparations is of fundamental importance in the treatment of diseases that affect low-income populations, given the deficiency of medical assistance, the influence of the oral transmission of cultural habits and the availability of flora.

The flora of the Caatinga has a rich plant diversity, with multiple potentials for economic exploitation in a sustainable way. However, in the case of medicinal or aromatic uses, this knowledge is still fragmented and tied to popular knowledge. Thus, studies on the popular knowledge about the native flora have deserved increasing attention, which contributes to the generation of knowledge and clarification of science, especially with regard to plants of the dry forests (Agra et al., 2008; Agra et al., 2007; Nunes et al., 2015; Pareyn, 2010).

5.3 AROMATIC AND MEDICINAL PLANTS FROM CAATINGA

Brazil stands out on the world stage as the fourth largest exporter of essential oils, after the United States, France and the United Kingdom (U.K.) (OEC, 2022).

According to Mattoso (2005), the essential oil from citrus leaves and its terpenic derivatives account for 90% of this export volume, and the rest is obtained from other species such as eucalyptus (*Eucalyptus citriodora*), pau-rosa (*Anibarosea odora* var. *amazonica* Ducke), limeira (*Citrus aurantifolia* Swingle) and lemongrass (*Cymbopogon citratus*).

Despite the peculiarity and potential of Brazilian biodiversity, the production of essential oils is still in its infancy to meet the world market (Amaral, 2010).

Prospective studies, sustainable use of Brazilian biodiversity, domestication of exotic species with commercial relevance, use of improvement techniques and development of new applications for essential oils are thematic lines, generally multidisciplinary, that have been driving the expansion of research on essential oils in Brazil (Bizzo, Hovell & Rezende, 2009).

In the Northeast, the study of odorous plants resulted in the discovery of several essential oils of potential economic importance (Craveiro et al., 1981).

Agra et al. (2007), Nunes et al. (2015) and Pareyn (2010) report that the flora of the Caatinga presents rich plant diversity, with multiple potentials for economic exploitation in a sustainable manner. However, in the case of medicinal or aromatic use, this knowledge is still fragmented, and much of it is linked to popular knowledge.

Several aromatic and/or medicinal species native to the Caatinga contain essential oils that are widely used as a source of raw material in the perfume and cosmetics, the pharmaceutical, hygiene and cleaning products and the food and paint industries as well as in agriculture for the biological control of diseases and pests (Biasi & Deschamps, 2009).

Despite the great economic importance of essential oils and their derivatives, until 1959 practically nothing was known about the chemical composition of the aromatic flora of the Brazilian northeast and its potentialities. The creation of the Program for the Chemical Study of Essential Oils from Plants of the Northeast (*Programa de Estudo Químico de Óleos Essenciais de Plantas do Nordeste*) in 1975 by the late pharmacist Dr. Francisco José de Abreu Matos and biologist Dr. Afrânio Gomes Fernandes, boosted research in this area.

In this study program, about 500 essential oils were obtained from approximately 150 different native and cultivated plant species, which resulted in the discovery of several essential oils of potential economic importance (unpublished report).

Since the 1980s, there has been an awakening to the richness of the Caatinga flora and the fascinating adaptations of plants to semi-arid habitats (Andrade-Lima, 1981, 1989; Giulietti et al., 2004). The selection of matrices, irrigation and mechanisation of the crop are practices that have been successfully applied in the northeast of the country for the production of essential oils (Bizzo, Hovell & Rezende, 2009).

The flora of the Caatinga has a rich plant diversity, with multiple potentials for economic exploitation in a sustainable manner. However, in the case of medicinal or aromatic use, this knowledge is still fragmented and tied to popular knowledge (Kiill, 2010).

The ten aromatic plant species described below were selected from the ethnopharmacopeia of Prof. Francisco José de Abreu Matos (Magalhães et al., 2019). The criteria used were species abundance, ease of cultivation, high yield of essential oil, chemical constituents of relevant economic importance and therapeutic potential. The scientific and popular names, synonyms, phytochemical composition, ethnobotanical and ethnopharmacological data are presented for each aromatic medicinal plant.

5.3.1 *Ageratum conyzoides* L.

- ***Taxonomic classification*** (APG IV/2016): Asterids – Asteracea – Asterales
- ***Botanical synonyms***: *Ageratum conyzoides* L. subsp. conyzoides
- Link to the taxon: http://servicos.jbrj.gov.br/flora/search/Ageratum_conyzoides
- ***Origin***: Native, non-endemic
- ***Popular names***: Mentrasto
- ***Phytochemistry***: The main composition of *A. conyzoides* is based on essential oils with terpenes (salinene, pinene, eugenol, cineol, felandren, limonene, linalool,

terpineol and caryophyllene), coumarin and benzofuran compounds, resins, alkaloids, flavones, flavonoids and chromones. There is also the presence of chromenes, especially precocenes I and II, which cause premature metamorphosis in several insect species, leading to the formation of sterile adults (Momesso, Moura & Constantino, 2009). The analysis of the essential oil from the leaves of *A. conyzoides* from the state of Ceará revealed the presence of chromene class substances such as precocene I (95.4%) and precocene II (4.5%) (Lima et al., 2005). Several studies point to precocene I as a component in higher concentration compared to precocene II in the essential oil of the species (Bayala et al., 2014; Menut et al., 1993; Patil et al., 2010). However, in some investigations carried out in other locations in Brazil, as well as other countries, a reversal in this proportion has been verified (Esper et al., 2015; Liu & Liu, 2014). In addition, the presence of β-caryophyllene (14.4%) and coumarin (5.05%) as significant volatile components of the leaves of the species have also been reported (Nogueira et al., 2010; Zoghbi et al., 2007). In a study conducted in the northern region of Brazil, it was found that the phytochemical composition of the stems, roots and flowers are similar to each other, presenting as major components β-caryophyllene (12.8%, 15.3% and 19.4%, respectively) and precocene I (71.6%, 67.4% and 55.5%, respectively) (Zoghbi et al., 2007).

- ***Part used/method of preparation/form of use*** (I or E), according to Magalhães et al. (2019): Root – tea (I), maceration in water (I); leaf – tea (I) its most commonly used extract forms are 2% and 5% infusion, 20% tincture and glycoholic extract.
- ***Life form and substratum***: Herb, sub-shrub/rupiculous, terrestrial
- ***Ethnobotanical data***: Painful ovulation (women's colic), depurative ("cleanses impurities from the blood"), flu, incomplete abortion, nervousness, abdominal cramps (stomachache), anti-inflammatory, muscle distention
- ***Ethnopharmacological/Biological data***: There is evidence supporting the use of the species as an antineuropathic pain agent (Sukmawan, Anggadiredja & Adnyana, 2021), as antibacterial against *Staphylococcus aureus* and *Enterococcus faecalis* (Kouame et al., 2018), antifungal, anti-inflammatory, anthelminthic and antitumor (Yadav et al., 2019).

5.3.2 *Cantinoa mutabilis* (Rich.) Harley & J.F.B. Pastore

- ***Taxonomic classification*** (APG IV/2016): Asterids – Lamiaceae – Lamiales
- ***Botanical synonyms***: *Hyptis polystachya, Hyptis rostrata, Hyptis singularis, Hyptisspicata, Hyptis tenuiflora, Hyptis trichocaly, Mesosphaerum barbatum, Teucrium rhombifolium, Hyptis mutabilis*
- ***Link to the taxon***: http://servicos.jbrj.gov.br/flora/search/Cantinoa_mutabilis
- ***Origin***: Native, non-endemic
- ***Phytochemistry***: A phytochemical study of the essential oil obtained from the aerial parts of the species (leaves and stem) was conducted in four different localities in the Amazon region. A specimen collected in the locality of Lago Grande, Amapá state, presented thymol (37.4%), p-cymene (19.3%) and γ-terpinene (16.6%) as major phytoconstituents. Phytochemistry analysis of an access from Retiro das Pedras municipality in Amapá state revealed σ-3-carene (25.5%), terpinolene (24.7%) and globulol (11.9%) as major components. The essential oil from a sample from the city of Belém, capital of

Pará state, showed β-caryophyllene (18.4%), 1.8-cineole (16.8%) and bicyclogermacrene as the most concentrated components (12.6%). The sample obtained from the Porto Almendrada locality in Peru was rich in (E)- (70.0%) and (Z)-methyl cinnamate (9.8%) (Aguiar et al., 2003). Analysis of the essential oil of the species occurring in southern Brazil showed that both leaves (26.61% and 24.23%) and inflorescences (11.6%) had globulol in their compositions. Other relevant constituents found in the leaves and flowers were β-pinene, germacrene D and E-caryophyllene (Silva et al., 2013). An investigation showed that the essential oil from *C. mutabilis* leaves had spathulenol (14.2%), E-caryophyllene (14.7%) and germacrene D (11.1%) as Eudicotyledoneae - Rosids - Fabids constituents (Oliva et al., 2006).

- ***Popular names***: Sambacuité
- ***Part used/method of preparation/form of use*** (I or E), according to Magalhães et al. (2019): Aerial part – juice (I, E); root – tea (I)
- ***Life form and substratum***: Shrub, herb, sub-shrub/terrestrial
- ***Ethnobotanical data***: Stomach ailments and analgesic
- ***Ethnopharmacological/Biological studies***: Evidence suggests that the species has antiulcerogenic (Barbosa & Ramos, 1992), sedative, anesthetic (Silva et al., 2013) and antimicrobial action against *Mucor* sp. (Oliva et al., 2006).

5.3.3 *Croton echioides* Baill.

- ***Taxonomic classification*** (APG IV/2016): major – Euphorbiaceae – Malpighiales
- ***Botanical synonyms***: *Croton macrobothrys* var. *Microbotrys, Oxydectes echioides*
- ***Link to the taxon***: http://servicos.jbrj.gov.br/flora/search/Croton_echioides
- ***Origin***: Native, endemic
- ***Popular names***: *Quebra-faca, caatinga-branca, canela-de-velho*
- ***Phytochemistry***: Phytochemical studies from the essential oil of the species are scarce. In a phytochemistry analysis of *C. Echioides* leaves conducted in the state of Alagoas, the following components were recorded: α-cubene (2), α-copaene (3), caryophyllene (4) and β-cubebene (5) (Craveiro, 1981).
- ***Part used/method of preparation/form of use*** (I or E), according to Magalhães et al. (2019): Bark (stem) – in natura/ scraping (I); whole plant – in natura (I)
- ***Life form and substratum***: Shrub, sub-shrub/terrestrial
- ***Ethnobotanical data***: Kidney problems, anticoagulant ("thinning of the blood").
- ***Ethnopharmacological/Biological studies***: Published data indicates that this species is considered promising for the development of antileishmaniasis drugs (Novello et al., 2020).

5.3.4 *Croton grewioides* Baill.

- ***Taxonomic classification*** (APG IV/2016): Eudicotyledoneae - Rosids - Fabids – Euphorbiaceae – Malpighiales
- ***Botanical synonyms***: *Croton zehntneri, Croton glycosmeus, Oxydectes grewioides*
- ***Link to the taxon***: http://servicos.jbrj.gov.br/flora/search/Croton_grewioides
- ***Origin***: Native, non-endemic
- ***Popular names***: Alecrim-de-cabocla, canelinha, canela do mato

- ***Phytochemistry***: In the state of Pernambuco, the essential oil from the leaves and stems of the species showed a predominance of phytoconstituents from the phenylpropanoid group. The volatile phytoconstituents found in the leaves were (E)-anethole (65.5%) and methyl eugenol (10.6%), while the essential oil analysis detected the presence of (E)-anethole (47.8%), (E)-methyl isoeugenol (30.0%) and cadalene (8.4%) (Silva et al., 2008). Analyses conducted in the state of Paraíba revealed that the essential oil from *C. grewioides* leaves was rich in mono- and sesquiterpenes, with α-pinene (47.43%) and sabinene (12.09%) as major components (Medeiros et al., 2017). The analysis of the essential oil of different genotypes of the species preserved in an active germplasm bank in the state of Sergipe proposed different chemotypes, which are (Z)-isoosmorhizole + (E)-isoosmorhizole (24.19% and 11.38%, respectively); methyl chavicol + eugenol (59.85% and 29.51%, respectively); methyl eugenol (85.58%) and eugenol + methyl eugenol (42.68% and 45.97%, respectively) (Oliveira et al., 2021).
- ***Life form and substratum***: Shrub, sub-shrub/terrestrial
- ***Part used/method of preparation/form of use*** (I or E), according to Magalhães et al. (2019): Twigs – tea (I, E)
- ***Ethnobotanical data***: Sinusitis
- ***Ethnopharmacological/Biological studies***: The volatile fractions of *C. grewioides* demonstrates antioxidant (Oliveira et al., 2021; Serra et al., 2019) and antispasmodic action on the trachea musculature (Lima et al., 2020). In addition, the essential oil showed low subacute oral toxicity (Coelho-de-Souza et al., 2019).

5.3.5 *Croton heliotropiifolius* Kunth.

- ***Taxonomic classification*** (APG IV/2016): Eudicotiledôneas – Rosídeas – Fabídeas – Euphorbiaceae – Malpighiales
- ***Botanical synonyms***: *Croton conduplicatus, Croton albicans, Croton lachnocladus, Croton moritibensis, Croton rhamnifolioides, Croton turnerifolius, Oxydectes lachnoclada*
- ***Link to the taxon***: http://servicos.jbrj.gov.br/flora/search/Croton_heliotropiifolius
- ***Origin***: Native, non-endemic
- ***Phytochemistry***: The literature reports that the essential oils of *Croton heliotropiifolius* showed bactericidal and fungicidal activity (Alencar-Filho et al, 2017). Essential oils obtained from the species are rich in sesquiterpenes, mainly hydrocarbon sesquiterpenes. The essential oil extracted from the aerial parts (leaves and stem), presents as main volatile components (E)-caryophyllene (23.85%), γ-muurolene (10.52%) and viridiflorene (8.08%). In a study conducted in the northeast of Brazil, it was observed that the chemical constituents of the essential oil of the species were the same in all seasons: β-caryophyllene, bicyclogermacrene, germacrene-D, limonene and 1,8-cineole. However, there was variation in the content of these, according to the season of collection (*Alencar-Filho et al,* 2017). Analysis of the essential oil obtained from hydrodistillation of the species stem revealed the predominance of oxygenated sesquiterpenes, including substances such as guaiol (18.38%) and valerianol (10.62%). However, in the volatile phytoconstituents profile of the leaves of the same specimen demonstrated the presence of hydrogenated sesquiterpenes as being prominent, encompassing β-caryophyllene (20.82%), spathulenol (16.37%) and β-elemene (6.81%). Analyses of

its essential oil identified a mixture of monoterpenes and sesquiterpenes, with the as main compounds being bicyclogermacrene (19.04%), Ecaryophyllene (18.51%), limonene (15.65%) and α-pinene (13.24%) (Torres et al., 2021). One of the major compounds described in three papers was α-pinene (Alencar-Filho et al, 2017; Brito et al, 2018; Oliveira et al, 2016).

- ***Popular names***: Velame
- ***Life form and substratum***: Shrub, sub-shrub/terrestrial
- ***Part used/method of preparation/form of use*** (I or E), according to Magalhães et al. (2019): Root – tea (I)
- ***Ethnobotanical data***: The Brazilian flora is well represented by this family, as 72 genera and about 1,100 known species occur (Júnior et al, 2022). One of the genera of the Euphorbiaceae family that has a wide variety of species rich in secondary metabolites is the genus *Croton* (Bezerra et al., 2020a). Brazil owns a good diversity of this genus; there are more than 350 native species (Trindade & Lameira, 2014). It is used as depurative ("cleanses impurities from the blood"), to treat eczema, cancer, thrush and for bladder ailments. The essential oil of the species is also used to combat *Aedes aegypti* mosquitoes (Alencar-Filho et al, 2017).
- ***Ethnopharmacological/Biological studies***: The investigations indicate that the volatile fraction of the species has antibacterial potential mainly against gram-positive bacteria (Araújo et al., 2017) and also demonstrates antinociceptive actions, (Oliveira Júnior et al., 2017), anxiolytic, sedative (Oliveira Júnior et al., 2018), anti-inflammatory (Martins et al., 2017), gastroprotective (Vidal et al., 2017) and antiparasitic actions (Alcântara et al., 2022).

5.3.6 *Croton jacobinensis* Baill.

- ***Taxonomic classification*** (APG IV/2016): Eudicots – Rosides – Fabids – Euphorbiaceae – Malpighiales
- ***Botanical Synonyms***: *Croton auriculatus, Croton sonderianus, Oxydectes sonderiana, Oxydectes jacobinensis*
- ***Link to the taxon***: http://servicos.jbrj.gov.br/flora/search/Croton_jacobinensis
- ***Origin***: Native, endemic
- ***Phytochemistry***: Analysis of the essential oils from the leaves and inflorescence allowed the identification of 32 compounds, representing more than 90% of the chemical composition of each oil. The main constituents of the leaf oil were bicyclogermacrene (34.0%), D-germacrene (19.0%) and trans-caryophyllene (17.8%), and the main components of the inflorescence oil were caryophyllene oxide (27.9%), spatulenol (16.7%) and 1,8-cineole (8.0%). *C. Sonderianus* has a high essential oil content that can vary from 0.5% to 1.5% (Santos, et al., 2005). Several studies indicate that the essential oil from the leaves of the species native to northeastern Brazil has a variable phytochemical composition, predominantly mono- and sesquiterpenes, according to the time and place of collection. The essential oil of the species showed as major components bicyclogermacrene (16.29%), β-phellandrene (15.42%), β-caryophyllene (13.82%) and α-pinene (9.87%) when collected 13:00 h, but the same individual plant showed the presence of spathulenol (18.32%), β-caryophyllene (14.58%), caryophyllene oxide (8.54%) and 1,8-cineole (8.38%) when collected at 21:00 h (Pinho et al., 2010). The essential oils from the different parts of the same specimen were also analysed. The

leaves (β-phellandrene, 20.4%), flowers (bicyclogermacrene, 29.1%), trunk bark (β-elemene, 22.8%) and roots (cyperene, 14.2%) showed distinct predominant components (Dourado & Silveira, 2005). Seasonality plays an important role on the phytoconstituents content of *C. sonderianus* essential oil occurring in Caatinga, as metabolites such as limonene (rainy season: 18.8%, dry season: 28.7%), spathulenol (rainy season: 10.3%, dry season: 31.5%) and 1,8-cineole (rainy season: 18.8%, dry season: 28.7%) show considerable phytochemical variation (Souza et al., 2017).

- ***Popular names***: *Marmeleiro preto, velame-de-nódea, marmeleiro-do-brejo*
- ***Life form and substratum***: Shrub, sub-shrub/terrestrial
- ***Part used/method of preparation/form of use*** (I or E), according to Magalhães et al. (2019): Bark (stem) – in natura (I); tea (I) – inner stem; bark – tea (I)
- ***Ethnobotanical data***: The leaves and bark are used as an infusion or simply chewed as a medicine for the treatment of gastrointestinal disorders, rheumatism and headaches (Santos et al., 2005). The volatile oils produced by species of the genus *Croton* have been the target of studies because of their biological activities, such as antimicrobial (Andrade et al., 2015), acaricidal (Câmara et al., 2017), antitumor (Araújo et al., 2017), insecticidal (Lima et al., 2013), among others. It is also used as an astringent, for liver diseases, diarrhea and intestinal ailments.
- ***Ethnopharmacological/Biological studies***: The essential oil of the species has antinociceptive (Santos, et al., 2005) and bronchodilatory properties (Pinho et al., 2010).

5.3.7 *Hymenaea courbaril* L.

- ***Taxonomic classification*** (APG IV/2016): Eudicots – Rosides – Fabids – Fabaceae – Fabales
- ***Botanical Synonyms***: *Hymenaea stilbocarpa*
- ***Link to the taxon***: http://servicos.jbrj.gov.br/flora/search/Hymenaea_courbaril
- ***Origin***: Native, non-endemic
- ***Phytochemistry:*** Analysis carried out on the fruits of the species from Ceará state of Brazil, demonstrated that the essential oil extracted by hydrodistillation was rich in sesquiterpenes. However, the phytochemical composition varied according to the stage of maturation. Phytoconstituents encountered were α-copaene (11.1%), spathulenol (10.0%), β-selinene (8.2%), γ-muurolene (7.9%), caryophyllene oxide (6.9%) and germacrene-D (17.61%) from the ripe fruits (Aguiar et al., 2010; Sales et al., 2014). The essential oil of unripened fruits contains as main substances germacrene D (31.9%), β-caryophyllene (27.1%), bicyclogermacrene (6.5%), α-humulene (4.2%) and α-copaene (4.2%) (Aguiar et al., 2010). The essential oil of the leaves of the species obtained by hydrodistillation in the state of Pernambuco, showed caryophyllene oxide (29.55%) and trans-caryophyllene (18.80%) as main components. (COSTA, 2017). In a study in the state of Maranhão, the bark of the tree was shown to be rich in monoterpenes, containing β-ocimene (23.33%), α-pinene (12.33%), β-pinene (11.79%) and β-myrcene (11,38) as the main volatile substances (Everton et al., 2021).
- ***Popular names***: Jatobá
- ***Part used/method of preparation/form of use*** (I or E), according to Magalhães et al. (2019): Bark (stem) – tea (I), maceration in liquor (I); lick (I), in natura/powder (I); "green" fruit – tea (I); bark – lick (I)
- ***Life form and substratum***: Tree/Terrestrial

- ***Ethnobotanical data***: Flu, convalescence (weakness), anemia, diarrhea (acute and chronic), flu, expectorant (full cough), tuberculosis, fortifier, cough, sexual impotence ("weakness in man"), bladder, lung, depurative ("cleans impurities from the blood") and convulsion.
- ***Ethnopharmacological/Biological studies***: The essential oil of the species demonstrates antimicrobial (Sales et al., 2014) and anti-inflammatory (Viveiros et al., 2017) potential.

5.3.8 *Lippia alba* (Mill.) N.E.Br

- ***Taxonomic classification*** (APG IV/2016): Asterids – Lamids – Verbenaceae - Lamiales
- ***Botanical synonyms***: *Camara alba, Lantana geminata, Lippia asperifólia, Lippia carterae, Lippia citrata, Lippia crenata, Lippia geminata, Lippia globiflora, Lippia havanensis, Lippia lantanifolia, Lippia lantanoides, Lippia lorentzii, Lippia obovata, Lippia panamensis, Lippia rondonensis, Lippia única, Verbena globiflora, Verbena lantanoides, Zappania geminata, Zappania lantanoides, Zappania odorata, Lantana alba, Zappania globiflora*
- ***Link to the taxon***: http://servicos.jbrj.gov.br/flora/search/Lippia_alba
- ***Origin***: Native, non-endemic
- ***Phytochemistry***: Several investigations point to distinct majority component profiles in *L. alba* essential oils, and it is possible to group this species into chemical classes or chemotypes. For example, linalool-chemotype, linalool + eucalyptol-chemotype, geranial + carvenone-chemotype, piperitone-chemotype, tagetenone-chemotype, citral-chemotype. However, in the Caatinga, three main chemotypes stand out: myrcene + citral (chemotype I), limonene + citral (chemotype II) and limonene + carvone (chemotype III) (Gomes et al., 2019; Blanco et al., 2013; Lorenzo et al., 2001; Montero-Villegas et al., 2018). Among the metabolites described for *L. alba* are essential oils, sulfated flavonoids, tannins, geniposides (iridoids), triterpenic saponins, resins and mucilages. The main constituents of *L. alba* volatiles are the monoterpenoids (borneol, camphor, 1,8-cineol, citronelol, geranial, linalol, myrcene, neral) and the sesquiterpenoids, such as b-caryophyllene and cadinene (Pascual et al., 2001). The composition of its essential oil presents quantitative and qualitative variation, leading to the separation into chemotypes (Matos et al., 1996; Frigheto et al., 1998; Zoghbi et al., 1998), which could present distinct pharmacological activities, as well as morphological differences (Matos et al., 1996; Corrêa, 1992; Vale et al., 1999). Based on chemical, organoleptic and morphological studies, Matos (1996) classified the plants known as "citron" in the Northeastern Region of Brazil into three fundamental chemotypes. The chemotype I, with high citral and myrcene values in the essential oil, whose leaves are rough and large; chemotype II, with high citral and limonene contents, in which delicate leaves and branches are observed; chemotype III, whose essential oil is rich in carvone and limonene, being morphologically similar to chemotype II, but with a sweet citrus odor.
- ***Popular names***: Erva-cidreira-de-Shrub, do-campo or brasileira, alecrim do campo or selvagem, cidreira-brava, falsa-melissa, cidró, cidrão, cidreira, carmelita (Biasi & Costa, 2003).
- ***Part used/method of preparation/form of use*** (I or E), according to Magalhães et al. (2019): Leaf – tea(I), decoction (I); twig – tea (I)

- ***Life form and substratum***: Shrub/Terrestrial
- ***Ethnobotanical data***: Treatments of anxiety, diarrhea, sleep disturbances, abdominal cramps, bloody stools, digestive diseases, gastroenteritis, liver disease and heart disease.
- ***Ethnopharmacological/biological data***: Vale et al. (1999) observed in animal tests anxiolytic, sedative, hypothermic and myorelaxant effects for chemotype II in behavioral evaluation tests. The sedative, myorelaxant and anxiolytic effects are similar to those obtained by the action of benzodiazepines on the GABA receptor, which suggests that both the components of the oils and the non-volatile fractions act on this receptor (Vale et al., 1999; Zétola et al., 2002). The *in vitro* antioxidant activity using linoleic acid oxidation to carbonyl compounds by the essential oil of *L. alba* exhibited similar effect as vitamin E and 2-(tert-butyl)-4- methoxyphenol (BHA) at concentrations of 5–20.0 g/L (Stashenko; Jaramillo & Martinez, 2004). Similarly, hydroalcoholic percolate showed considerable free radical scavenging potential according to the tested parameters: reduction of 1,1- diphenyl-2-picrylhydrazyl (DPPH) (IC50 < 30 μg/mL) and inhibition of lipid peroxidation *in vitro* (IC50 < 32 μg/mL). A methanolic extract of aerial parts showed IC50 values of 58.1μg/ml in *Entamoeba histolytica* (moderate) and 109.4 μg/ml in *Giardia lamblia* (weak). The positive controls were emetine (1.05 and 0.42 μg/ml) and metronidazole (0.04 and 0.21μg/ml) (Calzada, Yépez-Mulia & Aguilar, 2006).

5.3.9 *Lippia origanoides* Kunth

- ***Taxonomic classification*** (APG IV/2016): Asterids – Lamids – Verbenaceae – Lamiales
- ***Botanical synonyms***: *Lippia affinis, Lippia berteri, Lippia candicans, Lippia elegans, Lippia glandulosa, Lippia mattogrossensis, Lippia microphylla, Lippia obscura, Lippia pendula, Lippia polycephala, Lippia rígida, Lippia rubiginosa, Lippia salviaefolia, Lippia schomburgkian, Lippia sidoides, Lippia velutina*
- ***Link to the taxon***: http://servicos.jbrj.gov.br/flora/search/Lippia_origanoides
- ***Origin***: Native, Non-endemic
- ***Phytochemistry***: Several essential oil chemotypes are reported for the species which include, among others, carvacrol-chemotype, carvacrol + thymol-chemotype, cinnamate-chemotype, α- and β-phellandrenes + cymene + limonene-chemotype and ρ-cymene + 1,8-cineole-chemotype (Sarrazin et al., 2015; Ribeiro et al., 2014; Stashenko et al., 2010; Vicunã, Stashenko & Fuentes, 2010). However, several studies point out that the main chemotype found in Caatinga is the thymol-chemotype (Oliveira et al., 2018; Santos et al., 2015; Monteiro et al., 2007).
- ***Popular names***: alecrim-pimenta, alecrim-bravo, estrepa-cavalo e alecrim-grande (Soares et al., 2017; Magalhães et al., 2019)
- ***Part used/method of preparation/form of use*** (I or E), according to Magalhães et al. (2019): Leaf – tea (I, E)
- ***Life form and substratum***: Shrub/Terrestrial
- ***Ethnobotanical data***: Antiseptic for mucous membranes and skin (wound), mycosis of the skin, rhinitis (sore throat), throat problems, "athlete's foot" (chulé). Popularly, the leaves are used for sore throats, gum inflammations, skin lesions and for preparing cosmetics with the extract obtained from macerating the leaves with alcohol or from the essential oil (Veras et al., 2017).

- ***Ethnopharmacological data***: *Lippia sidoides* Cham (Verbenaceae) is an aromatic plant of interest to the pharmaceutical, food and cosmetic industries for its proven insecticidal, antibacterial, larvicidal, acaricidal and anti-inflammatory properties, among others (Baldim et al., 2019). Thymol and carvacrol are prominent compounds in the essential oil extracted from the leaves of this vegetable popularly called "rosemary pepper". With positive effects on skin wound treatment, this oil can be included in products as oral antiseptic and others with dermatological purposes (Betancourt et al., 2019).

5.3.10 *Mesosphaerum suaveolens* (L.) Kuntze.

- ***Taxonomic classification*** (APG IV/2016): Asterids – Lamids – Lamiaceae – Lamiales
- ***Botanical synonyms***: *Ballota suaveolens, Hyptis suaveolens, Bystropogon suaveolens, Bystropogon graveolens, Gnoteris cordata, Gnoteris villosa, Hyptis congesta, Hyptis ebracteata, Hyptis plumieri, Marrubium indicum, Schaueria graveolens*
- ***Link to the taxon***: http://servicos.jbrj.gov.br/flora/search/Mesosphaerum_suaveolens
- ***Origin:*** Native, Non-endemic
- ***Phytochemistry***: Analyses of the plant sample in the state of Maranhão in the dry, intermediate and rainy seasons, found that the essential oil from the aerial parts of the species had 1,8-cineole as the main phytoconstituent in all periods. However, the content of 1,8-cineole was higher in the intermediate period (64.44%) and dry season (46.31%) than in the rainy season (30.15%) (Luz, 2020). Occurrence of bicyclogermacrene (23.5%), germacrene D (17.2%) and (E)-caryophyllene (10.4%) were also recorded from another access collected in the same state (Lima, 2022). The phytochemical analysis of the essential oil from the leaves of the species collected in the state of Ceará detected β-caryophyllene (18.57%), sabinene (15.94%) and spathulenol (11.09%) as major phytoconstituents (Bezerra, 2020). The examination of the essential oil from the aerial parts of the species collected in the state of Piauí, showed the presence of 1,8-cineole (35.77%), sabinene (19.61%) and α-pinene (7.59%). *Mesosphaerum suaveolens* contains unique terpenoid metabolites, such as suaveolic acid, suaveolol, methyl suaveolate, beta-sitosterol and ursolic acid, and phenolic compounds, such as rosamarinic acid, methyl rosamarinate, that have potentiality to substitute the traditional drugs as therapeutic agent against the resistant and newly emerged bacterial and viral pathogens. Pentacyclic triterpenoid and ursolic acid have been reported to have effective antiviral response against the SARS-CoV2 responsible for the present COVID-19 pandemic and HIV virus (Mishra, Sohrab & Mishra, 2021).
- ***Popular names***: Bamburral
- ***Part used/method of preparation/form of use*** (I or E), according to Magalhães et al. (2019): Leaf – tea(I, E); whole plant – tea (I); root – tea (I)
- ***Life form and substratum***: Shrub, herb, sub-shrub/terrestrial
- ***Ethnobotanical data***: Amoebiasis, hoarseness, liver ailments, indigestion and throat problems.
- ***Ethnopharmacological/Biological data***: Antimicrobial, antioxidant and anthelmintic potential have been reported for extracts obtained from the species (Bezerra-Almeida et al., 2022; Bezerra et al., 2020c; Lima, Fernandes & Silva, 2022; Nantitanon, Chowwanapoonpohn & Okonogi, 2007).

5.4 CONCLUDING REMARKS

The pharmacotherapeutic potentials of aromatic medicinal plants native to the Caatinga biomes are highlighted in this chapter. However, the majority of these medicinal species are lacking experimental studies to verify their ethnomedicinal uses or even explore their potential pharmacological, cosmetological and biological potentials. Several important bioactive phytoconstituents are yet to be evaluated for pharmacological, biological and toxicological studies. This write up would serve as an eye opener for researchers that might be interested in further exploring these important aromatic medicinal plants and or their products.

REFERENCES

ABIFISA. (2022). Associação Brasileira das Empresas do Setor Fitoterápico, Suplemento Alimentar e de Promoção da Saúde. [WWW Document]. URL https://abifisa.org.br/ (accessed 4.10.22).

Agra, M.D.F., Baracho, G.S., Nurit, K., Basílio, I.J.L.D., and Coelho, V.P.M. (2007). Medicinal and poisonous diversity of the flora of "Cariri Paraibano", Brazil. *Journal of Ethnopharmacology*, 111(2), 383–395.

Agra, M.D.F., Silva, K.N., Basílio, I.J.L.D., Freitas, P.F.D., and Barbosa-Filho, J.M. (2008). Survey of medicinal plants used in the region Northeast of Brazil. *Revista Brasileira de Farmacognosia*, 18(3), 472-508.

Aguiar, E.H.A., Zoghbi, M.G.B., Silva, M.H.L., Maia, J.G.S., Amasifén, J.M.R., and Rojas, U.M. (2003). Chemical variation in the essential oils of Hyptis mutabilis (Rich.) Briq. *Journal of Essential Oil Research*, 15(2), 130–132. doi:10.1080/10412905.2003.9712089.

Aguiar, J.C.D., Santiago, G.M.P., Lavor, P.L., Veras, H.N.H., Ferreira, Y.S., Lima, M.A.A., Arriaga, Â.M.C., Lemos, T.L.G., Lima, J.Q., de Jesus, H.C.R., Alves, P.B., and Braz-Filho, R. (2010). Chemical constituents and larvicidal activity of hymenaea courbaril fruit peel. *Natural Product Communications*. doi:10.1177/1934578X1000501231.

Albuquerque, D.S., Souza, S.D.G., Souza, A.C.N., and Sousa, M.L.M. (2020). Cenário da desertificação no território brasileiro e ações de combate à problemática no estado do Ceará, Nordeste do Brasil. *Desenvolvimento e Meio Ambiente*, 55, 673–696. doi:10.5380/dma.v55i0.73214.

Albuquerque, S.G., and Bandeira, J.J.R.L. (1995). Effect of thinning and slashing on forage phytomass from a caatinga of Petrolina, Pernambuco, Brazil. *Pesquisa Agropecuária Brasileira*, 30(6), 885–891.

Albuquerque, U.P.D., and Andrade, L.D.H.C. (2002a). Uso de recursos vegetales de la Caatinga: el caso del campesino del estado de Pernambuco (Nordeste de Brasil). *Interciencia*, 27(7), 336–346.

Albuquerque, U.P.D., and Andrade, L.D.H.C. (2002b). Conhecimento botânico tradicional e conservação em uma área de caatinga no estado de Pernambuco, Nordeste do Brasil. *Acta Botanica Brasilica*, 16, 273–285. doi: 10.1590/S0102-33062002000300004.

Albuquerque, U.P.D., and Hanazaki, N. (2006). As pesquisas etnodirigidas na descoberta de novos fármacos de interesse médico e farmacêutico: fragilidades e pespectivas. *Revista Brasileira de Farmacognosia*, *16*, 678–689.

Albuquerque, U.P.D., and Oliveira, R.F. (2007). Is the use-impact on native caatinga species in Brazil reduced by the high species richness of medicinal plants? *Journal of Ethnopharmacology*, 113(1), 156–170.

Alcântara, I.S., Martins, A.O.B.P.B., de Oliveira, M.R.C., Coronel, C., Gomez, M.C.V., Rolón, M., Wanderley, A.G., Júnior, L.J.Q., de Souza Araújo, A.A., de Araújo, A.C.J., Freitas, P.R., Coutinho, H.D.M., and de Menezes, I.R.A. (2022). Cytotoxic potential and antiparasitic activity of the Croton rhamnifolioides Pax leaves; K. Hoffm essential oil and its inclusion complex (EOCr/β-CD). *Polymer Bulletin*, 79(2), 1175–1185. doi:10.1007/s00289-021-03556-6.

Alcântara-Júnior, J.P., Ayala-Osuna, J.T., Queiroz, S.R.O.D., and Rios, A.P. (2005). Levantamento etnobotânico e etnofarmacológico de plantas medicinais do município de Itaberaba-BA para cultivo e preservação. *Sitientibus Serie Ciencias Biologicas*, 5(1), 39-44.

Alencar-Filho, J.M.T., Araújo, L.C., Oliveira, A.P. et al. (2017). Chemical composition and antibacterial activity of essential oil from leaves of Croton heliotropiifolius in different seasons of the year. *Revista Brasileira de Farmacognosia*, 27(4), 440–444. doi:10.1016/j.bjp.2017.02.004.

Almeida, C.F.C.B.R., Silva, T.C.L., Amorim, E.L.C., Maia, M.B.S., and Albuquerque, U.P. (2005). Lifestrategy and chemical composition as predictors of the selection of medicinal plants from the caatinga (Northeast Brazil). *Journal of Arid Environments*, 62(1), 127–142. doi: 10.1016/j.jaridenv.2004.09.020.

Almeida-Bezerra, J.W., Rodrigues, F.C., Lima Bezerra, J.J., Vieira Pinheiro, A.A., Almeida de Menezes, S., Tavares, A.B., Costa, A.R., Augusta de Sousa Fernandes, P., Bezerra da Silva, V., Martins da Costa, J.G., Pereira da Cruz, R., Bezerra Morais-Braga, M.F., Melo Coutinho, H.D., Teixeira de Albergaria, E., Meiado, M.V., Siyadatpanah, A., Kim, B., and Morais de Oliveira, A.F. (2022). Traditional uses, phytochemistry, and bioactivities of Mesosphaerum suaveolens (L.) Kuntze. *Evidence-Based Complementary and Alternative Medicine*, 2022, 1–28. doi:10.1155/2022/3829180.

Amaral, W. (2010). *Prospecção da flora aromática de um segmento de campos gerais da floresta atlântica no Estado do Paraná*. MSc thesis, Universidade Federal do Paraná, Curitiba.

Amorozo, M.C.M, and Gély, A. (1988). Uso de plantas medicinais por caboclos do baixo Amazonas Barcarena, PA, Brasil. *Bol do Mus Para Emílio Goeldi*, 4(1), 47–131.

Andrade-Lima, D.A., and Zappi, D.C. (1989). *Plantas das caatingas*. Academia Brasileira de Ciências.

Andrade-Lima, D.A. (1981). The caatinga dominium. *Revista Brasileira de Botânica*, 4, 149–163.

Anon. (1995). International Conference and Programme for Plant Genetic Resources (ICPPGR). [WWW Document] URL https://www.embrapa.br/recursos-geneticos-e-biotecnologia. (accessed 4.11.22).

APG. (2016). An update of the Angiosperm Phylogeny Group classification for the orders and families of flowering plants: APG IV. *Botanical Journal of the Linnean Society*, 181(1), 1–20. doi:10.1111/boj.12385.

Araújo, C.D.S.F., and Sousa, A.N.D. (2011). Estudo do processo de desertificação na caatinga: uma proposta de educação ambiental. *Ciência & Educação*, 17(4), 975–986.

Araújo, F.M., Dantas, M.C.S.M., e Silva, L.S., Aona, L.Y.S., Tavares, I.F., and de Souza-Neta, L.C. (2017). Antibacterial activity and chemical composition of the essential oil of Croton heliotropiifolius Kunth from Amargosa, Bahia, Brazil. *Industrial Crops and Products*, 105, 203–206. doi:10.1016/j.indcrop.2017.05.016.

Badke, M.R., Budó, M.D.L.D., Alvim, N.A.T., Zanetti, G.D., and Heisler, E.V. (2012). Saberes e práticas populares de cuidado em saúde com o uso de plantas medicinais. *Texto & Contexto-Enfermagem*, 21, 363–370.

Baldim, I., Tonani, L., von Zeska Kress, M.R., and Pereira Oliveira, W. (2019). Lippia sidoides essential oil encapsulated in lipid nanosystem as an anti-Candida agent. *Industrial Crops and Products, 127*, 73–81. doi: 10.1016/j.indcrop.2018.10.064.

Barbosa, P.P.P., and Ramos, C.P. (1992). Studies on the antiulcerogenic activity of the essential oil of Hyptis mutabilis Briq. in Rats. *Phytotherapy Research*, 6(2), 114–115. doi:10.1002/ptr.2650060214.

Barros, M.L.B. (2004). Prefacio. In: Leal, I.R., Tabarelli, M., Silva, J.M.C. (eds.). *Ecologia e Conservação da Caatinga*. Editora Universitária da UFPE.

Bayala B., Bassole, I.H.N., Gnoula C. et al. (2014). Chemical composition, antioxidant, anti-inflammatory and anti-proliferative activities of essential oils of plants from Burkina Faso. *PLoS ONE*, 9(3), e92122. doi:10.1371/journal.pone.0092122.

Betancourt, L., Hume, M., Rodríguez, F., Nisbet, D., Sohail, M.U., and Afanador-Tellez, G. (2019). Effects of Colombian oregano essential oil (Lippia origanoides Kunth) and Eimeria species on broiler production and cecal microbiota. *Poultry Science*, 98(10), 4777–4786. doi:10.3382/ps/pez193.

Bezerra, F.W., Bezerra, P.N., Oliveira, M.S., Costa, W.A., Ferreira, G.C., and Carvalho, R.N. (2020a). Bioactive compounds and biological activity of croton species (Euphorbiaceae): An overview. *Current Bioactive Compounds, 16*(4), 383–393. doi: 10.2174/1573407215666181122103511.

Bezerra, J.W.A., Grangeiro-Neto, F.A., Pereira-Filho, J.T. et al. (2020b). Chemical composition and insecticide activity of essential oil of Mesosphaerum suaveolens against Nauphoeta cinerea. *Journal of Agricultural Studies*, 8(2), 352–361. doi:10.5296/jas.v8i2.15737.

Bezerra, J.W.A., Rodrigues, F.C., Gonçalo, M.A.B.F., dos Santos, M.A.F., Macedo, G.F., de Souza Bezerra, J., and Coutinho, H.D.M. (2020c). Chemical composition and antibacterial activity of the essential oil of Mesosphaerum suaveolens (Lamiaceae). In *Essential Oils-Bioactive Compounds, New Perspectives and Applications*. IntechOpen.

Biasi, L.A. and Deschamps, C. (2009). *Plantas aromáticas do cultivo à produção de óleo essencial.* Curitiba: Layer Studio Gráfico e Editora Ltda.

Biasi, L.A., and Costa, G. (2003). Propagação vegetativa de Lippia alba. *Ciência Rural*, 33(3), 455–459. doi:10.1590/S0103-84782003000300010.

Bizzo, H.R., Hovell, A.M.C., and Rezende, C.M. (2009). Óleos essenciais no Brasil: aspectos gerais, desenvolvimento e perspectivas. *Química Nova*, 32, 588–594. doi:10.1590/S0100-40422009000300005.

Blanco, M.A., Colareda, G.A., van Baren, C., Bandoni, A.L., Ringuelet, J., and Consolini, A.E. (2013). Antispasmodic effects and composition of the essential oils from two South American chemotypes of Lippia alba. *Journal of ethnopharmacology*, 149(3), 803–809. doi:10.1016/j.jep.2013.08.007.

Brasil. (1998). *Primeiro relatório nacional para a conservação sobre diversidade biológica.* Brasília.

Brasil. (2002a). *Biodiversidade Brasileira: avaliação e identificação de áreas e ações prioritárias para conservação, utilização sustentável e repartição dos benefícios da biodiversidade nos biomas brasileiros.* Ministério do Meio Ambiente.

Brasil. (2002b). Caatinga. [WWW Document]. URL https://antigo.mma.gov.br/biomas/caatinga.html (accessed 4.11.22).

Brasil. (2008a). Instrucao Normativa no 6 de 23 de setembro de 2008. *Diário Oficial da União*, 185(Secao 1), 75–83.

Brasil. (2008b). Manejo Florestal da Caatinga: uma alternativa de desenvolvimento sustentável em projetos de assentamentos rurais do semi-árido em Pernambuco. *Estatística Florestal da Caatinga.* Natal, 6–17.

Brasil. (2008c). *Unidades de conservação e terras indígenas do bioma caatinga.* The Nature Conservancy e Ministério do Meio Ambiente.

Brasil. (2011). Plano de Divulgação do Bioma Caatinga. Ministério do Meio Ambiente Secretaria de Biodiversidade *e* Florestas Departamento de Conservação da Biodiversidade Núcleo do Bioma, Brasília.

Brasil. (2017). Portaria n° 350, de 08 de setembro de 2017. *Aprova os instrumentos de Termos de Compromisso a serem firmados entre o usuário e a União, para fins de regularização do acesso ao patrimônio genético e ao conhecimento tradicional associado, nos termos da Lei nº 13.123, de 20 de maio de 2015.* Diário Oficial da União.

Brasil. (2018). *Espécies nativas da flora brasileira de valor econômico atual or potencial: plantas para o futuro: região Nordeste.* Ministério do Meio Ambiente.

Brito, S.S.S., Silva, F., Malheiro, R., Baptista, P., and Pereira, J.A. (2018). Croton argyrophyllus Kunth and Croton heliotropiifolius Kunth: Phytochemical characterization and bioactive properties. *Industrial Crops and Products*, 113, 308–315. doi:10.1016/J.INDCROP.2018.01.044.

Calzada, F., Yépez-Mulia, L., and Aguilar, A. (2006). In vitro susceptibility of Entamoeba histolytica and Giardia lamblia to plants used in Mexican traditional medicine for the treatment of gastrointestinal disorders. *Journal of Ethnopharmacology*, 108(3), 367–370.

Camara, C.A.G., Moraes, M.M., Melo, J.P.R., and da Silva, M.M.C. (2017). Chemical composition and acaricidal activity of essential oils from *Croton rhamnifolioides* Pax and Hoffm. in different regions of a Caatinga Biome in Northeastern Brazil. *Journal of Essential Oil Bearing Plants*, *20*(6), 1434–1449. doi: 10.1080/0972060X.2017.1416677.

Cartaxo, S.L., Souza, M.M.A., and Albuquerque, U.P. (2009). Medicinal plants with bioprospecting potential used in semi-arid northeastern Brazil. *Journal of Ethnopharmacology*, 131(2), 326–342. doi: 10.1016/j.jep.2010.07.003.

Castelletti, C.H.M., Silva, J.M.C.D., Tabarelli, M., and Santos, A.M.M. (2005). *Quanto ainda resta da Caatinga?* Uma estimativa preliminar. Ecologia e conservação da caatinga. Recife, pp 719–34.

Coelho-de-Souza, A.N., Rocha, M.V.A.P., Oliveira, K.A., Vasconcelos, Y.A.G., Santos, E.C., Silva-Alves, K.S., Diniz, L.R.L., Ferreira-da-Silva, F.W., Oliveira, A.C., Ponte, E.L., Evangelista, J.S.-A.M., Assreuy, A.M.S., and Leal-Cardoso, J.H. (2019). Volatile oil of Croton zehntneri per oral sub-acute treatment offers small toxicity: perspective of therapeutic use. *Revista Brasileira de Farmacognosia*, 29(2), 228–233. doi:10.1016/j.bjp.2018.11.005.

Corrêa, C.B.V. (1992). Contribuição ao estudo de Lippia alba (Mill.) NE Br. ex Britt & Wilson-erva-cidreira. *Revista Brasileira de Farmácia*, 73(3), 57–64.

Costa, C.C.D.A., Camacho, R.G.V., Macedo, I.D.D., and Silva, P.C.M.D. (2010). Análise comparativa da produção de serapilheira em fragmentos arbóreos e arbustivos em área de caatinga na FLONA de Açu-RN. *Revista Árvore*, 34, 259–265.

Costa, M., Silva, A., Silva, A., Lima, V., Bezerra-Silva, P., Rocha, S., Navarro, D., Correia, M., Napoleão, T., Silva, M., and Paiva, P. (2017). Essential oils from leaves of medicinal plants of Brazilian flora: Chemical composition and activity against Candida species. *Medicines*, 4(2), 27. doi:10.3390/medicines4020027.

Coutinho,L.M.(2006).Oconceitodebioma.*ActabotBras*,20(1),13–23.doi:10.1590/S0102-33062006000100002.

Craveiro, A.A., Fernandes A.G., Andrade C.H.S., Matos F.J.A., Alencar J.W., and Machado, M.I.L. (1981). *Óleos Essenciais de Plantas do Nordeste*. UFC.

Craveiro, A.A., Machado, M.I.L., Alencar, J.W., and Matos, F.J.A. (1994). Natural product chemistry in north-eastern Brazil. In: Prance, G.T., Chadwick, D.J, and March, J. Symposium On Ethnobotany And Search For New Drugs. Ciba Foundation Symposium 185. Fortaleza, Brasil, p. 95–102.

Dantas, J. (2020, November 10). Photo leafless tree on a brown grass field during the day. *Unsplash*. URL https://unsplash.com/pt-br/fotografias/v4thvMdfuSw (accessed 3.3.22).

Diegues, A.C. (2000). Etnoconservação da natureza: enfoques alternativos. *Etnoconservação: novos rumos para a proteção da natureza nos trópicos*, 2, 1–46.

Dourado, R.C., and Silveira, E.R. (2005). Preliminary investigation on the volatile constituents of Croton sonderianus Muell. Arg.: Habitat, plant part and harvest time variation. *Journal of Essential Oil Research*, 17(1), 36–40. doi:10.1080/10412905.2005.9698823.

Esper R.H., Gonçalez E., Felicio R.C., Felicio J.D. (2015). Fungicidal activity and constituents of Ageratum conyzoides essential oil from three regions in São Paulo state, Brazil. *Arquivos Do Instituto Biológico*, 82. doi:10.1590/1808-1657000482013.

Everton, G.O., Pereira, A.P.M., Rosa, P.V.S., Mafra, N.S.C., Santos Júnior, P.S., Souza, F.S., Mendonça, C.D.J.S., Silva, F.C., Gomes, P.R.B., and Mouchrek Filho, V.E. (2021). Chemical characterization, toxicity, antioxidant and antimicrobial activity of the essential oils of Hymenaea courbaril L. and Syzygium cumini (L.) Skeels. *Ciência e Natura*, *43*, e11. doi:10.5902/2179460X43819.

Flora do Brasil. (2022). Jardim Botânico do Rio de Janeiro. [WWW Document]. URL https://floradobrasil.jbrj.gov.br (accessed 4.11.22).

Forzza, R.C., Peixoto, A.L., Pirani, J.R., Queiroz, L.P., Stehmann, J.R., Walter, B.M.T., and Coelho, N. (2010). As angiospermas do Brasil (Vol. 2). In *Catálogo de Fungos e Plantas do Brasil*, 78–89. Instituto de Pesquisas Jardim Botânico do Rio de Janeiro.

Forzza, R.C., Baumgratz, J.F.A., Bicudo, C.E.M. et al. (2012). New Brazilian floristic list highlights conservation challenges. *BioScience*, 62(1), 39–45. doi:10.1525/bio.2012.62.1.8.

Frighetto, N., Oliveira, J.G., Siani, A.C., and Chagas, K.C. (1998). *Lippia alba* Mill N.E. Br. (Verbenaceae) as a Source of Linalool. *Journal of Essential Oil Research*, 10(5), 578–580. doi:10.1080/10412905.1998.9700976.

Gera, M., Blsht, N.S., and Rana, A.K. (2003). Market information system for sustainable management of medicinal plants. *Indian Forester*, 129(1), 102–108.

Giulietti, A.M., Bocage Neta, A.L., Castro, A.A.J.F., Gamarra-Rojas, C.F., Sampaio, E.V.S.B., Virgínio, J.F., and Harley, R.M. (2004). Diagnóstico da vegetação nativa do bioma Caatinga. *Biodiversidade da Caatinga: áreas e ações prioritárias para a conservação*, 48–90.

Giulietti, A., Rapini, A., Andrade, M., and Queiroz, L.E.S. (2009). *Plantas Raras do Brasil*. Conservação Internacional. Belo Horizonte.

Gomes, A.F., Almeida, M.P., Leite, M.F., Schwaiger, S., Stuppner, H., Halabalaki, M., and David, J.M. (2019). Seasonal variation in the chemical composition of two chemotypes of Lippia alba. *Food Chemistry*, 273, 186-193. doi:10.1016/j.foodchem.2017.11.089.

Gomes, E.C., Barbosa, J., Vilar, F.C., Perez, J., Vilar, R., and Dias, T. (2008). Plantas da caatinga de uso terapeutico:levantamento etnobotanico. *Engenharia Ambiental: pesquisa e tecnologia*, 5(2), 74–85.

Gomes, T.B., and Bandeira, F.P.S.D.F. (2012). Uso e diversidade de plantas medicinais em uma comunidade quilombola no Raso da Catarina, Bahia. *Acta Botanica Brasilica*, 26(4), 796–809.

Hauff, S.N. (2010). *Representatividade do Sistema Nacional de Unidades de Conservação na Caatinga*. Programa das Nações Unidas para o Desenvolvimento (PNUD).

Holanda, A.C.D., Lima, F.T.D., Silva, B.M., Dourado, R.G., and Alves, A.R. (2015). Estrutura da vegetação em remanescentes de caatinga com diferentes históricos de perturbação em Cajazeirinhas (PB). *Revista Caatinga*, 28, 142–150.

Jha, A.K. (1995). Medicinal plants: Poor regulation blocks conservation. *Economic and Political Weekly*, 3270–3270.

Junior, J.I.G., Ferreira, M.R.A., de Oliveira, A.M., and Soares, L.A.L. (2022). Croton sp.: a review about popular uses, biological activities and chemical composition. *Research, Society and Development*, 11(2), e57311225306–e57311225306.doi:10.33448/rsd-v11i2.25306.

Kiill, L.H.P. (2010). Caatinga: patrimônio brasileiro ameaçado. Embrapa Semiárido.

Kouame, B.K.F.P., Toure, D., Kablan, L., Bedi, G., Tea, I., Robins, R., Chalchat, J.C., and Tonzibo, F. (2018). Chemical constituents and antibacterial activity of essential oils from flowers and stems of Ageratum conyzoides from Ivory Coast. *Records of Natural Products, 12*(2), 160–168. doi:10.25135/rnp.22.17.06.040.

Leal, I.R., Tabarelli, M., and Silva, M.J.C. (2003). *Ecologia e conservação da Caatinga*. UFPE.

Lima, G.P.G., Souza, T.M., Freire, G.P., Farias, D.F., Cunha, A.P., Ricardo, N.M.P.S., Morais, S.M., and Carvalho, A.F.U. (2013). Further insecticidal activities of essential oils from Lippia sidoides and Croton species against Aedes aegypti L. *Parasitology Research*, 112, 1953–1958. doi:10.1007/s00436-013-3351-1.

Lima, A.S., Fernandes, Y.M.L., Silva, C.R. (2022). Anthelmintic evaluation and essential oils composition of Hyptis dilatata Benth. and Mesosphaerum suaveolens Kuntze from the Brazilian Amazon. *Acta Tropica*, 228, 106321. doi: 10.1016/j.actatropica.2022.106321.

Lima, C.C., de Holanda-Angelin-Alves, C.M., Pereira-Gonçalves, Á., Kennedy-Feitosa, E., Evangelista-Costa, E., Bezerra, M.A.C., Coelho-de-Souza, A.N., and Leal-Cardoso, J.H. (2020). Antispasmodic effects of the essential oil of Croton zehnteneri, anethole, and estragole, on tracheal smooth muscle. *Heliyon, 6*(11), e05445. doi:10.1016/j.heliyon. 2020.e05445.

Lima, M.A.S., Barros, M.C.P., Pinheiro, S.M., Nascimento, R.F., Matos, F.J.A., and Silveira, E.R. (2005). Volatile compositions of two Asteraceae from the northeast of Brazil: Ageratum conyzoides and Acritopappus confertus (Eupatorieae). *Flavour and Fragrance Journal*, 20(6), 559–561. doi:10.1002/ffj.1483.

Liu, X.C., and Liu, Z.L. (2014). Evaluation of larvicidal activity of the essential oil of Ageratum conyzoides L. aerial parts and its major constituents against Aedes albopictus. *Journal of Entomology and Zoology Studies*, 2(4), 345–350.

Lorenzo, D., Paz, D., Davies, P., Vila, R., Cañigueral, S., and Dellacassa, E. (2001). Composition of a new essential oil type of Lippia alba (Mill.) NE Brown from Uruguay. *Flavour and Fragrance Journal*, 16(5), 356–359. doi:10.1002/ffj.1011.

Luz, T.R.S.A., Leite, J.A.C., and Mesquita, L.S.S. (2020). Seasonal variation in the chemical composition and biological activity of the essential oil of Mesosphaerum suaveolens (L.) Kuntze. *Industrial Crops and Products*, 153, 112600. doi:10.1016/j.indcrop.2020.112600.

Magalhães, K.N., Guarniz, W.A.S., Sá, K.M., Freire, A.B., Monteiro, M.P., Nojosa, R.T., and Bandeira, M.A.M. (2019). Medicinal plants of the Caatinga, northeastern Brazil: Ethnopharmacopeia (1980–1990) of the late professor Francisco José de Abreu Matos. *Journal of Ethnopharmacology*, 237, 314–353.

Martins, A.O.B.P.B., Rodrigues, L.B., Cesário, F.R.A.S., de Oliveira, M.R.C., Tintino, C.D.M., Castro, F.F. e, Alcântara, I.S., Fernandes, M.N.M., de Albuquerque, T.R., da Silva, M.S.A., de Sousa Araújo, A.A., Júniur, L.J.Q., da Costa, J.G.M., de Menezes, I.R.A., and Wanderley, A.G. (2017). Anti-edematogenic and anti-inflammatory activity of the essential oil from Croton rhamnifolioides leaves and its major constituent 1,8-cineole (eucalyptol). *Biomedicine & Pharmacotherapy*, 96, 384–395. doi:10.1016/j.biopha.2017.10.005.

Martins, E.R., Castro, D.M., Castellani, D.C. and Dias, J.E. (2000). *Plantas Medicinais*. Editora UFV: Viçosa, 220 p.

Matos, F.J.A. (1999). *Plantas da medicina popular do Nordeste: propriedades atribuídas e confirmadas*. UFC.

Matos, F.J.A., Machado, M.I.L., Craveiro, A.A., and Alencar, J.W. (1996). Essential oil composition of two chemotypes of *Lippia alba* Grown in Northeast Brazil. *Journal of Essential Oil Research, 8*(6), 695–698. doi:10.1080/10412905.1996.9701047.

Mattoso, E. (2005). *Estudo de fragrâncias amadeiradas da Amazônia*. MSc thesis, Universidade Estadual de Campinas, Campinas.

Medeiros, V.M., Nascimento, Y.M., Souto, A.L. et al. (2017). Chemical composition and modulation of bacterial drug resistance of the essential oil from leaves of Croton grewioides. *Microbial Pathogenesis*, 111, 468–471.doi: 10.1016/j.micpath.2017.09.034.

Menut, C., Lamaty, G., Zollo, P.H.A., Kuiate, J.R., and Bessière, J.M. (1993). Aromatic plants of tropical central Africa. Part X Chemical composition of the essential oils of Ageratum houstonianum Mill. and Ageratum conyzoides L. from Cameroon. *Flavour and Fragrance Journal*, 8(1), 1–4. doi:10.1002/ffj.2730080102.

Mishra, P., Sohrab, S., and Mishra, S.K. (2021). A review on the phytochemical and pharmacological properties of Hyptis suaveolens (L.) Poit. *Future Journal of Pharmaceutical Sciences*, 7(1), 65. doi:10.1186/s43094-021-00219-1.

Momesso, L.S., Moura, R.M.X., and Constantino, D.H.J. (2009). Atividade antitumoral do Ageratum conyzoides L. (Asteraceae). *Revista Brasileira de Farmacognosia*, 19(3), 660–663. doi:10.1590/S0102-695X2009000500002.

Monteiro, M.V.B. de Melo Leite, A.K.R Bertini, L.M. de Morais, S.M., and Nunes-Pinheiro, D.C.S. (2007). Topical anti-inflammatory, gastroprotective and antioxidant effects of the essential oil of Lippia sidoides Cham. leaves. *Journal of Ethnopharmacology*, 111(2), 378–382. doi:10.1016/j.jep.2006.11.036.

Montero-Villegas, S., Crespo, R., Rodenak-Kladniew, B., Castro, M.A., Galle, M., Cicció, J.F., and Polo, M. (2018). Cytotoxic effects of essential oils from four Lippia alba chemotypes in human liver and lung cancer cell lines. *Journal of Essential Oil Research*, 30(3), 167–181. doi:10.1080/10412905.2018.1431966.

Morais, S.M.D., Dantas, J., Silva, A.R.A.D., and Magalhães, E.F. (2005). Plantas medicinais usadas pelos índios Tapebas do Ceará. *Revista Brasileira de Farmacognosia*, 15, 169–177.

Nantitanon, W., Chowwanapoonpohn S., and Okonogi S. (2007). Antioxidant and antimicrobial activities of hyptis suaveolens essential oil. *Scientia Pharmaceutica*, 75(1), 35–46. doi:10.3797/scipharm.2007.75.35.

Nascimento, D.T.F. and Ribeiro, S.A. (2017). *Os biomas brasileiros e a defesa da vida*. Kelps.

Nogueira, J.H.C., Gonçalez, E., Galleti, S.R., Facanali, R., Marques, M.O.M., and Felício, J.D. (2010). Ageratum conyzoides essential oil as aflatoxin suppressor of Aspergillus flavus. *International Journal of Food Microbiology*, 137(1), 55–60. doi: 10.1016/j.ijfoodmicro.2009.10.017.

Novello, C.R., Düsman, E., Balbinot, R.B., de Paula, J.C., Nakamura, C.V., de Mello, J.C.P., Sarragiotto, M.H. (2020). Antileishmanial activity of neo-clerodane diterpenes from Croton echioides. *Natural Product Research*, 36(4), 925–931. doi:10.1080/14786419.2020.1851221.

Nunes, A.T., Lucena, R.F.P., Santos, M.V.F., and Albuquerque, U.P. (2015). Local knowledge about fodder plants in the semi-arid region of Northeastern Brazil. *Journal of Ethnobiology and Ethnomedicine*, 11(1), 1–12. doi:10.1186/1746-4269-11-12.

OEC. (2022). Essential oils (HS: 3301) Product trade, exporters and importers | OEC - The Observatory of Economic Complexity [WWW Document]. URL https://oec.world/en/profile/hs92/essential-oils (accessed 4.11.22).

Oliva, M.M., Demo, M.S., Lopez, A.G., Lopez, M.L., and Zygadlo, J.A. (2006). Antimicrobial activity and composition of hyptis mutabilis essential oil. *Journal of Herbs, Spices & Medicinal Plants*, 11(4), 57–63. doi:10.1300/J044v11n04_07.

Oliveira, A.P., Santos, A.A., Santana, A.S., Lima, A.P.S., Melo, C.R., Santana, E.D., and Bacci, L. (2018). Essential oil of Lippia sidoides and its major compound thymol: Toxicity and walking res ponse of populations of Sitophilus zeamais (Coleoptera: Curculionidae). *Crop Protection*, 112, 33–38.

Oliveira, D.D., Silva, C.V., Guedes, M.L.S., and Velozo, E.S. (2016). Fixed and volatile constituents of *Croton heliotropiifolius* Kunth from Bahia-Brazil. *African Journal of Pharmacy and Pharmacology*, 10(26), 540–545. doi:10.5897/AJPP2015.4509.

Oliveira Júnior, R., Ferraz, C., Silva, J., de Oliveira, A., Diniz, T., e Silva, M., Quintans Júnior, L., de Souza, A., dos Santos, U., Turatti, I., Lopes, N., Lorenzo, V., and Almeida, J. (2017). Antinociceptive effect of the essential oil from Croton conduplicatus Kunth (Euphorbiaceae). *Molecules*, 22(6), 900. doi:10.3390/molecules22060900.

Oliveira Júnior, R.G. de, Ferraz, C.A.A., Silva, J.C., de Andrade Teles, R.B., Silva, M.G., Diniz, T.C., dos Santos, U.S., de Souza, A.V.V., Nunes, C.E.P., Salvador, M.J., Lorenzo, V.P., Quintans Júnior, L.J., and Almeida, J.R.G.daS. (2018). Neuropharmacological effects of essential oil from the leaves of Croton conduplicatus Kunth and possible mechanisms of action involved. *Journal of Ethnopharmacology*, 221, 65–76. doi:10.1016/j.jep.2018.04.009.

Oliveira, S.D.D.S., Oliveira, A.M., Blank, A.F. et al. (2021). Radical scavenging activity of the essential oils from Croton grewioides Baill accessions and the major compounds eugenol, methyl eugenol and methyl chavicol. *Journal of Essential Oil Research*, 33(1), 94–103. doi:10.1080/10412905.2020.1779139.

Pareyn, F.G.C. (2010). A importância da produção não madeireira na Caatinga. In: Gariglio, M.A., Sampaio, E.V.S.B., Cestaro, L.A., and Kageyama, P.Y. *Uso sustentável e conservação dos recursos Florestais da Caatinga*. Serviço Florestal Brasileiro.

Pascual, M.E., Slowing, K., Carretero, M.E., and Villar, Á. (2001). Antiulcerogenic activity of Lippia alba (Mill.) N. E. Brown (Verbenaceae). *Il Farmaco*, 56(5–7), 501–504. doi:10.1016/S0014-827X(01)01086-2.

Patil, R.P., Nimbalkar, M.S., Jadhav, U.U., Dawkar, V.V., and Govindwar, S.P. (2010). Antiaflatoxigenic and antioxidant activity of an essential oil from *Ageratum conyzoides* L. *Journal of the Science of Food and Agriculture*, 90(4), 608–614. doi:10.1002/jsfa.3857.

Pereira-Júnior, L.R., Andrade, A.P.D., Araújo, K.D., Barbosa, A.D.S., and Barbosa, F.M. (2014). Espécies da caatinga como alternativa para o desenvolvimento de novos fitofármacos. *Floresta e Ambiente*, 21(4), 509–520.

Pinho, L.S, Maia, P.V.M, Garcia, T.M.N., Cruz, J.S., Morais, S.M., Souza, A.N.C., and Cardoso, J.H.L. (2010). Croton sonderianus essential oil samples distinctly affect rat airway smooth muscle. *Phytomedicine*, 17(10), 721–725. doi:10.1016/j.phymed.2010.01.015.

Posey, D.A. (1983). Folk Apiculture of the Kayapo Indians of Brazil. *Biotropica*, *15*(2), 154–158. doi:0.2307/2387963.

Queiroz, L.P., Conceição, A.A, and Giulietti, A.M. (2006). Nordeste semi-árido: Caracterização geral e lista das fanerógamas. In: Giulietti, A.M., Conceição, A.A. and Queiroz, L.P. (eds.). *Diversidade e caracterização das fanerógamas do semi-árido brasileiro*. MCT, pp.15–359.

Oliveira, A.P., Santos, A.A., Santana, A.S., Lima, A.P.S., Melo, C.R., Santana, E.D., ... and Bacci, L. (2018). Essential oil of Lippia sidoides and its major compound thymol: Toxicity and walking response of populations of Sitophilus zeamais (Coleoptera: Curculionidae). *Crop Protection*, *112*, 33–38. doi:10.1016/j.cropro.2018.05.011.

Ribeiro, A.F., Andrade, E.H.A., Salimena, F.R.G., and Maia, J.G.S. (2014). Circadian and seasonal study of the cinnamate chemotype from *Lippia origanoides* Kunth. *Biochemical Systematics and Ecology*, 55, 249–259.

Rocha, M.S., Islam, T., Silva, T.G., and Militão, G.C.G. (2015). Isolation, characterization and evaluation of antimicrobial and cytotoxic activity of estragole, obtained from the essential oil of croton zehntneri (Euphorbiaceae). *Anais da Academia Brasileira de Ciências*, 87(1), 173–182. doi:10.1590/0001-3765201520140111.

Sales, G.W.P., Batista, A.H.M., Rocha, L.Q, and Nogueira, N.A.P. (2014). Efeito antimicrobiano e modulador do óleo essencial extraído da casca de frutos da Hymenaea courbaril. *Rev. Ciênc. Farm. Básica Apl.*, 35, pp. 709–715.

Sampaio, E.V.S.B. (1995). Overview of the Brazilian caatinga. *Seasonally Dry Tropical Forests*, 1, 35–63.

Sampaio, E.V.S.B., and Rodal, M.J. (2002). Uso das plantas da caatinga. Sampaio, E.V.S.B., Giulietti, A.M., Virgínio, J. and Gamarra-Rojas, C.F.L. *Vegetação e flora da caatinga*. Recife.

Santos, C.P. de Oliveira, T.C., Pinto, J.A.O. Fontes, S.S., Cruz, E.M.O., de Fátima Arrigoni-Blank, M., and Blank, A.F. (2015). Chemical diversity and influence of plant age on the essential oil from Lippia sidoides Cham. germplasm. *Industrial Crops and Products*, *76*, 416–421. doi:10.1016/j.indcrop.2015.07.017.

Santos, F.A., Jeferson, F.A., Santos, C.C., Silveira, E.R., and Rao, V.S.N. (2005). Antinociceptive effect of leaf essential oil from Croton sonderianus in mice. *Life Sciences*, *77*(23), 2953–2963. doi:10.1016/j.lfs.2005.05.032.

Sarrazin S.L.F., Da Silva L.A., De Assunção A.P.F., Oliveira R.B., Calao V.Y.P., Da Silva R., Stashenko E.E., Maia J.G., and Mourão R.H.V. (2015). Antimicrobial and seasonal evaluation of the Carvacrol-chemotype oil from *Lippia origanoides* Kunth. *Molecules*, 20(2), 1860–1871. doi:10.3390/molecules20021860.

Serra, D.S., Gomes, M.D.M., Cavalcante, F.S.Á., and Leal-Cardoso, J.H. (2019). Essential oil of Croton Zehntneri attenuates lung injury in the OVA-induced asthma model. *Journal of Asthma*, *56*(1), 1–10. doi:10.1080/02770903.2018.1430828.

Silva, J.M.C. (2003). Caatinga. In: Maury, C.M. (eds.). *Biodiversidade brasileira: Avaliação e identificação de áreas e ações prioritárias para conservação, utilização sustentável e repartição dos benefícios da biodiversidade nos biomas brasileiros*. Ministério do Meio Ambiente.

Silva, C.G., Zago, H.B., Júnior, H.J.G.S. et al. (2008). Composition and insecticidal activity of the essential oil of Croton grewioides Baill. against Mexican Bean Weevil (Zabrotes subfasciatus Boheman). *Journal of Essential Oil Research*, 20(2), 179–182. doi:10.1080/10412905.2008.9699985.

Silva, L.L, Garlet, Q.I., Benovit, S.C. et al. (2013). Sedative and anesthetic activities of the essential oils of Hyptis mutabilis (Rich.) Briq. and their isolated components in silver catfish (Rhamdia quelen). *Brazilian Journal of Medical and Biological Research*, 46(9), 771–779. doi: 10.1590/1414-431X2 0133013.

Silva, J.D., Tabarelli, M., Fonseca, M.D., and Lins, L.V. (2004). *Biodiversidade da Caatinga: áreas e ações prioritárias para a conservação.* Ministério do Meio Ambiente.

Silva, J.M.C. (2003). Caatinga. In: Maury, C.M. *Biodiversidade brasileira: Avaliação e identificação de áreas e ações prioritárias para conservação, utilização sustentável e repartição dos benefícios da biodiversidade nos biomas brasileiros.*

Silva, V.A., Andrade, L.D.H.C., and Albuquerque, U.P. (2005). Revising the cultural significance index: the case of the Fulni-ô in northeastern Brazil. *Field Methods*, 18(1), 98–108.

Soares, B.V., Neves, L.R., Ferreira, D.O., Oliveira, M.S.B., Chaves, F.C.M., Chagas, E.C., Gonçalves, R.A., and Tavares-Dias, M. (2017). Antiparasitic activity, histopathology and physiology of Colossoma macropomum (tambaqui) exposed to the essential oil of Lippia sidoides (Verbenaceae). *Veterinary Parasitology*, 234, 49–56. doi:10.1016/j.vetpar.2016.12.012.

Souza, A.V.V., Britto, D., Santos, U.S., Bispo, L.P., Turatti, I.C.C., Lopes, N.P., Oliveira, A.P., and Almeida, J.R.G.S. (2017). Influence of season, drying temperature and extraction time on the yield and chemical composition of 'marmeleiro' (*Croton sonderianus*) essential oil. *Journal of Essential Oil Research*, 29(1), 76–84. doi:10.1080/10412905.2016.1178183

Stashenko, E.E., Jaramillo, B.E., and Martínez, J.R. (2004). Comparison of different extraction methods for the analysis of volatile secondary metabolites of Lippia alba (Mill.) N.E. Brown, grown in Colombia, and evaluation of its in vitro antioxidant activity. *Journal of Chromatography A*, 1025(1), 93–103. doi:10.1016/j.chroma.2003.10.058.

Stashenko, E.E., Martínez, J.R., Ruíz, C.A., Arias, G., Durán, C., Salgar, W., and Cala, M. (2010). Lippia origanoides chemotype differentiation based on essential oil GC-MS and principal component analysis. *Journal of Separation Science*, 33(1), 93–103. doi: 10.1002/jssc.200900452.

Sukmawan, Y.P., Anggadiredja, K., and Adnyana, I.K. (2021). Anti-neuropathic pain activity of Ageratum conyzoides L due to the essential oil components. *CNS & Neurological Disorders - Drug Targets*, 20(2), 181–189. doi:10.2174/1871527319666201120144228.

Tabarelli, M., and Vicente, A. (2002). Lacunas de conhecimento sobre as plantas da Caatinga. In: Sampaio, E.V.S.B., Giulietti, A.M., Virgírio, J., Gamarrarojas, C.F.L. *Vegetação e flora da Caatinga*. Associação Plantas do Nordeste e Centro Nordestino de Informações sobre Plantas: Recife, pp. 25–40.

Tabarelli, M., Leal I.R., Scarano, F.R., and Silva J.M.C. (2018), Caatinga: legado, trajetória e desafios rumo à sustentabilidade. *Ciência e Cultura*, 70(4), 25–29. doi: 10.21800/2317-66602018000400009.

Teixeira, L.P. (2018). *Análise da Distribuição Espacial e Representatividade Geográfica das Unidades de Conservação do Domínio Fitogeográfico da Caatinga*. Universidade Federal do Ceará.

Torres M.C.M., Luz M.A., Oliveira F.B., Barbosa A.J.C., and Araújo L.G. (2021). Composição química do óleo essencial das folhas de Croton heliotropiifolius Kunth (Euphorbiaceae). *Brazilian Journal of Development*, 7(2), 15862–15872. doi:10.34117/bjdv7n2-284.

Trindade, M.J.S., and Lameira, O.A. (2014). Species from the Euphorbiaceae family used for medicinal purposes in Brazil. *Revista Cubana de Plantas Medicinales*, 19(4), 292–309.

Trombulak, S.C., and Frissell, C.A. (2000). Review of ecological effects of roads on terrestrial and aquatic communities. *Conservation Biology*, 14(1), 18–30.

Ulloa, C.U, Acevedo-Rodríguez, P., Beck, S. et al. (2017). An integrated assessment of the vascular plant species of the Americas. *Science*, 358(6370), 1614–1617. doi:10.1126/science.aao0398.

Vale, T.G., Matos, F.J.A., de Lima, T.C.M., and Viana, G.S.B. (1999). Behavioral effects of essential oils from Lippia alba (Mill.) N.E. Brown chemotypes. *Journal of Ethnopharmacology*, 67(2), 127–133. doi:10.1016/S0378-8741(98)00215-3.

Varjabedian, R. (2010). Lei da Mata Atlântica: Retrocesso ambiental. *Estudos Avançados*, 24(68), 147–160. https://doi.org/10.1590/S0103-40142010000100013.

Veras, H.N., Rodrigues, F.F., Botelho, M.A., Menezes, I.R., Coutinho, H.D., and Costa, J.G. (2017). Enhancement of aminoglycosides and β-lactams antibiotic activity by essential oil of Lippia sidoides Cham. and the thymol. *Arabian Journal of Chemistry*, 10, S2790-S2795.

Vicuña, G.C., Stashenko, E.E., and Fuentes, J.L. (2010). Chemical composition of the Lippia origanoides essential oils and their antigenotoxicity against bleomycin-induced DNA damage. *Fitoterapia*, 81(5), 343–349. doi:10.1016/j.fitote.2009.10.008.

Vidal, C.S., Oliveira Brito Pereira Bezerra Martins, A., de Alencar Silva, A., de Oliveira, M.R.C., Ribeiro-Filho, J., de Albuquerque, T.R., Coutinho, H.D.M., da Silva Almeida, J.R.G., Quintans, L.J., and de Menezes, I.R.A. (2017). Gastroprotective effect and mechanism of action of Croton rhamnifolioides essential oil in mice. *Biomedicine & Pharmacotherapy*, *89*, 47–55. doi: 10.1016/j.biopha.2017.02.005.

Vieira, R.F., and Martins, M.D.M. (1996, 24–29 March). Estudos etnobotânicos de especies medicinais de uso popular no Cerrado. VIII Symposium about the Cerrado (Brazilian Savannah), Brasília, Brazil.

Vieira, R.F., and Martins, M.D.M. (1998). Estudos etnobotânicos de espécies medicinais de uso popular no Cerrado. In *VIII Simpósio sobre o Cerrado*. Brasília.

Viveiros, M.M.H.H., Rainho, C.A., Silva, M.G., da Costa, J.G.M., de Oliveira, A.G., Padovani, C.R., and Schellini, S.A. (2017). Anti-inflammatory effects of Hymenaea courbaril essential oil compounds on pterygium fibroblasts. *Investigative Ophthalmology & Visual Science*, 58(8), 1090-1090.

Zappi, D. (2009). Fitofisionomia da Caatinga associada à Cadeia do Espinhaço. *Megadiversidade*, 4(1–2), 34–38.

WHO. (2012). *International classification of primary care*. 2nd ed. (ICPC-2). World Health Organization [WWW Document]. URL https://www.who.int/standards/classifications/other-classifications/international-classification-of-primary-care (accessed 4.11.22).

Yadav, N., Ganie, S.A., Singh, B., Chhillar, A.K., and Yadav, S.S. (2019). Phytochemical constituents and ethnopharmacological properties of *Ageratum conyzoides* L. *Phytotherapy Research*, 33(9), 2163–2178. doi:10.1002/ptr.6405.

Zappi, D. (2009). Fitofisionomia da Caatinga associada à Cadeia do Espinhaço. *Megadiversidade*, 4(1–2), 34–38.

Zétola, M., de Lima, T.C.M., Sonaglio, D., González-Ortega, G., Limberger, R.P., Petrovick, P.R., and Bassani, V.L. (2002). CNS activities of liquid and spray-dried extracts from Lippia alba—Verbenaceae (Brazilian false melissa). *Journal of Ethnopharmacology*, 82(2–3), 207–215. doi:10.1016/S0378-8741(02)00187-3.

Zoghbi, M.G.B., Andrade, E.H.A., Santos, A.S., Silva, M.H.L., and Maia, J.G.S. (1998). Essential oils of Lippia alba (Mill.) N. E. Br growing wild in the Brazilian Amazon. *Flavour and Fragrance Journal*, 13(1), 47–48. doi:10.1002/(SICI)10991026(199801/02)13:1<47::AID-FFJ690>3.0.CO;2-0.

Zoghbi, M.G.B., Bastos, M.deN.doC., Jardim, M.A.G., and Trigo, J.R. (2007). Volatiles of inflorescences, leaves, stems and roots of Ageratum conyzoides L. growing wild in the North of Brazil. *Journal of Essential Oil Bearing Plants*, 10(4), 297–303. doi:10.1080/0972060X.2007.10643558.

Chapter 6

Plants of the Chihuahuan Semi-desert for the Control of Phytopathogens

Claudio Alexis Candido-del Toro, Roberto Arredondo-Valdés, Mayela Govea-Salas, Julia Cecilia Anguiano-Cabello, Elda Patricia Segura-Ceniceros, Rodolfo Ramos-González, Juan Alberto Ascacio-Valdés, Elan Iñaky Laredo-Alcalá, and Anna Iliná

CONTENTS

6.1 INTRODUCTION

There is a sizeable territorial extension with extreme climates, water scarcity, and nutrient-poor soils called deserts. Days last on average 14 hours and reach temperatures of 43°C (Castillo-Quiroz et al., 2014; Granados-Sánchez et al., 2011). Vegetation that manages to develop within these territories suffers from desert stress. Desert stress refers to all the changes that plants undergo to develop and grow in this environment. Extreme climates, intense heat during the day, and low temperatures at night cause plants to develop changes to resist this environment and retain and store water due to the great scarcity (Castillo-Quiroz et al., 2014).

The technique plants use to survive this type of desert stress is the segregation of chemical compounds that help them develop. These compounds, also known as secondary metabolites or bioactive compounds, help plants with self-care and preservation; tannins, flavonoids, phenols, saponins, among others, are the compounds that have been identified as secondary metabolites (Gonzalez & Arias, 2009; Tawaha et al., 2007). These compounds are of great importance because

DOI: 10.1201/9781003251255-6

they have a great variety of activities that can be used as antioxidants, antimicrobial, anticancer, antidiabetic, among others (Abdel-Zaher et al., 2005; Dudai et al., 2003; Harlev et al., 2012; Liu et al., 2004; Sabu & Kuttan, 2002; Tawaha et al., 2007; Yamamoto et al., 2011).

These compounds have activities, but they have been shown to help in cardiovascular and neurodegenerative diseases (Finkel & Holbrook, 2000; Harlev et al., 2012). However, in plants, it happens differently. Besides helping self-preservation and care, they are also used to control phytopathogens that reach the plant (Sampietro et al., 2016). The most common pathogens are bacteria and fungi; however, they can also be attacked by nematodes or viruses. In addition, phytopathogens take advantage of mechanical damage, or some insects penetrate the plant cells and, thus, starting infection (Tournas, 2005). This is coupled with the fact that phytopathogens are not only specific to desert plants but are also capable of infecting crop plants, as annual crop losses of around 30% due to some of the phytopathogens mentioned above have been reported (Hilder & Boulter, 1999). Therefore, one of the most practical solutions is the use of herbicides; however, herbicides contain chemical compounds such as glyphosate (Woodburn, 2000), which helps to control weeds, bacteria, fungi, among others, but can also cause harm to humans because sometimes traces of herbicides can be found in vegetables (Böcker et al., 2019; Hynes & Boyetchko, 2006). Therefore, an alternative is being sought to solve the problem, one option being plant extracts. For example, it has been reported that aqueous extracts of *Agave lechuguilla* and *Larrea tridentata* are practical and valuable for controlling phytopathogens (Abutbul et al., 2005; Harlev et al., 2012; Terán Baptista et al., 2020).

6.2 SEMI-DESERT PLANTS

Deserts are significant territorial regions in which the environment and ecosystems can become very dry, so the soil decreases vegetation because of the high altitudes and freezing nights that can last 10 and a half hours (Granados-Sánchez et al., 2011). For such reasons, desert soils are very difficult when it comes to plant growth and adaptability because of the desert stress caused by the ecosystem, a consequence of growing in soils poor in nutrients and harsh climates, such as, extreme temperatures that reach no lower than 43°C during the middle of the day, water shortage due to the length of the day (13 and a half hours on average), which causes shortages in cloud production and water evaporation (Castillo-Quiroz et al., 2014; Castillo et al., 2011), intense solar radiation due to the sun hitting the soil surface, causing heating and radiation on vegetation, among other factors. This causes plants that manage to develop in this environment to be susceptible to various attacks, both from animals and from the same reactive oxygen species, which leads them to evolve by developing natural defenses with secondary metabolites that are responsible for defending the plant, such as phenols, flavonoids, tannins, saponins, terpenes, among others (Gonzalez & Arias, 2009; Tawaha et al. , 2007). These metabolites contribute to their protection, but they can also have a curative role, even achieving resistance to the following diseases, giving them great value in the medical, agricultural, and other industries; these metabolites can be used for the treatment and/or prevention of cancer, diabetes, neurodegenerative and cardiovascular diseases, and carcinogenesis, among others (Finkel & Holbrook, 2000; Harlev et al., 2012).

Secondary plant metabolites are chemical compounds produced by plants at risk, i.e., when they suffer mechanical damage or infection by phytopathogens, including damage caused by animals such as insects and mammals. Because of this, metabolites must constantly evolve to avoid, prevent, or repair the damage caused by any of these factors; however, genetic modification also

TABLE 6.1 PATHOGEN INHIBITION RELATIONSHIP WITH DESERT PLANT EXTRACTS

Type of Extract	Part of the Plant	Pathogen	Reference
Aqueous	Flower and stem	*Aeromonas hydrophila*	(Harlev et al., 2012; Terán Baptista et al., 2020)
Aqueous	Stem	*Photobacterium damselae*	(Abutbul et al., 2005; Harlev et al., 2012)
Aqueous	Flower and stem	*Streptococcus iniae*	(Abutbul et al., 2005; Terán Baptista et al., 2020)
Aqueous	Flower	*Vibrio Alginolyticus*	(Abutbul et al., 2005; Harlev et al., 2012)
Ethyl acetate	Flower and stem	*Varthemia iphinoides*	(Al-Dabbas et al., 2005)
Aqueous	Flower, stem, and root	*Fusarium solani*	(Afifi et al., 1991)
Aqueous	Flower and stem	*Candida tropicalis*	
Ethanolic	Flower and stem	*Aspergilus parasiticus*	

has much to do when the environment induces changes. These metabolites, being in constant desert stress, cause a high concentration of compounds, which gives them the ability to have antibacterial, antidiabetic (Abdel-Zaher et al., 2005; Sabu & Kuttan, 2002), antifungal, antioxidant (Liu et al., 2004; Tawaha et al., 2007), antiviral, and anticancer properties (Dudai et al., 2003; Harlev et al., 2012; Yamamoto et al., 2011). Due to the incredible capabilities of desert plant compounds and their high adaptability, current science is focused on obtaining extracts and essential oils. These oils and extracts represent only 1% of the dried plant; however, despite their low production, they are believed to inhibit specific pathogens, as shown in Table 6.1.

Mexico has arid and semi-arid regions with a large population of plants with excellent resource potential (Alcaraz Meléndez & Véliz Murillo, 2006). Within desert plants, *A. lechuguilla* (**lechuguilla**) (Castillo-Quiroz et al., 2014; Kaab et al., 2019), *Glycine max* (**Soybean**), *L. tridentata* (**Gobernadora**) (Arteaga et al., 2005; Kaab et al., 2019), *Schizandra chinensis* (**Bayas de cinco sabores**), *Trichosanthes kirilowii Maxim* (**Tian hua fen**), and *Yucca schidigera* (**Yuca de Mojave**) (Harlev et al., 2012) can contribute bioactive compounds for help in different areas. *L. tridentata* and *A. lechuguilla* are the most used plants in studies. Among the best-known compounds is nordihydroguaiaretic acid, produced mainly by the desert shrub *L. tridentata*; this compound is popular because of its high capacity for the treatment of diseases, as it has demonstrated effects against tumor cells, which shows high cytotoxicity against these cells, which is of benefit for treatment alternatives (Harlev et al., 2012).

The Chihuahuan Desert has high, dry, and hot climates where the plant population is divided into three types: Microphyll desert scrub (M.D.M.), rosetophytic desert scrub (M.D.R.), and crassicaule desert scrub (M.D.C.), which are characterised by the place where they are found, alluvial soils, limestone, and rocks, respectively (Granados-Sánchez et al., 2011). M.D.M are characterised by shrubs with small leaves, not always with thorns, and are found more in flat terrain, such as valleys, where the soils are deep and have a layer of rocks. M.D.R. are a predominantly shrubby and sub-shrubby species and have longer and narrower leaves. They are found in areas called calcic desert scrublands and are also known as *izotales magueyales* and *lechuguillas*.

TABLE 6.2 DESERT CACTI

Genre	Species	Reference
Ariocarpus	*Agavoides*	(Granados-Sánchez et al., 2011)
	Retusus	
	Fissuratus	
Coryphanta	*Radians*	
	Gladuligera	
	Speciosa	
Echinocactus	*Horizontalonius*	
	Parryi	
	Platyacanthus	
Ferocactus	*Hamatacanthus*	
	Histrix	
	Pilosus	
Neobuxbaumia	*Macrocephala*	(Miguel-Talonia et al., 2014)
	Mezcalaensis	
	Tetetzo	
Pachycereus	*Fulviceps*	
	Weberi	
Stenocereus	*Stellatus*	
	Dumortieri	

Finally, M.D.C., also known as *mesquite cactus* and or *nopalera* are found in nopal areas and are found in mesquite grassland (Granados-Sánchez et al., 2011; Aguero & Carranza Gonzalez, 2019). Within this vegetation, the most predominant shrubs are *L. tridentata*, *A. lechuguilla*, *Dasylirion* spp. and *Yucca* sp. (Granados-Sánchez et al., 2011; Sánchez Escalante, n.d.).

However, we find not only this type of vegetation but also, because of the extreme climates and water scarcity, a great diversity of cacti. In Table 6.2, some of the many species are mentioned.

6.2.1 Generalities of Bioactive Compounds (Metabolites)

The bioactive compounds of plants are an essential source of research. Generally, bioactive compounds are found in the aerial parts of plants, such as the flower, leaves, and the stem; there are some reports of finding them in roots. The bioactive compounds from desert plants are potent because they can survive and adapt to desert stress. After all, these compounds should be more potent because they can survive and adapt to desert stress, which, in terms of their activities for use, make them potent. Table 6.3 shows the compounds that have been identified in a general way.

These compounds not only have a healing capacity in plants but also a wide range of activities against pathogens, as well as activities against diseases such as cancer, and it has even been demonstrated how these compounds can have therapeutic capacities in different treatments and/or as medicines (Harlev et al., 2012).

TABLE 6.3 BIOACTIVE COMPOUNDS OF DESERT PLANTS

Bioactive Compound	Activity	Description	Reference
Phenols	Antioxidant	Catalyzes lipid peroxidation	(Al-Mustafa & Al-Thenibat, 2008; Schroeter et al., 2002)
Flavonoids	Antioxidant, although it has not been as widely reported as phenols.	Eliminates free radicals	(Pier-Giorgio, 2000; Rice-Evans et al., 1996)
Saponins	Anticancer, cardiac immunostimulant	They have natural detergent properties, which help plants adapt to desert climatic conditions (water conservation and resistance to extreme temperatures, among others).	(Gurfinkel & Rao, 2003; Harmatha, 2000; Sen et al., 1998)

6.2.2 *Agave Lechuguilla*

A. lechuguilla belongs to the Agavaceae family and is one of more than 300 non-timber forest species in this family. It is generally found in about 100,000 km^2 in arid and semi-arid regions, in limestone and rocky terrain of northeastern Mexico in the states of Coahuila, Chihuahua, Nuevo Leon, Durango, San Luis Potosi and Zacatecas (Castillo-Quiroz et al., 2014; Castillo et al., 2012; Hernandez et al., 2005; Reyes-Aguero et al., 2017) as well as in part of Central America. Insects, bats, and birds attack it because of its high concentration of nutrients in the leaves (Castillas et al., 2012; Castillo-Quiroz et al., 2014). This plant can retain up to 60% of the water it receives; has, on average, 11–30 leaves; can measure up to 40 cm wide and 70 cm long; and has long, thin roots, which contain a substantial amount of fiber and can expand up to 12 cm below the soil. The leaves have a light green and milky yellow color, can measure up to 50 cm high and 6 cm wide, and, on average, have eight to 20 thorns per leaf. Since ancient times, this plant has been used medicinally and for everyday objects such as making sandals, ropes, baskets, nets, bags, and blankets (Reyes-Agüero et al., 2017) as well as in the automotive industry, among other uses (Grove, 2016). *A. lechuguilla* extract has been shown to have potential in several areas, including bioethanol production (Castillo-Quiroz et al., 2014), inhibition of weed growth and germination (Kaab et al., 2019), and poison, among others, because of its high content of flavonoids and steroids (Hernández et al., 2005; Reyes-Agüero et al., 2017). In addition, this plant has specific properties that allow it to adapt to climatic changes and infertile soils and grow with little water (Castillo et al., 2008, 2011, 2012).

6.2.3 *Larrea Tridentata*

L. tridentata, also known as creosote bush, governess, chaparral, or greasewood, belongs to the Zygophyllaceae family and is one of the 250 species of this family (Jones, 1987; Skouta et al., 2018). It covers about 19 million hectares, most commonly in North American deserts (Chihuahuan semi-desert) (Arteaga et al., 2005). The thickets of this species are always green; however, in extreme drought, the foliage turns yellowish (Granados-Sánchez et al., 2011). It can reach up to 3 meters tall in its branching stage, has leaves that are 1 cm long, and flowers with yellow petals and thorns. Since ancient times, it has been used as an infusion made with leaves and branches. Although pharmaceuticals have already been presented in capsules and tablets, the extracts have great potential for industries (Abou-Gazar et al., 2004; Skouta et al., 2018). By the potential to inhibit the growth and germination of weeds (Kaab et al., 2019), antioxidant activity, and soap making, among others. Other ways Mexican ancestors have used it as a medicinal plant, but are not all scientifically proven, include for infertility, rheumatism, arthritis, diabetes, pain, inflammation (Arteaga et al., 2005), cancer, tuberculosis, menstrual pain (Lambert et al., 2005), obesity (Del Vecchyo-Tenorio et al., 2016), infections, kidney problems, and tumor treatment (Martins et al., 2012). This is because of the plant's yield, as it can have up to 50% of extractable matter and only 0.1% of essential oils, from which compounds such as flavonoids (Arteaga et al., 2005; Lambert et al., 2005; Martins et al., 2012), saponins, triterpenes, and triterpenoids (Jitsuno & Mimaki, 2010; Martins et al., 2013) can be obtained. It is believed to have all these qualities because it has been shown to have a wide variety of activities, such as antiherpes, antioxidant, antifungal, anti-inflammatory (Arteaga et al., 2005; Bashyal et al., 2017; Gnabre et al., 2015), antiparasitic, antibacterial (Martins et al., 2012, 2013) and antiviral (Brent, 1999) activity.

6.3 PHYTOPATHOGENIC BACTERIA

One of the biggest problems encountered are diseases caused by phytopathogenic bacteria, which, being crop limiting, cause a tremendous annual economic loss worldwide (Sampietro et al., 2016). Most of these diseases are caused by bacteria of the Gram-negative *Bacillus* genus, which infects plants through leaves, stems, or roots that have suffered some mechanical damage, which allows the bacteria to invade the plant tissue more quickly or, to the contrary, by insects that carry the bacteria and are phytophagous borers (Tournas, 2005). Once the bacteria have penetrated the plant and infected it, they can cause a wide range of plant damage, among which the most common are wilting, necrosis, softening, and tumors, among others, caused by bacteria such as *Pseudomonas*, *Xanthomonas*, *Erwinia*, *Agrobacterium*, respectively (Sampietro et al., 2016).

For disease control treatment, chemical products containing copper, as well as some antibiotics, are constantly used. Today, scientists are interested in bioactives from plants to confront the phytopathogenic bacteria (Kumar et al., 2005), as several bacteria have developed resistance to traditional treatments, requiring increased doses and that can produce a high impact on the environment (McLeod et al., 2017). Furthermore, as mentioned above, the use of bioactive compounds from desert plants gives even greater viability for use in order to replace or reduce the doses of commercial biocides (Sampietro et al., 2016). Plant extracts from desert plants, leaves, and stems have been developed because of their high antimicrobial capacities, in which compounds with this capacity have been found, such as quercetin, kaempferol, and this flavone, from leaves (Terán Baptista et al., 2020).

Within phytopathogenic bacteria, there is a classification usually for phytopathogenic fungi that has also been implemented in bacteria. They can be classified as biotrophs and necrotrophs. Biotrophs feed on nutrients from living cells, while necrotrophs feed on dead cells. There are times when necotrophic bacteria can have a biotrophic part; such bacteria are called hemibiotrophs, which have a biotrophic part at the beginning of the infection. Table 6.4 shows some examples of biotrophic and necrotrophic bacteria.

However, it has been demonstrated that it does not matter what type of characteristic the bacteria have; at the moment of invading the plant tissue and suppressing the plant defenses, a

TABLE 6.4 CLASSIFICATION OF PHYTOPATHOGENIC BACTERIA		
Bacteria	**Classification**	**Reference**
Pseaudomonas syringae	Biotrophic/hemibiotrophic	(Kraepiel & Barny, 2016; Szczesny et al., 2010)
Agrobacterium	Biotrophic	
Xanthomonas spp.	Necrotrophic	
Erwinia amylovora	Necrotrophic/hemibiotrophic	
Pectobacterium spp.	Necrotrophic	
Dickeya spp.	Necrotrophic	
Ralstoniaa solanacearum	Biotrophic/necrotrophic	

necrotrophic phase occurs, which is when the bacteria enter to infect the plants, killing the plant cells. In the same way, one phase depends on the other to induce the disease, regardless of how long each bacterium lasts in the different phases (Kraepiel & Barny, 2016).

Once the bacteria manage to infect the plant, they prevent the passage of water from the root to the leaves, which is commonly known as wilting; the degree of severity depends on factors such as environmental temperature, soil moisture, bacterial load, the virulence of the strain, and even the susceptibility of the plants (Gonzalez & Arias, 2009; Momol et al., 2008).

6.4 PHYTOPATHOGENIC FUNGI

Another of the most common problems affecting plants is phytopathogenic fungi, which, in addition to invading plants, can cause crop losses between 20% and 30% (Hilder & Boulter, 1999; Santamarina et al., 2017; Wang et al., 2014), which is a substantial economic loss worldwide (Pilar Santamarina et al., 2015). For this reason, desert plants are of great importance, as they have been shown to have biocidal activity, and extracts and essential oils of these plants have been used to attack this problem (Santamarina et al., 2017). Among the applications of these oils and extracts are antifungal activity (Bouabidi et al., 2015; Javed et al., 2012; Moghaddam et al., 2015; Petretto et al., 2013). In addition, several fungi cause diseases in plants (see Table 6.5), in which

TABLE 6.5 CHARACTERISTIC SYMPTOMS OF FUNGAL DISEASES		
Phytopathogenic Fungi	**Symptom**	**Reference**
Colletotrichum	Circular lesions with black spots	(De Silva et al., 2017; Perfect et al., 1999)
Fusarium	Fusarium blight and wilt	(Gale et al., 2005; Zhang et al., 2020)
Phoma	Large spots on leaves	(Bennett et al., 2018)
Botrytis	Leaf necrosis	(AbuQamar et al., 2017)
Rhizoctonia	Brown stain, bare spot, rot	(Keijer et al., 1997)
Sclerotinia	White filaments, water-filled lesions on stems	(Smolińska & Kowalska, 2018)
Botryodiplodia	Chlorosis	(Moreira-Morrillo et al., 2021)
Alternaria	Small green/brown spots	(Santamarina et al., 2017)
Penicillium	Brown, white, and wet spots	(Santamarina et al., 2017)

wilting and softening have been demonstrated as the most characteristic (Wang et al., 2014). In addition, the fungus infection cause deformities in plants and can affect the yield of plants and lower their production of secondary metabolites, or in the case of harvest, decrease the nutritional value of crops (Duniway, 2002; Liu et al., 2008). To combat this, many have resorted to using of chemical fungicides that attack the fungus but, at the same time, damage the crops (Agathokleous & Calabrese, 2021). Currently, the aim is to avoid this type of infection so that yields increase in production and reduce annual losses due to phytopathogenic fungi, which account for approximately 14% of losses (Ranjekar et al., 2003).

Plants can be infected by several factors, among which the main cause of infection is due to poor handling of food at the time of preservation for sale (Sumalan et al., 2013).

6.5 PHYTOPATHOGENIC VIRUSES

Viruses are a significant problem in agriculture because there are some that attack crops, damage production, and create a tremendous economic loss worldwide. However, unlike past pathogens, this one represents a significant challenge because the virus can quickly mutate, so there is a great diversity among viruses that can attack plants, making the damage more extensive (Jacquemond, 2012).

Infection can occur through different factors, among which are transmission by aphids (aphids) and through infected plants, seeds, and fungi (Andika et al., 2017; Jacquemond, 2012), among others. Through aphids, there are two main ways (Ng & Perry, 2004). As for infection by plants, although it is a severe problem, unstable viruses may not remain viable when the infected seed is sown (O'Keefe et al., 2007).

These viruses not only can attack crops but also desert plants and weeds, where a latent infection occurs. In the case of weeds and desert plants, despite being infected by the virus, they do not develop apparent symptoms; however, they cause the highest percentage of crop infections (Jones, 2004).

However, for viruses to replicate optimally, the plants must have what is needed so that, it can attack the crops and plant shoots. For this reason, the relationship between transmission efficiency and virus accumulation in plants has been studied (Ali et al., 2006; Betancourt et al., 2008).

In terms of research, one of the most widely used viruses is the cucumber mosaic virus because it is easy to handle and its infection has been shown to occur through fungi, plants, and insects; therefore, it allows studies to acquire more knowledge about how viruses attack plants (Andika et al., 2017; Jacquemond, 2012).

6.6 PHYTOPATHOGENIC NEMATODES

Nematodes are another global problem in agriculture because crop losses can become very large. However, solutions, such as biocides, represent a problem because they attack the nematodes and the nutrients in the same crops and plants. Furthermore, this type of nematode directly interacts with phytopathogenic fungi, which infects the plant and transmits it to other plants (Zhang et al., 2020). There are about 500,000 species, of which some are useful as food for fungi (Neher & Weicht, 1998). However, 90% of these species help the environment with the nitrogen cycle, generally inhabiting the rhizosphere of plants (Zhang et al., 2020). Unlike the others, this

TABLE 6.6 CROPS AFFECTED BY NEMATODES

Nematode	Cultivation	Reference
Meloidogyne spp.	Bean	(Carneiro et al., 2010)
	Potato	(Khalifa et al., 2012)
	Beans	(Al-Hazmi & Al-Nadary, 2015)
	Tomato	(Hajji-Hedfi et al., 2018; Wanjohi et al., 2018)
	Watermelon	(Keinath et al., 2019)
	Eggplant	(Khan & Siddiqui, 2017)
	Carrot	(Ahmad et al., 2019)
Pratylenchus	Potato	(Björsell et al., 2017)

type of pest does not decompose organic material but feeds on living cells. Two factors cause this type of infection: Through the soil and through infected plants. Once the nematodes reach the plant, it must adapt to infect it so that the plant facilitates entry to the infection, which causes deficiency in resistance to pathogens and susceptibility to infections (Zhang et al., 2020). Infection by nematodes is due to their interaction with other plant pathogens. Some of the examples of crops attacked by nematodes are shown in Table 6.6.

6.7 HERBICIDES

Herbicides, being molecules of reduced size, target specific/selective processes in plants and can be found in different presentations, solutions, powders, and emulsions, among others (Gómez, 1995), which represent about 60% in volume, concerning all pesticides used worldwide (Dayan, 2007). In recent years, pests have been creating resistance to the chemicals used for this type of problem (Ahmed, 2018; Jabran et al., 2015; Liu et al., 2021), which is more significant once the weeds begin the flowering process (Montoya et al., 2008). The actual resistance by the herbicides is because the continuous use of chemical products in crops over the years, which causes the doses to increase, also creating a tolerance to it (Papa et al., 2004), so it has become of greater importance to start looking for new, green alternatives, such as extracts from plants (Kaab et al., 2019; Morra et al., 2018; Sbai et al., 2016; Wang et al., 2011). According to Jones and Burges (1998), herbicides must have specific characteristics: Be stable for storage and processing, easy to apply, have protection against environmental conditions, and have herbicidal activity.

6.8 CONCLUSION

Desert plants are of great importance because they must go through many processes to survive in desert stress, making them a rich source of chemical compounds used to treat and diagnose diseases, both human and plants, as well as for crop care. In addition, because plants can adapt to extreme climates and water scarcity, they must protect themselves from these factors and diseases and/or infections by phytopathogens, such as bacteria, fungi, or nematodes.

6.9 ACKNOWLEDGMENTS

Thanks go to the National Science and Technology Council (CONACYT) for supporting the development of the document by financing the **'Nano and microencapsulated bioherbicides loaded with plant extracts from the Chihuahuan semi-desert for the control of plant development' (320692)** project.

REFERENCES

Abdel-Zaher, A.O., Salim, S.Y., Assaf, M.H., & Abdel-Hady, R.H. (2005). Anti-diabetic activity and toxicity of Zizyphus spina-christi leaves. *Journal of Ethnopharmacology, 101*(1–3), 129–138. https://doi.org/10.1016/j.jep.2005.04.007

Abou-Gazar, H., Bedir, E., Takamatsu, S., Ferreira, D., & Khan, I.A. (2004). Antioxidant lignans from *Larrea tridentata*. *Phytochemistry, 65*(17), 2499–2505. https://doi.org/10.1016/j.phytochem.2004.07.009

AbuQamar, S., Moustafa, K., & Tran, L.S.P. (2017). Mechanisms and strategies of plant defense against Botrytis cinerea. *Critical Reviews in Biotechnology, 37*(2), 262–274. https://doi.org/10.1080/07388551.2016.1271767

Abutbul, S., Golan, A., Barazani, O., Ofir, R., & Zilberg, D. (2005). Screening of desert plants for use against bacterial pathogens in fish. *The Israeli Journal of Aquaculture = Bamidgeh, 57*(2), 71–80. https://doi.org/10.46989/001c.20405

Afifi, F., Al-Khalil, S., Abdul-Haq, B.K., Mahasneh, A., Al-Eisawi, D.M., Sharaf, M., Wong, L.K., & Schiff, P.L. (1991). Antifungal flavonoids from Varthemia iphionoides. *Phytotherapy Research, 5*(4), 173–175. https://doi.org/10.1002/ptr.2650050407

Agathokleous, E., & Calabrese, E.J. (2021). Fungicide-induced hormesis in phytopathogenic fungi: A critical determinant of successful agriculture and environmental sustainability. *Journal of Agricultural and Food Chemistry, 69*(16), 4561–4563. https://doi.org/10.1021/acs.jafc.1c01824

Ahmad, L., Siddiqui, Z.A., & Abd_Allah, E.F. (2019). Effects of interaction of Meloidogyne incognita, Alternaria dauci and Rhizoctonia solani on the growth, chlorophyll, carotenoid and proline contents of carrot in three types of soil. *Acta Agriculturae Scandinavica Section B: Soil and Plant Science, 69*(4), 324–331. https://doi.org/10.1080/09064710.2019.1568541

Ahmed, H.M. (2018). Phytochemical screening, total phenolic content and phytotoxic activity of corn (Zea mays) extracts against some indicator species. *Natural Product Research, 32*(6), 714–718. https://doi.org/10.1080/14786419.2017.1333992

Al-Dabbas, M.M., Hashinaga, F., Abdelgaleil, S.A.M., Suganuma, T., Akiyama, K., & Hayashi, H. (2005). Antibacterial activity of an eudesmane sesquiterpene isolated from common Varthemia, Varthemia iphionoides. *Journal of Ethnopharmacology, 97*(2), 237–240. https://doi.org/10.1016/j.jep.2004.11.007

Al-Hazmi, A.S., & Al-Nadary, S.N. (2015). Interaction between Meloidogyne incognita and Rhizoctonia solani on green beans. *Saudi J Biol Sci, 22*(5), 570–574. https://doi.org/10.1016/j.sjbs.2015.04.008

Al-Mustafa, A.H., & Al-Thenibat, O.Y. (2008). Antioxidant activity of some Jordanian medicinal plants used traditionally for treatment of diabetes. *Pakistan Journal of Biological Sciences, 11*(3), 351–358. https://doi.org/10.3923/pjbs.2008.351.358

Alcaraz Meléndez, L., & Véliz Murillo, M.G. (2006). Comercialización de una planta del desierto : Damiana (Turnera diffusa). *Revista Mexicana de Agronegocios Comercialización, 10*, 1405–9282.

Ali, A., Li, H., Schneider, W.L., Sherman, D.J., Gray, S., Smith, D., and Roossinck, M.J. (2006). Analysis of genetic bottlenecks during horizontal transmission of cucumber mosaic virus. *Journal of Virology, 80*(17), 8345–8350. https://doi.org/10.1128/jvi.00568-06

Andika, I.B., Wei, S., Cao, C., Salaipeth, L., Kondo, H., & Sun, L. (2017). Phytopathogenic fungus hosts a plant virus: A naturally occurring cross-kingdom viral infection. *Proceedings of the National Academy of Sciences of the United States of America, 114*(46), 12267–12272. https://doi.org/10.1073/pnas.1714916114

Arteaga, S., Andrade-Cetto, A., & Cárdenas, R. (2005). *Larrea tridentata* (Creosote bush), an abundant plant of Mexican and US-American deserts and its metabolite nordihydroguaiaretic acid. *Journal of Ethnopharmacology, 98*(3), 231–239. https://doi.org/10.1016/j.jep.2005.02.002

Bashyal, B., Li, L., Bains, T., Debnath, A., & LaBarbera, D.V. (2017). *Larrea tridentata*: A novel source for anti-parasitic agents active against Entamoeba histolytica, Giardia lamblia and Naegleria fowleri. *PLoS Neglected Tropical Diseases, 11*(8), 1–19. https://doi.org/10.1371/journal.pntd.0005832

Bennett, A., Ponder, M.M., & Garcia-Diaz, J. (2018). Phoma infections: Classification, potential food sources, and their clinical impact. *Microorganisms, 6*(3), 2–13. https://doi.org/10.3390/microorganisms6030058

Betancourt, M., Fereres, A., Fraile, A., & García-Arenal, F. (2008). Estimation of the effective number of founders that initiate an infection after Aphid transmission of a multipartite plant virus. *Journal of Virology, 82*(24), 12416–12421. https://doi.org/10.1128/jvi.01542-08

Björsell, P., Edin, E., & Viketoft, M. (2017). Interactions between some plant-parasitic nematodes and Rhizoctonia solani in potato fields. *Applied Soil Ecology, 113*, 151–154. https://doi.org/10.1016/j.apsoil.2017.02.010

Böcker, T., Möhring, N., & Finger, R. (2019). Herbicide free agriculture? A bio-economic modelling application to Swiss wheat production. *Agricultural Systems, 173*(August 2018), 378–392. https://doi.org/10.1016/j.agsy.2019.03.001

Bouabidi, W., Hanana, M., Gargouri, S., Amri, I., Fezzani, T., Ksontini, M., Jamoussi, B., & Hamrouni, L. (2015). Chemical composition, phytotoxic and antifungal properties of Ruta chalepensis L. essential oils. *Natural Product Research, 29*(9), 864–868. https://doi.org/10.1080/14786419.2014.980246

Brent, J. (1999). Three new herbal hepatotoxic syndromes. *Journal of Toxicology - Clinical Toxicology, 37*(6), 715–719. https://doi.org/10.1081/CLT-100102449

Carneiro, F.F., Ramalho, M.A.P., & Pereira, M.J.Z. (2010). Fusarium oxysporum f. sp. phaseoli and Meloidogyne incognita interaction in common bean. *Crop Breeding and Applied Biotechnology, 10*(3), 271–274. https://doi.org/10.1590/s1984-70332010000300014

Castillas, F., Cardenas, A., Morales, C., Verde, M., & Cruz-Vega, D. (2012). Cytotoxic activity of *Agave lechuguilla* Torr. *African Journal of Biotechnology, 11*(58), 12229–12231. https://doi.org/10.5897/ajb-12-1123

Castillo, Q., Berlanga, R., Pando, M., & Cano, P. (2008). Regeneración del cogollo de *Agave lechuguilla* Torr., de cinco procedencias bajo cultivo. *Rev. Cien. For. En México., 33* (103), 27–40.

Castillo, Q., Cano, P., & Berlanga, R. (2012). Establecimiento y aprovechamiento de lechuguilla. Comisión Nacional Forestal-Instituto Nacional de Investigaciones Forestales, Agrícolas y Pecuarias. México, 33.

Castillo, Q., Mares, A., & Villavicencio, G. (2011). Lechuguilla (*Agave lechuguilla* Torr.) planta suculenta de importancia económica y social de las zonas áridas y semiáridas de México. *Boletín de La Sociedad Latinoamericana y Del Caribe de Cactáceas y Otras Suculentas, 8*(2), 6–9.

Castillo-Quiroz, D., Martinez-Burciaga, O.U., Ríos-González, L.J., Rodríguez-de la Garza, J.A., Morales-artínez, T.K., Castillo-Reyes, F., & Avila-Flores, D.Y. (2014). Determinación de áreas potenciales para plantaciones de Agave lechuguilla Torr. para la producción de etanol. *Acta Química Mexicana, 6*(12), 5–11.

Dayan, F.E. (2007). Current status and future prospects in herbicide discovery. *Weed Management Handbook*, 93–113. https://doi.org/10.1002/9780470751039.ch6

De Silva, D.D., Crous, P.W., Ades, P.K., Hyde, K.D., & Taylor, P.W.J. (2017). Life styles of Colletotrichum species and implications for plant biosecurity. *Fungal Biology Reviews, 31*(3), 155–168. https://doi.org/10.1016/j.fbr.2017.05.001

Del Vecchyo-Tenorio, G., Rodríguez-Cruz, M., Andrade-Cetto, A., & Cárdenas-Vázquez, R. (2016). Creosote bush (*Larrea tridentata*) improves insulin sensitivity and reduces plasma and hepatic lipids in hamsters fed a high fat and cholesterol diet. *Frontiers in Pharmacology, 7*(JUN), 1–10. https://doi.org/10.3389/fphar.2016.00194

Dudai, N., Larkov, O., Chaimovitsh, D., Lewinsohn, E., Freiman, L., & Ravid, U. (2003). Essential oil compounds of Origanum dayi post. *Flavour and Fragrance Journal, 18*(4), 334–337. https://doi.org/10.1002/ffj.1237

Duniway, J.M. (2002). Status of chemical alternatives to methyl bromide for pre-plant fumigation of soil. *Phytopathology, 92*(12), 1337–1343. https://doi.org/10.1094/PHYTO.2002.92.12.1337

Finkel, T., & Holbrook, N.J. (2000). Oxidants, oxidative stress and the biology of ageing. *Nature, 408*(6809), 239–247. https://doi.org/10.1038/35041687

Gale, L.R., Bryant, J.D., Calvo, S., Giese, H., Katan, T., O'Donnell, K., Suga, H., Taga, M., Usgaard, T.R., Ward, T.J., & Kistler, H.C. (2005). Chromosome complement of the fungal plant pathogen Fusarium graminearum based on genetic and physical mapping and cytological observations. *Genetics, 171*(3), 985–1001. https://doi.org/10.1534/genetics.105.044842

Gnabre, J., Bates, R., & Huang, R.C. (2015). Creosote bush lignans for human disease treatment and prevention: Perspectives on combination therapy. *Journal of Traditional and Complementary Medicine, 5*(3), 119–126. https://doi.org/10.1016/j.jtcme.2014.11.024

Gómez, J.F. (1995). Control de malezas 7. *CENICAÑA. El Cultivo de La Caña En La Zona Azucarera de Colombia* (pp. 143–152). Cali: CENICAÑA.

González, I., & Y. Arias. 2009. Interacción planta-bacterias fitopatógenas: Caso de estudio *Ralstonia solanacearum*- plantas hospedantes plant-phytopathogen bacteria interaction: Case study *Ralstonia. Revista de protección vegetal* 24 (2): 69–80.

Granados-Sánchez, D., Sánchez-González, A., Granados Victorino, R.L., & Borja de la Rosa, A. (2011). Ecología de la vegetación del desierto chihuahuense. *Revista de Chapingo Serie Ciencias Forestales y Del Ambiente, 17*, 111–130. https://doi.org/10.5154/r.rchscfa

Grove, A.R. (2016). Morphological study of *agave lechuguilla. Contributions from the Hull Botanical Laboratory, 534*, 103. http://www.journals.uchicago.edu/t-and-c.

Gurfinkel, D.M., & Rao, A.V. (2003). Soyasaponins: The relationship between chemical structure and colon anticarcinogenic activity. *Nutrition and Cancer, 47*(1), 24–33. https://doi.org/10.1207/s15327914nc4701_3

Hajji-Hedfi, L., Regaieg, H., Larayedh, A., Chihani, N., & Horrigue-Raouani, N. (2018). Biological control of wilt disease complex on tomato crop caused by Meloidogyne javanica and Fusarium oxysporum f.sp. lycopersici by Verticillium leptobactrum. *Environmental Science and Pollution Research, 25*(19), 18297–18302. https://doi.org/10.1007/s11356-017-0233-6

Harlev, E., Nevo, E., Lansky, E.P., & Lansky, S. (2012). Anticancer attributes of desert plants : a review. *Anticancer Drugs, 23*(3), 255–271. https://doi.org/10.1097/CAD.0b013e32834f968c

Harmatha, J. (2000). Chapter 14. Chemo-ecological role of spirostanol saponins in the interaction between plants and insects. *Saponins in Food, Feedstuffs and Medicinal Plants, 45*, 129–141. https://doi.org/10.1007/978-94-015-9339-7_14

Hernández, R., Lugo, E.C., Díaz, L., & Villanueva, S. (2005). Extracción y cuantificación indirecta de las saponinas de agave lechuguilla torrey. *E- Gnosis Revista Digital Cientifica y Tecnologica, 3*, 3–12.

Hilder, V.A., & Boulter, D. (1999). Genetic engineering of crop plants for insect resistance - A critical review. *Crop Protection, 18*(3), 177–191. https://doi.org/10.1016/S0261-2194(99)00028-9

Hynes, R.K., & Boyetchko, S.M. (2006). Research initiatives in the art and science of biopesticide formulations. *Soil Biology and Biochemistry, 38*(4), 845–849. https://doi.org/10.1016/j.soilbio.2005.07.003

Jabran, K., Mahajan, G., Sardana, V., & Chauhan, B.S. (2015). Allelopathy for weed control in agricultural systems. *Crop Protection, 72*, 57–65. https://doi.org/10.1016/j.cropro.2015.03.004

Jacquemond, M. (2012). Chapter 13: Cucumber mosaic virus. In *Advances in Virus Research* (1st ed., Vol. 84). Elsevier Inc. https://doi.org/10.1016/B978-0-12-394314-9.00013-0

Javed, S., Shoaib, A., Mahmood, Z., Mushtaq, S., & Iftikhar, S. (2012). Analysis of phytochemical constituents of Eucalyptus citriodora L. responsible for antifungal activity against post-harvest fungi Analysis of phytochemical constituents of Eucalyptus citriodora L. *Natural Product Research, 26*(18), 1732–1736. https://doi.org/10.1080/14786419.2011.607451

Jitsuno, M., & Mimaki, Y. (2010). Triterpene glycosides from the aerial parts of *Larrea tridentata. Phytochemistry, 71*(17–18), 2157–2167. https://doi.org/10.1016/j.phytochem.2010.09.012

Jones, K.A., & Burges, H.. (1998). Formulation of microbial biopesticides: Beneficial microorganisms, nematodes and seed treatments. 1 st edition, In: Burges, H.D. (Ed.), 7–30. DOI 10.1007/978-94-011-4926-6

Jones, R.A.C. (2004). Using epidemiological information to develop effective integrated virus disease management strategies. *Virus Research, 100*(1), 5–30. https://doi.org/10.1016/j.virusres.2003.12.011

Jones, S. (1987). *Sistematica Vegetal* (pp. 405–406). Mexico: McGraw-Hill.

Kaab, S.B., I. B. Rebey, M. Hanafi, K. MkadminiHammi, A. Smaoui, M. L. Fauconnier, C. Declerck, H. Jijakli and R. Ksouri. 2019. Screening of Tunisian plant extracts for herbicidal activity and formulation of bioherbicide based on *Cynara cardunculus. South African Journal of Botany* 128 (2020): 67–76. https://doi.org/10.1016/j.sajb.2019.10.018

Keijer, J., Korsman, M.G., Dullemans, A.M., Houterman, P.M., De Bree, J., & Van Silfhout, C.H. (1997). In vitro analysis of host plant specificity in Rhizoctonia solani. *Plant Pathology, 46*(5), 659–669. https://doi.org/10.1046/j.1365-3059.1997.d01-61.x

Keinath, A.P., Wechter, W.P., Rutter, W.B., & Agudelo, P.A. (2019). Cucurbit Rootstocks Resistant to Fusarium oxysporum f. sp. niveum Remain Resistant When Co-Infected by Meloidogyne incognita in the Field. *Plant Disease, 103*(6), 1383–1390. https://doi.org/10.1094/PDIS-10-18-1869-RE

Khalifa, E.Z., Ammar, M.M., Mousa, E.M., & Hafez, S.L. (2012). Biological control of the disease complex on potato caused by root-knot nematode and Fusarium wilt fungus. *Nematologia Mediterranea, 40*(2), 169–172.

Khan, M., & Siddiqui, Z.A. (2017). Interactions of Meloidogyne incognita, Ralstonia solanacearum and Phomopsis vexans on eggplant in sand mix and fly ash mix soils. *Scientia Horticulturae, 225*, 177–184. https://doi.org/10.1016/j.scienta.2017.06.016

Kraepiel, Y., & Barny, M.A. (2016). Gram-negative phytopathogenic bacteria, all hemibiotrophs after all? *Molecular Plant Pathology, 17*(3), 313–316. https://doi.org/10.1111/mpp.12345

Kumar, K., Gupta, S.C., Chander, Y., & Singh, A.K. (2005). Antibiotic use in agriculture and its impact on the terrestrial environment. *Advances in Agronomy, 87*(05), 1–54. https://doi.org/10.1016/S0065-2113(05)87001-4

Lambert, J.D., Sang, S., Dougherty, A., Caldwell, C.G., Meyers, R.O., Dorr, R.T., & Timmermann, B.N. (2005). Cytotoxic lignans from *Larrea tridentata*. *Phytochemistry, 66*(7), 811–815. https://doi.org/10.1016/j.phytochem.2005.02.007

Liu, C.Z., Murch, S.J., El-Demerdash, M., & Saxena, P.K. (2004). Artemisia judaica L.: Micropropagation and antioxidant activity. *Journal of Biotechnology, 110*(1), 63–71. https://doi.org/10.1016/j.jbiotec.2004.01.011

Liu, R., Kumar, V., Jhala, A., Jha, P., & Stahlman, P.W. (2021). Control of glyphosate- and mesotrione-resistant Palmer amaranth in glyphosate- and glufosinate-resistant corn. Agronomy Journal, June, 1–11. https://doi.org/10.1002/agj2.20770

Liu, S., Ruan, W., Li, J., Xu, H., Wang, J., Gao, Y., & Wang, J. (2008). Biological control of phytopathogenic fungi by fatty acids. *Mycopathologia, 166*(2), 93–102. https://doi.org/10.1007/s11046-008-9124-1

Martins, S., Aguilar, C.N., Teixeira, J.A., & Mussatto, S.I. (2012). Bioactive compounds (phytoestrogens) recovery from *Larrea tridentata* leaves by solvents extraction. *Separation and Purification Technology, 88*, 163–167. https://doi.org/10.1016/j.seppur.2011.12.020

Martins, S., Amorim, E.L.C., Sobrinho, T.J.S.P., Saraiva, A.M., Pisciottano, M.N.C., Aguilar, C.N., Teixeira, J.A., & Mussatto, S.I. (2013). Antibacterial activity of crude methanolic extract and fractions obtained from *Larrea tridentata* leaves. *Industrial Crops and Products, 41*(1), 306–311. https://doi.org/10.1016/j.indcrop.2012.04.037

McLeod, A., Masimba, T., Jensen, T., Serfontein, K., & Coertze, S. (2017). Evaluating spray programs for managing copper resistant Pseudomonas syringae pv. tomato populations on tomato in the Limpopo region of South Africa. *Crop Protection, 102*, 32–42. https://doi.org/10.1016/j.cropro.2017.08.005

Miguel-Talonia, C., Téllez-Valdés, O., & Murguía-Romero, M. (2014). Las cactáceas del Valle de Tehuacán-Cuicatlán, México: estimación de la calidad del muestreo. *Revista Mexicana de Biodiversidad, 85*(2), 436–444. https://doi.org/10.7550/rmb.31390

Moghaddam, M., Taheri, P., Pirbalouti, A.G., & Mehdizadeh, L. (2015). Chemical composition and antifungal activity of essential oil from the seed of Echinophora platyloba DC. against phytopathogens fungi by two different screening methods. *Lwt, 61*(2), 536–542. https://doi.org/10.1016/j.lwt.2014.12.008

Momol, T., Ji, P., Pernezny, K., McGovern, R., & Olson, S. (2008). Three Soilborne Tomato Diseases Caused by Ralstonia and Fusarium Species and their Field Diagnostics. *Edis, 2*, 205–210. https://doi.org/10.32473/edis-pp127-2005

Montoya, J. C., C. Porfiri, N. Romano and N. Rodríguez. 2008. Cultivo de girasol en la región semiárida pampeana. *Manejo de las malezas en el cultivo de girasol* 6: 49–64.

Moreira-Morrillo, A.A., Cedeño-Moreira, Á.V., Canchignia-Martínez, F., & Garcés-Fiallos, F.R. (2021). Lasiodiplodia theobromae (Pat.) Griffon & Maul [(sin.) Botryodiplodia theobromae Pat] en el cultivo de cacao: síntomas, ciclo biológico y estrategias de manejo. *Scientia Agropecuaria, 14*. https://doi.org/10.17268/sci.agropecu.2021.068

Morra, M.J., Popova, I.E., & Boydston, R.A. (2018). Bioherbicidal activity of Sinapis alba seed meal extracts. *Industrial Crops and Products, 115*(February), 174–181. https://doi.org/10.1016/j.indcrop.2018.02.027

Neher, D.A., & Weicht, T.R. (1998). Functional diversity of nematodes. *Journal of Nematology, 10*, 239–251. https://doi.org/10.1016/S0929-1393(98)00123-1

Ng, J.C.K., & Perry, K.L. (2004). Transmission of plant viruses by aphid vectors. *Molecular Plant Pathology, 5*(5), 505–511. https://doi.org/10.1111/J.1364-3703.2004.00240.X

O'Keefe, D.C., Berryman, D.I., Coutts, B.A., & Jones, R.A.C. (2007). Lack of seed coat contamination with Cucumber mosaic virus in lupin permits reliable, large-scale detection of seed transmission in seed samples. *Plant Disease, 91*(5), 504–508. https://doi.org/10.1094/PDIS-91-5-0504

Papa, J.C.M., Felizia, J.C., & Esteban, A.J. (2004). Tolerancia y resistencia a herbicidas. *Sitio Argentino de Produción Animal, 1*, 1–6. http://www.produccion-animal.com.ar/produccion_y_manejo_pasturas/pasturas_combate_de_plagas_y_malezas/25-tolerancia_y_resistencia_a_herbicidas.pdf

Perfect, S.E., Hughes, H.B., O'Connell, R.J., & Green, J.R. (1999). Colletotrichum: A model genus for studies on pathology and fungal–plant interactions. *Fungal Genetics and Biology, 27*(2–3), 186–198. https://doi.org/10.1006/fgbi.1999.1143

Petretto, G.L., Chessa, M., Piana, A., Masia, M.D., Foddai, M., Mangano, G., Culeddu, N., Afifi, F.U., & Pintore, G. (2013). Chemical and biological study on the essential oil of Artemisia caerulescens L. ssp. densiflora (Viv.). *Natural Product Research, 27*(19), 1709–1715. https://doi.org/10.1080/14786419.2012.749471

Pier-Giorgio, P. (2000). Flavonoids as antioxidants. *Journal of Natural Products, 63*(7), 1035–1042. https://doi.org/10.1021/np9904509

Ranjekar, P.K., Patankar, A., Gupta, V., Bhatnagar, R., Bentur, J., & Kumar, P.A. (2003). Genetic engineering of crop plants for insect resistance. *Current Science, 84*(3), 321–329. https://doi.org/10.1007/978-94-010-9646-1_4

Reyes-Agüero, J.A., Aguirre-Rivera, J.R., & Peña-Valdivia, C.B. (2017). Biología y aprovechamiento de *Agave lechuguilla* Torrey. *Botanical Sciences, 88*(67), 75. https://doi.org/10.17129/botsci.1626

Reyes Aguero, J. A. and E. Carranza Gonzalez. 2019. Los botánicos Rzedowski-Calderón en San Luis Potosí. *Universitarios Postosinos* 13 (209): 10–15. ISSN-1870–1698

Rice-Evans, C.A., Miller, N.J., & Paganga, G. (1996). Structure-antioxidant activity relationships of flavonoids and phenolic acids. *Free Radical Biology and Medicine, 20*(7), 933–956. https://doi.org/10.1016/0891-5849(95)02227-9

Sabu, M.C., & Kuttan, R. (2002). Anti-diabetic activity of medicinal plants and its relationship with their antioxidant property. *Journal of Ethnopharmacology, 81*(2), 155–160. https://doi.org/10.1016/S0378-8741(02)00034-X

Sampietro, D.A., Lizarraga, E.F., Ibatayev, Z.A., Omarova, A.B., Suleimen, Y.M., & Catalán, C.A.N. (2016). Chemical composition and antimicrobial activity of essential oils from Acantholippia deserticola, Artemisia proceriformis, Achillea micrantha and Libanotis buchtormensis against phytopathogenic bacteria and fungi. *Natural Product Research, 30*(17), 1950–1955. https://doi.org/10.1080/14786419.2015.1091453

Sánchez Escalante, J. J. 2007. Plantas nativas de Sonora: Las plantas del desierto sonorense. *Revista Universitaria de Sonora* 19: 20–22.

Santamarina, M.P., Roselló, J., Sempere, F., Giménez, S., & Blázquez, M.A. (2015). Commercial Origanum compactum Benth. and Cinnamomum zeylanicum Blume essential oils against natural mycoflora in Valencia rice. *Natural Product Research, 29*(23), 2215–2218. https://doi.org/10.1080/14786419.2014.1002406

Santamarina, M.P., Ibáñez, M.D., Marqués, M., Roselló, J., Giménez, S., & Blázquez, M.A. (2017). Bioactivity of essential oils in phytopathogenic and post-harvest fungi control. *Natural Product Research, 31*(22), 2675–2679. https://doi.org/10.1080/14786419.2017.1286479

Sbai, H., Saad, I., Ghezal, N., Greca, M.D., & Haouala, R. (2016). Bioactive compounds isolated from Petroselinum crispum L. leaves using bioguided fractionation. *Industrial Crops and Products, 89*, 207–214. https://doi.org/10.1016/j.indcrop.2016.05.020

Schroeter, H., Boyd, C., Spencer, J.P.E., Williams, R.J., Cadenas, E., & Rice-Evans, C. (2002). MAPK signaling in neurodegeneration: Influences of flavonoids and of nitric oxide. *Neurobiology of Aging*, *23*(5), 861–880. https://doi.org/10.1016/S0197-4580(02)00075-1

Sen, S., Makkar, H.P.S., Muetzel, S., & Becker, K. (1998). Effect of Quillaja saponaria saponins and Yucca schidigera plant extract on growth of Escherichia coli. *Letters in Applied Microbiology*, *27*(1), 35–38. https://doi.org/10.1046/j.1472-765X.1998.00379.x

Skouta, R., Morán-Santibañez, K., Valenzuela, C.A., Vasquez, A.H., & Fenelon, K. (2018). Assessing the antioxidant properties of *Larrea tridentata* extract as a potential molecular therapy against oxidative stress. *Molecules*, *23*(7), 1–17. https://doi.org/10.3390/molecules23071826

Smolińska, U., & Kowalska, B. (2018). Biological control of the soil-borne fungal pathogen Sclerotinia sclerotiorum. *Journal of Plant Pathology*, *100*, 1–12. https://doi.org/10.1007/s42161-018-0023-0

Sumalan, R.M., Alexa, E., & Poiana, M.A. (2013). Assessment of inhibitory potential of essential oils on natural mycoflora and Fusarium mycotoxins production in wheat. *Chemistry Central Journal*, *7*(1), 1–12. https://doi.org/10.1186/1752-153X-7-32

Szczesny, R., Jordan, M., Schramm, C., Schulz, S., Cogez, V., Bonas, U., & Büttner, D. (2010). Functional characterization of the Xcs and Xps type II secretion systems from the plant pathogenic bacterium Xanthomonas campestris pv vesicatoria. *New Phytologist*, *187*(4), 983–1002. https://doi.org/10.1111/j.1469-8137.2010.03312.x

Tawaha, K., Alali, F.Q., Gharaibeh, M., Mohammad, M., & El-Elimat, T. (2007). Antioxidant activity and total phenolic content of selected Jordanian plant species. *Food Chemistry*, *104*(4), 1372–1378. https://doi.org/10.1016/j.foodchem.2007.01.064

Terán Baptista, Z.P., de los Angeles Gómez, A., Kritsanida, M., Grougnet, R., Mandova, T., Aredes Fernandez, P.A., & Sampietro, D.A. (2020). Antibacterial activity of native plants from Northwest Argentina against phytopathogenic bacteria. *Natural Product Research*, *34*(12), 1782–1785. https://doi.org/10.1080/14786419.2018.1525716

Tournas, V.H. (2005). Spoilage of vegetable crops by bacteria and fungi and related health hazards. *Critical Reviews in Microbiology*, *31*(1), 33–44. https://doi.org/10.1080/10408410590886024

Wang, C., Zhu, M., Chen, X., & Bo, Q. (2011). Review on allelopathy of exotic invasive plants. *Procedia Engineering*, *18*, 240–246. https://doi.org/10.1016/j.proeng.2011.11.038

Wang, L., Zhang, Y., Wang, L., Liu, F., Cao, L., Yang, J., Qiao, C., & Ye, Y. (2014). European journal of medicinal chemistry benzofurazan derivatives as antifungal agents against phytopathogenic fungi. *European Journal of Medicinal Chemistry*, *80*, 535–542. https://doi.org/10.1016/j.ejmech.2014.04.058

Wanjohi, W.J., Wafula, G.O., & Macharia, C.M. (2018). Integrated management of Fusarium Wilt-root knot nematode complex on tomato in Central Highlands of Kenya. *Sustainable Agriculture Research*, *7*(2), 8. https://doi.org/10.5539/sar.v7n2p8

Woodburn, A.T. (2000). Glyphosate: Production, pricing and use worldwide. *Pest Management Science*, *56*(4), 309–312. https://doi.org/10.1002/(SICI)1526-4998(200004)56:4<309::AID-PS143>3.0.CO;2-C

Yamamoto, N., Kanemoto, Y., Ueda, M., Kawasaki, K., Fukuda, I., & Ashida, H. (2011). Anti-obesity and anti-diabetic effects of ethanol extract of Artemisia princeps in C57BL/6 mice fed a high-fat diet. *Food and Function*, *2*(1), 45–52. https://doi.org/10.1039/c0fo00129e

Zhang, Y., Li, S., Li, H., Wang, R., Zhang, K.Q., & Xu, J. (2020). Fungi–nematode interactions: Diversity, ecology, and biocontrol prospects in agriculture. *Journal of Fungi*, *6*(4), 1–24. https://doi.org/10.3390/jof6040206

Chapter 7

Phytochemical Compounds from Desert Plants to Management of Plant-parasitic Nematodes

Marco Antonio Tucuch-Pérez, Roberto Arredondo-Valdés, Francisco Daniel Hernández-Castillo, Yisa María Ochoa-Fuentes, Elan Iñaky Laredo-Alcalá, and Julia Cecilia Anguiano-Cabello

CONTENTS

7.1 INTRODUCTION

Agriculture is a pillar of human civilization that provides essential products such as food, fuel, energy, wood, and others; this activity has allowed developers to establish current human societies (Tucuch-Pérez et al., 2021). Research and processes have been developed to enhance food production to satisfy the food necessity to over 7.700 million people. Nevertheless, although food production has been increasing, it is affected by phytosanitary problems such as plagues, weeds, and diseases that put the food security of many countries at risk. There are four main types of phytopathogenic organisms: Bacteria, fungus, viruses, and nematodes. The plant-parasitic nematodes (PPNs) are animals with a worm form that are generally microscopic that have a structure known as a stylet that allows them to penetrate the plant tissue to obtain nutrients (Agrios, 1997).

The P.P.N.s are common in all cultivated soils worldwide. This group of phytopathogenic organisms can devastate entire plantations when there is no efficient management; it has been

DOI: 10.1201/9781003251255-7

estimated that the losses caused by PPNs around $157 billion per year (Ferreira-Barros et al., 2021). This problem grows with increasing temperatures due to climate change and because once PPNs are established, their eradication is complicated. Therefore, the primary strategy for management is to reduce the population of nematodes to an economically acceptable level with chemical nematicides (Maleita et al., 2017). However, due to the chemical products' undesired effect on environmental and human health, several nematicides have been removed from the market, limiting the options to control PPNs (Burns et al., 2017).

Therefore, it is necessary to develop less toxic and sustainable alternatives, and plants with a high number of phytochemical compounds present in their metabolism are an attractive option. Desert plants have the potential to control plant diseases, which contain various phytochemical compounds (PCs) with nematicidal activity such as terpenoids, alkaloids, flavonoids, and tannins (Al-Saleem et al., 2018). These plants have developed different physiological processes due to the extreme conditions that the desert presents, allowing the production of secondary metabolites with high nematicidal effect, such as the species *Argemone mexicana* and *Citrullus colocynthis*, which have PCs that inhibit the activity of PPNs such as *Meloidogyne incognita* (Khan et al., 2017). Therefore, this research focused on collecting information about PCs from desert plants and their nematicidal activity and the activity of some PCs against PPNs.

7.2 NEMATODES

Nematodes are a phylum of pseudocelomate organisms with long, narrow bodies that are threadlike without segments; they are round or cylindrical worms due to the form of their body (Kiontke and Fitch, 2013). Nematodes are aquatic organisms, although they also proliferate in terrestrial environments where they represent 80% of pluricellular organisms and up to 90% in marine environments. Many species of nematodes exist and the free-living species are the most numerous, ubiquitous component with ecological importance in ecosystems because they interfere in degradation and recycle organic matter for compliance of biologic cycles in ecosystems. On the other hand, the parasitic species of nematode causes diseases to animals, including humans, and plants. Nevertheless, the parasitic species are considered a minor group compared to the free-living species (Danovaro et al., 2008).

7.2.1 Plant-parasitic Nematodes

Although the number of parasitic nematodes species is minor, the damage that they cause to animals, humans, and plants can be devastating. The significant damage and adverse effects caused by PPNs are related to agricultural production. PPNs are specialised parasites that infect plants through several strategies, with their principal objective being the plant's roots, although they can also be present on other plant organs. PPNs have a needle-like structure called a stylet, which allows them to damage the plant's cells as they feed. However, the stylet is also used to introduce substances such as proteins and metabolites to host tissue that allows the parasitism of plants (Haegeman et al., 2012). The stylet is connected to the pharynx, a specialised muscle that allows expansion and contraction of the esophagus. Muscular movements transfer the food of PPNs toward their intestine through the stylet and expel substances from their salivary glands to the plant cells. After moving through the intestine, the food ends in the rectum in the female and cloaca in male nematodes (Lambert and Bekal, 2002).

Usually, male and female PPNs have different characteristics, although species with asexual reproduction by parthenogenesis can exist. In females, the main visible reproductive organs are the ovaries and the vulva, which can be used for species identification under the microscope, in contrast, male nematodes reproductive structures include one or two testes and a structure called a spicule that is a reproductive organ that guides the sperm into the vagina of females (Lambert and Bekal, 2002).

The life cycle of PPNs species is similar; the cycle starts with the egg stage in which the embryo develops until the first juvenile stage, or J1; depending on the specie, the J1 can hatch from the egg or develop in the egg until the J2 stage and hatch. P.P.N.s usually wait to hatch until J2 stage; the other stages are J3, J4, and adult, at which point they molt at each one (Lawrence and Lawrence, 2020). Most P.P.N.s develop their life cycle in plant roots, although species grow and develop in other parts of the plant, such as leaves and other aerial parts, depending on the type of feeding and life cycle.

There are four ways for PPNs to infect and develop their life cycle in plants; the first is the ectoparasites. This class of nematodes grows outside of the roots and uses the stylet to damage and penetrate the plant's roots to obtain nutrients; one example is the genus *Xiphinema*. The second class is the semi-endoparasites; the PPNs with this feeding type grow outside of the plant's roots, and when they reach the infective stage, they penetrate the root cells and form a permanent feeding cell where the nematodes will swell, losing their capacity to move (i.e., *Rotylenchulus reniformis, Tylenchulus semipenetrans*). The third feeding type is the migratory endoparasites. This class of PPNs hatch outside, and they can penetrate the root tissue in all developing stages; once inside the roots, they migrate through them, destroying them and getting nutrients from the plant cells, causing tissue necrosis. Some genus with this feeding type are *Pratylenchus, Radopholus,* and *Hirschmanniella*. The fourth feeding type is the sedentary endoparasite. The nematodes with this feeding class are embedded in roots where they hatch and develop to the first stage (J1), at which point they leave the roots and molt in the soil, thus reaching the infective stage (J2). They inject secretions into the roots to form a syncytial feeding cell where they continue in their life cycle development stages until they become adults; the adult males leave the roots again, but the females stay in roots where they take a saclike shape and lay a large number of eggs. The PPNs with this feeding type are considered to be the most destructive to plants and include genus such as *Meloidogyne, Heterodera,* and *Globodera* (Lambert and Bekal, 2002)

7.2.2 Main Species of PPNs in Agriculture

PPNs are a significant phytosanitary problem in agriculture that is present in all soils globally; it is estimated that PPNs causes losses of $157 billion per year, affecting the yield and quality of agricultural products (Ferreira-Barros et al., 2021). There are currently reports of more than to 4,100 species of PPNs; nevertheless, some species are considered the main causes of economic losses on crops. According to experts from around the world, the following species are the most critical in agriculture around the world: *Meloidogyne* spp., *Heterodera* spp., *Globodera* spp., *Pratylenchus* spp., *Radopholus similis, Ditylenchus dipsaci, R. reniformis,* and *Aphelenchoides besseyi* (Jones et al., 2013).

Meloidogyne spp., known as root-knot nematodes, are found worldwide. The genus consists of 98 described species; the hosts of these PPNs include a vast number of plant species, such as avocado, alfalfa, cotton, amaranth, peanut, pumpkin, coffee tree, onion, chili, cabbage, peach, strawberry, bean, chickpea, corn, apple tree, melon, banana, potato, papaya, watermelon, tobacco, tomato, and vine. The females of these species lay the eggs in a gelatinous matrix into the roots galls or around the roots. The main symptoms of PPNs in affected plants are the formation of giant cells, galls, necrosis, a decrease of root growth, and decreased plant development (Cepeda-Siller, 1996; Jones et al., 2013).

Heterodera spp. and *Globodera* spp. are another group of PPNs that affect many crops worldwide; these genera are known as cyst nematodes, and their main characteristic is the capacity to form cysts on the roots. Cyst formation occurs when the females insert their heads into cells and produce a syncytial feeding cell. Once fixed, the females swell, and the males fertilise them to produce viable eggs; finally, the females die and suffer a process that changes their cuticle. making it hard, forming the cysts that can be viable for 10 years. The symptoms generally appear on plantations, and the affected plants suffer chlorosis and dwarfism (Cepeda-Siller, 1996; Ibrahim et al., 2017).

Pratylenchus spp., *R. similis,* and *D. dipsaci* cause severe damage in plants, specifically on roots. *Pratylenchus* spp., known as the root-lesion nematode, is a cosmopolite genus that attacks a vast variety of plant species; juveniles and adults can enter and leave the roots damaging them. The females lay the eggs into the roots or in the rhizosphere. the main symptoms of infection are necrotic roots and reduced root growth (Cepeda-Siller, 1996). *R. similis* is an endoparasitic nematode that parasitises peppers, citrus and banana crops, causing significant losses. All stages of development for this species occurs within the roots. The females lay eggs in the roots or in the soil if the roots are overcrowded. Usually, the reproduction is sexual, but in the absence of males or females, nematodes can be hermaphrodites. The symptoms produced in roots consist of dark lesions, and trees can fall due to the loss of their root system (Jones et al., 2013).

D. dipsaci, classified as stem and bulb nematode, can affect more than 500 plant species. Nevertheless, the crops studied for economic problems are onion, garlic, potato, strawberry, oat, and alfalfa. These PPNs are migratory endoparasites and are considered quarantined organisms by several territories in the world. *D. dipsaci* develop their first stage within the egg, hatching in the J2 stage; the infective of this specie is the adult stage. In the adult stage, this species secretes enzymes that smooth cells allowing feeding. The main symptoms of plant infection are discolored bulbs, twisted stems, stunted growth, and rot of bulbs (Cepeda-Siller, 1996; Jones et al., 2013; Mimee et al., 2019).

R. reniformis and *Xiphinema index* are other species of PPNs that considered economically important in agriculture because of the damage and losses they cause to crops. *R. reniformis,* also called the reniform nematode, is a sedentary semi-endoparasite nematode that affects important vegetables, ornamentals, and fruits. The reniform nematodes hatch from the eggs in the J2 stage; once hatched, the juveniles suffer three molts before reaching the J3 stage, at which point they are smaller than when they entered into the J2 stage. Infection is caused only by juvenile females, which penetrate the roots and insert their head in a root cell, forming a feeding site similar to the syncytial site of cyst nematodes; subsequently, the posterior part of the body is swelled and stays outside of the roots, assuming a kidney shape. The males fertilise the eggs, and they are trapped into the female. The symptoms caused by these PPNs are root necrosis, decreased root growth, chlorotic leaves, and egg sacs that may be on the root surface (Jones et al., 2013). On the other hand, *A. besseyi* is another essential species of nematode that causes white tip disease on rice crops; this species is seed-borne and can survive on stored seeds in anhydrobiosis for many years, reactivating when the seeds are rehydrated. This species' reproduction is amphimictic because the males are more numerous than the females, although they can reproduce through parthenogenesis. An attractive characteristic of these PPNs is that they can feed on fungi when their host is not present; the main symptom is chlorotic tips on leaves and reduced grain production (Jones et al., 2013; Wang et al., 2014).

7.3 DESERT PLANTS

The desert is one of the more extensive ecosystems globally, accounting for one-third of the global land surface. Although it is a biome with dry weather and scarce precipitation, around the world, 4% is considered highly arid, 15% arid, and 14.6% semi-arid. This classification may be derived through several approaches including the aridity index, which is a widely used classification developed by the United Nations Environmental Program (Laity, 2009; Pointing and Belnap, 2012). The aridity present in deserts causes extreme temperatures, where the diurnal temperature can exceed 45 °C but fall below 0 °C at night; such temperature variations have a destructive effect on the desert soil, inducing an effect known as weathering, through which the rocks in the soil rupture, creating small particles that form the desert dust and sand. These environmental factors significantly influence desert organisms such as microorganisms, animals, and plants.

Desert plants present several strategies to adapt to their environment due to the problematic conditions these plants have to survive. One of these strategies is known as drought escaping; plants with this strategy generally are annual plants that are ephemeral and grow when there is sufficient environmental water or moisture. Another strategy desert plants implement is drought evading. Plants with this strategy modified their anatomical structures and physiological responses to avoid periods of limited moisture. These plants, such as riparian trees (e.g., *Acacia erioloba* and *Boscia albitrunca*) and stem succulents (e.g., Cactaceae and Agavaceae), develop deep roots and modify lost water by stomatal control and crassulacean acid metabolism photosynthesis (Ward, 2016). The other strategy that desert plants use is drought enduring. These plants make a rapid gas exchange and shed their leaves when drought occurs. Drought resisting plants with this feature modulate their gas exchange depending on water availability, small vertical leaves, and low hydraulic conductivity in xylem (e.g., *Larrea tridentate, Zygophyllum dumosum,* and *Anabasis articulate)* (Smith et al., 1997).

Desert plants structural and physiological adaptations allow them to survive and produce secondary metabolites known as PCs. These compounds carry out critical cellular functions in physiological processes and signal defense and stress responses (Isah, 2019). Desert plants, such as those mentioned previously, offer a considerable number of PCs due to the challenging environmental conditions that allow the biosynthesis of unique compounds that can be used in agricultural diseases management caused by microorganisms such as PPNs (Ofir, 2020).

7.4 PHYTOCHEMICAL COMPOUNDS FROM DESERT PLANTS WITH NEMATICIDAL ACTIVITY

Around the world, several studies have demonstrated desert plants' capacity to produce a considerable number of PCs as part of their metabolisms; these PCs are present when the desert plants are processed and submitted to an extraction process to obtain plant extracts (Ofir, 2020). Recently, the PCs have taken relevance in agriculture, as many studies have demonstrated their biological activity, which may offer an alternative for organic farming.

In the study of PCs, the biosynthesis of compounds is an essential process due to existing pathways through which the PCs can be produced. Phenols, tannins. and aromatic alkaloids are produced by the shikimic acid pathway, while terpenes, steroids, and some alkaloids are generated through the mevalonic acid pathway (Dewick, 2002). PCs can be classified by their structure in terpenoids (e.g., gibberellic acid, steroids, and carotenoids), which are derivate compounds from isoprene and have insecticidal, allelopathy, and hormones functions in plants. Alkaloids are nitrogen compounds that can act as enzyme inhibitors, ion channel blockers, or neurotransmission interference. Phenolic compounds, which are the largest group of PCs synthesised by plants, present diverse structures and have hydroxylated aromatic rings in common. Flavonoids which can be monomers, dimmers, and higher oligomers, are compounds that can be divided into three classes: Anthocyanins, flavones, and flavonols, and their main functions are to protect the plants from UV irradiation, chemical messengers, physiological regulators, and act as pigments in diverse organs of the plants. Tannins are phenolics formed by oligomers and polymers and are divided into two groups: Hydrolyzable tannins (e.g., ellagitannins, gallotannins, and complex tannins) and condensed tannins. Saponins are compounds whose aglycone is a triterpenoid or steroidal structure; because of the lipophilic sugars in their structure, the saponins can act as a detergent on membranes. Within their biological properties are the capacity to act as antioxidants, antimicrobials, and insecticides. Finally, glycosides, can be sulfur compounds, alcohol, or phenol. Many plants that produce glycosides store inactive P.C.s and activate them when interacting with enzymes (Kabera et al., 2014; Tiwari and Rana, 2015).

PCs can be obtained from desert plants using extraction techniques obtaining natural products with biological activities. For example, in the case of agriculture, plant extracts and essential oils that contain concentrated PCs can be obtained (Tucuch et al., 2021); in this context, many researchers have identified PCs from desert plants with antimicrobial activity capable of controlling PPNs that cause diseases and problems in many crops around the world. Table 7.1 presents some desert plants and the phytochemical with nematicidal activity.

TABLE 7.1 PHYTOCHEMICAL COMPOUNDS FROM DESERT PLANTS WITH NEMATICIDAL ACTIVITY

Source of Phytochemical Compound	Compound	Reference
Abutilon pannosum	Alkaloids, phenolic compounds, flavonoids, glycosides, and saponins	Deora and Bano (2019)
Achillea fragrantissima	Flavonoids	Elmann et al. (2015)
A. fragrantissima	Flavonoids, alkaloids, tannins, saponins, phenolic compounds	El-Sherbiny et al. (2010)
Aloe vera	Tannins, flavonoids, terpenoids, and alkaloids	Saadoon et al. (2019)
A. mexicana	Alkaloids, flavonoids, saponins, and tannins	Khan et al. (2017)
Atriplex halimus	Alkaloids, saponins, glycosides, and flavonoids	Abdelnabby and Abdelrahman (2012)
Brassica sinaica	Glycosides and phenolic compounds	El-Sherbiny et al. (2010)
Carduus getulus	Alkaloids, saponins, glycosides, and flavonoids	Abdelnabby and Abdelrahman (2012)
C. colocynthis	Phenolic compounds	Rizvi and Shahina (2014)
Larrea divaricata	Phenolic compounds	Gómez et al. (2021)
Origanum vulgare	Phenolic compounds	Oka et al. (2000)
Phoenix dactylifera	Glycosides, tannins, flavonoids, alkaloids, and saponins	Alam and El-Nuby (2019)
Quillaja saponari	Saponins, glycosides, phenolic compounds, and tannins.	Insunza et al. (2001)
Thymus vulgari	Phenolic compounds	Oka et al. (2000)
Trichodesma africanum	Phenolic compounds, flavonoids, and alkaloids	El-Sherbiny et al. (2010)

7.4.1 Biological Effectiveness Studies of Phytochemical Compounds from Desert Plants with Nematicidal Activity

Biological effectiveness studies are necessary to develop alternatives for managing diseases caused by PPNs to understand and determine how the tested substances act against the P.P.N.s. Biological effectiveness studies determine how effective PCs can be, using many methods and evaluating variables such as mortality percentage, hatching percentage, mobility of nematodes, and nematostatic effect. The bioactive compounds from desert plants have a considerable potential to be an alternative against PPNs; therefore, researchers worldwide have carried out biological effectiveness studies with PCs from desert plants against PPNs (see Table 7.1). For example, El-Sherbiny et al. (2010) tested P.C.s obtained from *A. fragrantissima, B. sinaica*, and *T. africanum* against *Meloidogyne javanica* on common bean plants and observed reduction of disease percentage of 68%, 61%, and 96%, respectively. In another study, Saadoon et al. (2019) used an extract of *A. vera* to control *M. incognita* in cucumber plants; the result obtained demonstrated a significant reduction of the number of nematodes. On the other hand, Khan et al. (2017) made an extract from *A. mexicana*, and they applied it on *M. incognita in vitro*, determining the potential of this extract to control *M. incognita* due to the high nematicidal effect and the effect on the hatching of eggs.

In another study, Abdelnabby and Abdelrahman (2012) tested a methanolic extract from *A. halimus* and *C. getulus in vitro* against *M. incognita* in the J2 stage using the microwell process and observed nematicidal activity of 20% and 34%, respectively. In Africa, specifically in Egypt, there are many plants of interest, such as *C. colocynthis*, which was used by Rizvi and Shahina (2014) to obtain a methanolic extract that was tested on *Meloidogyne* spp.; results reported mortalities of 60%, 80%, and 100% with the extract from the pulp, fruit, and creeper of *C. colocynthis*, respectively. The resins from the plants can also be used in products with P.C.s. Gómez et al. (2021) produced resin from *L. divaricate* and applied it on *M. incognita* and observed the nematicidal effect of the resin. The *P. dactylifera*, known as date palm, is a common plant in the desert that has been used to obtain PCs. Alam and El-Nuby (2019) obtained an aqueous extract from this plant and tested it against *M. incognita*; the result was interesting as a mortality of 90% of *M. incognita* in J2 stage was reported.

7.4.2 Action Mode of Phytochemical Compounds with Nematicidal Activity

The action mode of substances refers to how the molecules interact inside the organism with receptors, enzymes, neurotransmitters, or other molecules that, when they are in contact with PCs of interest, allow affecting vital physiological processes for the life of organisms. Although elucidating the action mode of substances is complicated, as researching proteomic, genetic, enzymatic, and other areas is necessary, many research centers and companies invest in these studies because understanding the action mode of molecules in organisms allows the development of new products and alternatives, such as PCs. In the case of PCs with nematicidal activity from desert plants, there are not enough studies that elucidate the action mode of PPNs; nevertheless, some studies try to elucidate the action mode of other promising PCs with nematicidal activity (see Table 7.2). For example, α-terthienyl is considered a promising control of PPNs, suggesting that it can reduce oxidative stress inhibiting detox enzymes such as glutathione-s-transferase and superoxide dismutase in the hypodermis of PPNs (Hamaguchi et al., 2019). In several studies, carvacrol and thymol demonstrated an effect on tyramine receptors in PPNs—explicitly interacting with the SER-2

TABLE 7.2 ACTION MODE OF PHYTOCHEMICAL COMPOUNDS WITH NEMATICIDAL ACTIVITY

Phytochemical Compound	Action Mode	Reference
α-terthienyl	Induction of oxidative stress by inhibition of detox enzymes	Hamaguchi et al. (2019)
Carvacrol and thymol	Affectation of tyramine receptors	Andrés et al. (2012)
Geraniol and citronellol	Affectation of the cellular membrane and ionic channels	Andrés et al. (2012)
Eugenol	Inhibition of free radicals and reactive species involved in toxic effects, inhibition of acetylcholinesterase, and affect octopamine receptors	D'Addabbo and Avato (2021)
Cinnamaldehyde	Ovicidal effect and in the hatching of eggs.	Ferreira-Barros et al. (2021)
Phenylpropanoids	Induction of oxidative stress	Eloh et al. (2020)
Tannins, flavonoid, alkaloids, and saponins	Dissolution of the cytoplasmic membrane, interference with the structure of enzymes and electron flow of respiratory chain and the ADP phosphorylation, and affectation of embryonic development.	Saadoon et al. (2019); Asif et al. (2014)
Glucosinolates and isothiocyanates	Inhibition of enzymes	Avato et al. (2013)
Naphthoquinones	Induction of formation of reactive oxygen species with toxic effects	Maleita et al. (2017)
Ricin and ricinus agglutinin	Modification of behavior chemotactic due to the adhering of phasmids	Arboleda et al. (2012)
Isothiocyanates	Affectation of ADN by oxidative damage	Wu et al. (2011)

receptor. In another way, geraniol and citronellol can affect the cellular membrane and ionic channels, disrupting intracellular signaling pathways and altering proteins from the membrane (Andrés et al., 2012). The eugenol and the cinnamaldehyde are also PCs with biological activity. In the case of eugenol, the nematicidal effect is due to the allylic double bond and the phenolic proton that interact with cellular systems as an antioxidant, inhibiting free radicals and reactive species involved in toxic effects of acetylcholinesterase as a blocker of octopamine receptors affecting the nervous system of PPNs. In addition, cinnamaldehyde has an ovicidal effect and stops egg from hatching due to its effect on the permeability of the membrane (D'Addabbo and Avato, 2021; Ferreira-Barros et al., 2021).

Other PCs are the phenylpropanoids synthesised from phenylalanine and tyrosine amino acids. These compounds induce oxidative stress in PPNs, increasing lipid peroxidation levels and reducing antioxidant activity (Eloh et al., 2020). Furthermore, the nematicidal effect of PCs is attributed to lipophilic properties of oxygenated compounds such as tannins, flavonoids, alkaloids, and saponins, which are capable of dissolving the cytoplasmic membrane of PPNs, interfering with the enzyme protein structure and with the electron flow of the respiratory chain and the ADP phosphorylation. Furthermore, these compounds destroy the egg masses, killing the eggs and affecting the embryonic development, preventing them from hatching (Asif et al., 2014; Saadoon et al., 2019). Finally, the glucosinolates and isothiocyanates are PCs standard in the secondary metabolism of plants from the Brassicaceae family; the action mode proposed to these compounds is that they can react with biological nucleophiles essential to PPNs, provoking the inhibition of enzymes (Avato et al., 2013).

7.5 BENEFITS OF PHYTOCHEMICAL COMPOUNDS TO THE MANAGEMENT OF PLANT-PARASITIC NEMATODES

PPNs are microorganisms that affect all crops around the world. Although an infestation is rarely fatal for a crop, the presence of these organism reduces the yield and quality of agricultural products. Because of the aforementioned reasons, farmers need alternative solutions to manage PPNs, as the chemical products primarily in use are toxic to human health and cause environmental damage. Therefore, the farmers and consumers are conscious of changing from chemical to sustainable agriculture (Clemensen et al., 2020; Desmedt et al., 2020). This change of paradigm can be achieved through many alternatives—the use of PCs to control PPNs, being an ecofriendly and efficient option. Furthermore, the use of PCs to manage PPNs has benefits that can improve the agroecosystem (Godlewska et al., 2021).

One of the main components in agricultural production is the soil; soil health is essential because the balance of the soil components allows the plant to have optimal requirements for them to develop. However, nematicides are applied directly in the soil, which is one cause of component soil losses. Because of these problems, it is necessary to develop alternatives that do not affect soil components and can improve their structure. Thus, PCs arise as alternatives with benefits to the soil; an example is the flavonoids from *Medicago sativa* that induce the growth of N-fixing bacterium as *Sinorhizobium meliloti*; also, PCs as tannins increase the movilization of N in the soil or the alkaloids that induce resistance against water stress in plants that develop in drought conditions (Clemensen et al., 2020; Sarker and Oba, 2018; Yang et al., 2018). Plants treated with PCs obtain benefits such as the attraction of pollinators, induction of resistance or direct protection against other phytopathogens, and protection against UV irradiation (Clemensen et al., 2020).

All the benefits obtained from using PCs in agriculture allow food production to focus on organic and sustainable agriculture, allowing healthy production without affecting the environment or human health. Furthermore, the resistance of PPNs to active ingredients is null when they are treated with PCs, in that the control is equal or better than the chemical nematicides. Therefore, several companies and research centers focus on resources to obtain a new alternative based on PCs each year in sustainable agriculture. Moreover, there is high demand from the farmers around the world for organic products; also, many governments are enacting legislation to lead to sustainable food production to obtain all the benefits of this kind of agriculture for the human population (Dhakal and Singh, 2019).

7.6 CONCLUSION

Without good management, PPNs are a phytosanitary problem that can cause severe problems. One problem with PPNs is the lack of information because most studies focused on controlling other phytopathogenic microorganisms. Furthermore, the majority of farmers use chemical products to control the diseases caused by PPNs, but the problems of chemical agriculture have caused them to opt for sustainable products, such as PCs from plants. The desert plants are a source of many PCs due to the environmental conditions these plants tolerate, which has allowed the development of several PCs as part of their secondary metabolism.

Thus, the exploitation of this kind of PCs is an option to develop new alternatives against PPNs; likewise, elucidating the action mode of these PCs is also essential because it helps us to understand how the bioactive compounds are acting in the PPNs, so it is necessary to continue researching the action mode of PCs. Currently, organic farming is growing because of the environmental problems caused by chemical products; this has brought a paradigm change that involves a change in production systems to green technologies, which have less impact on the environment. The natural and biological product market is increasing. It is estimated that the biopesticides market has reached US$4 billion, Principal reasons for selecting organic farming include getting a better yield and better resistance management, managing residues safety, biodegradability, reentry of workers in short times (Marrone, 2019). Hence, it is essential to continue studying plants from the desert because these plants can be the alternatives that help in the move to sustainable agriculture.

REFERENCES LIST

Abdelnabby, H.M., & Abdelrahman, S.M. (2012). Nematicidal activity of selected flora of Egypt. *Egyptian Journal of Agronematology, 11*(1), 106–124.

Agrios, G.N. (1997). *Plant Pathology*. San Diego. Academic Press.

Alam, E.A., & El-Nuby, A.S.M. (2019). Phytochemical and antinematodal screening on water extracts of some plant wastes against *Meloidogyne incognita*. *International Journal of Chemical and Pharmaceutical Sciences, 10*, 1–16. http://www.ijcps.com/sub_pages/vol10issue4.html

Al-Saleem, M.S., Awaad, A.S., Alothman, M.R., & Alqasoumi, S.I. (2018). Phytochemical standardization and biological activities of certain desert plants growing in Saudi Arabia. *Saudi Pharmaceutical Journal, 26*(2), 198–204. doi: 10.1016/j.jsps.2017.12.011

Andrés, M.F., González-Coloma, A., Sanz, J., Burillo, J., & Sainz, P. (2012). Nematicidal activity of essential oils: A review. *Phytochemistry Reviews, 11*(4), 371–390. doi: 10.1007/s11101-012-9263-3

Arboleda, F.D.J., Guzmán, Ó.A., & Mejía, L.F. (2012). Efecto de extractos cetónicos de higuerilla (*ricinus communis*) sobre el nematodo barrenador *Radopholus similis* cobb. en condiciones *in vitro*. *Luna Azul, 35*, 28-47. http://www.scielo.org.co/scielo.php?pid=S1909-24742012000200003andscript=sci_abstractandtlng=es

Asif, M., Parihar, K., Rehman, B., Ashraf Ganai, M., Usman, A., & Siddiqui, M.A. (2014). Bio-efficacy of some leaf extracts on the inhibition of egg hatching and mortality of *Meloidogyne incognita*. *Archives of Phytopathology and Plant Protection, 47*(8), 1015–1021. doi: 10.1080/03235408.2013.829626

Avato, P., D'Addabbo, T., Leonetti, P., & Argentieri, M.P. (2013). Nematicidal potential of Brassicaceae. *Phytochemistry Reviews, 12*(4), 791-802. doi: 10.1007/s11101-013-9303-7

Burns, A.R., Bagg, R., Yeo, M., Luciani, G.M., Schertzberg, M., Fraser, A.G., & Roy, P.J. (2017). The novel nematicide wact-86 Interacts with aldicarb to kill nematodes. *PLoS Neglected Tropical Diseases, 11*, e0005502. doi: 10.1371/journ al.pntd.0005502

Cepeda-Siller, M. (1996). *Nematologia Agricola*. Mexico: Trillas.

Clemensen, A.K., Provenza, F.D., Hendrickson, J.R., & Grusak, M.A. (2020). Ecological implications of plant secondary metabolites-phytochemical diversity can enhance agricultural sustainability. *Frontiers in Sustainable Food Systems*, 4, 547826. doi: 10.3389/fsufs.2020.547826

D'Addabbo, T., & Avato, P. (2021). Chemical composition and nematicidal properties of sixteen essential oils: A review. *Plants, 10*(7), 1368. doi: 10.3390/plants10071368

Danovaro, R., Gambi, C., Dell'Anno, A., Corinaldesi, C., Fraschetti, S., Vanreusel, A., & Gooday, A.J. (2008). Exponential decline of deep-sea ecosystem functioning linked to benthic biodiversity loss. *Current Biology, 18*(1), 1–8. doi:10.1016/j.cub.2007.11.056

Deora, G.S., & Bano, I. (2019). Preliminary phytochemical screening and GC-MS analysis of methanolic leaf extract of *Abutilon pannosum* (Forst. F.) Schlect. from Indian Thar desert. *Journal of Pharmacognosy and Phytochemistry, 8*(1), 894-899. https://www.phytojournal.com/archives?year=2019andvol=8andissue=1andArticleId=6866

Desmedt, W., Mangelinckx, S., Kyndt, T., & Vanholme, B. (2020). A phytochemical perspective on plant defense against nematodes. *Frontiers in Plant Science*, 11, 602079. doi:10.3389/fpls.2020.602079

Dewick, P.M. (2002). *Medicinal Natural Products: A Biosynthetic Approach*. New York: John Wiley & Sons.

Dhakal, R., & Singh, D.N. (2019). Biopesticides: A key to sustainable agriculture. *International Journal of Pure and Applied Bioscience, 7*(3), 391–396. doi: 10.18782/2320-7051.7034

Elmann, A., Telerman, A., Mordechay, S., Erlank, H., Rindner, M., Kashman, Y., & Ofir, R. (2015). Downregulation of microglial activation by achillolide A. *Planta Medica, 81*(03), 215–221. doi: 10.1055/s-0034-1396204

Eloh, K., Kpegba, K., Sasanelli, N., Koumaglo, H.K., & Caboni, P. (2020). Nematicidal activity of some essential plant oils from tropical West Africa. *International Journal of Pest Management, 66*(2), 131–141. doi: 10.1080/09670874.2019.1576950

El-Sherbiny, A.A., & Al-Yahya, F.A. (2010). Efficacy of thirteen species of wild flora as soil amendments in the control of the root-knot nematode, *Meloidogyne javanica* on common bean in Saudi Arabia. *Alexandria Science Exchange Journal, 31*, 371–379. doi: 10.21608/ASEJAIQJSAE.2010.2333

Ferreira-Barros, A., Paulo Campos, V., Lopes de Paula, L., Alais Pedroso, L., de Jesus Silva, F., Carlos Pereira da Silva, J. and Humberto Silva, G. (2021). The role of *Cinnamomum zeylanicum* essential oil, (E)-cinnamaldehyde and (E)-cinnamaldehyde oxime in the control of *Meloidogyne incognita*. *Journal of Phytopathology, 169*(4), 229–238. doi: 10.1111/jph.12979

Godlewska, K., Ronga, D., & Michalak, I. (2021). Plant extracts-importance in sustainable agriculture. *Italian Journal of Agronomy, 16*(2). doi: 10.4081/ija.2021.1851

Gómez, J., Simirgiotis, M.J., Manrique, S., Piñeiro, M., Lima, B., Bórquez, J., & Tapia, A. (2021). UHPLC-ESI-OT-MS phenolics profiling, free radical scavenging, antibacterial and nematicidal activities of "Yellow-Brown Resins" from Larrea spp. *Antioxidants, 10*(2), 185. doi: 10.3390/antiox10020185

Haegeman, A., Mantelin, S., Jones, J.T., & Gheysen, G. (2012). Functional roles of effectors of plant-parasitic nematodes. *Gene, 492*(1), 19-31. doi: 10.1016/j.gene.2011.10.040

Hamaguchi, T., Sato, K., Vicente, C.S., & Hasegawa, K. (2019). Nematicidal actions of the marigold exudate α-terthienyl: oxidative stress-inducing compound penetrates nematode hypodermis. *Biology Open, 8*(4), bio038646. doi: 10.1242/bio.038646

Ibrahim, I.K.A., Handoo, Z.A., & Basyony, A.B.A. (2017). The cyst nematodes *Heterodera* and *Globodera* species in Egypt. *Pakistan Journal of Nematology, 35*(2), 151–154. doi:10.18681/pjn.v35.i02.p151-154

Insunza, V.I.O.L.E.T.A., Aballay, E., & Macaya, J. (2001). In vitro nematicidal activity of aqueous plant extracts on Chilean populations of Xiphinema americanum sensu lato. *Nematropica*, 47–54. https://journals.flvc.org/nematropica/article/view/69612

Isah, T. (2019). Stress and defense responses in plant secondary metabolites production. *Biological Research, 52*. doi: 10.1186/s40659-019-0246-3

Jones, J.T., Haegeman, A., Danchin, E.G., Gaur, H.S., Helder, J., Jones, M.G., & Perry, R.N. (2013). Top 10 plant-parasitic nematodes in molecular plant pathology. *Molecular Plant Pathology, 14*(9), 946–961. doi: 10.1111/mpp.12057

Kabera, J.N., Semana, E., Mussa, A.R., & He, X. (2014). Plant secondary metabolites: biosynthesis, classification, function and pharmacological properties. *Journal of Pharmacy and Pharmacology, 2*(7), 377–392.

Khan, A., Asif, M., Tariq, M., Rehman, B., Parihar, K., & Siddiqui, M.A. (2017). Phytochemical investigation, nematostatic and nematicidal potential of weeds extract against the root-knot nematode, *Meloidogyne incognita in vitro. Asian Journal of Biological Sciences, 10*, 38–46. doi: 10.3923/ajbs.2017.38.46

Kiontke, K., & Fitch, D.H. (2013). Nematodes. *Current Biology, 23*(19), R862–R864. doi: 10.1016/j.cub.2013.08.009

Laity, J. (2009). *Deserts and Desert Environments*. Chichester, UK. Wiley-Blackwell.

Lambert, K., & Bekal, S. (2002). Introduction to plant-parasitic nematodes. *The plant Health instructor*, Chichester: *10*, 1094-1218. doi: 10.1094/PHI-I-2002-1218-01

Lawrence, K.S., & Lawrence, G.W. (2020). Plant-parasitic nematode management. In J. Bergtold and M. Sailus (Eds.), *Conservation tillage systems in the southeast* (164–180 p). SARE. www.sare.org/conservation-tillage-in-the-southeast.

Maleita, C., Esteves, I., Chim, R., Fonseca, L., Braga, M.E., Abrantes, I., & de Sousa, H.C. (2017). Naphthoquinones from walnut husk residues show strong nematicidal activities against the root-knot nematode *Meloidogyne hispanica*. *ACS Sustainable Chemistry and Engineering, 5*(4), 3390–3398. doi: 10.1021/acssuschemeng.7b00039

Marrone, P.G. (2019). Pesticidal natural products–status and future potential. *Pest Management Science, 75*(9), 2325–2340. doi: 10.1002/ps.5433

Mimee, B., Lord, E., Véronneau, P.Y., Masonbrink, R., Yu, Q., & Eves-van den Akker, S. (2019). The draft genome of *Ditylenchus dipsaci*. *Journal of Nematology, 51*. doi: 10.21307/jofnem-2019-027

Ofir, R. (2020). "Desert Chemotypes": The potential of desert plants-derived metabolome to become a sustainable resource for drug leads. *Medical Research Archives, 8*(7), 1–15. doi: 10.18103/mra.v8i7.2169

Oka, Y., Nacar, S., Putievsky, E., Ravid, U., Yaniv, Z., & Spiegel, Y. (2000). Nematicidal activity of essential oils and their components against the root-knot nematode. *Phytopathology, 90*(7), 710–715. doi: 10.1094/PHYTO.2000.90.7.710

Pointing, S.B., & Belnap, J. (2012). Microbial colonization and controls in dryland systems. *Nature Reviews Microbiology, 10*(8), 551–562. doi: 10.1038/nrmicro2831

Rizvi, T.S., & Shahina, F. (2014). Nematicidal activity of *Citrullus colocynthis* extracts against root-knot nematodes. *Pakistan Journal of Nematology, 32*(1), 101–112. http://www.pjn.com.pk/papers/1508494058.pdf

Saadoon, S.M., Jabbar, A.S., & Gad, S.B. (2019). Efficiency of using magnetized water in improving *Meloidogyne incognita* control by three concentration of *Aloe vera* extract on cucumber plant. *Plant Archives, 19*(1), 721–727. http://plantarchives.org/PDF%2019-1/721-727%20(4758).pdf

Sarker, U., & Oba, S. (2018). Drought stress enhances nutritional and bioactive compounds, phenolic acids and antioxidant capacity of *Amaranthus* leafy vegetable. *BMC Plant Biology, 18*, 258. doi: 10.1186/s12870-018-1484-1

Smith, S.D., Anderson, J.E., & Monson, R.K. (1997). *Physiological Ecology of North American Desert Plants*. Berlin: Springer. doi: 10.1007/978-3-642-59212-6

Tiwari, R., & Rana, C.S. (2015). Plant secondary metabolites: A review. *International Journal of Engineering Research and General Science, 3*(5), 661–670. http://pnrsolution.org/Datacenter/Vol3/Issue5/82.pdf

Tucuch-Pérez, M.A., Arredondo-Valdés, R., Laredo-Alcalá, E.I., Alvarado-Canche, C.N., & Castillo, F.D.H. (2021). A review of nano and microencapsulate of phytochemical compounds for diseases management in agriculture. *Acta Agrícola y Pecuaria, 7*(1). doi: 10.30973/aap/2021.7.0071005

Wang, F., Li, D., Wang, Z., Dong, A., Liu, L., Wang, B., & Liu, X. (2014). Transcriptomic analysis of the rice white tip nematode, *Aphelenchoides besseyi* (Nematoda: Aphelenchoididae). *PloS One, 9*(3), e91591. doi: 10.1371/journal.pone.0091591

Ward, D. (2016). *The biology of deserts*. Oxford: Oxford University Press.

Wu, H., Wang, C.J., Bian, X.W., Zeng, S.Y., Lin, K.C., Wu, B., ... & Zhang, X. (2011). Nematicidal efficacy of isothiocyanates against root-knot nematode *Meloidogyne javanica* in cucumber. *Crop Protection, 30*(1), 33–37.doi: 10.1016/j.cropro.2010.09.004

Yang, L.,Wen, K.-S., Ruan, X., Zhao, Y.-X.,Wei, F., & Wang, Q. (2018). Response of plant secondary metabolites to environmental factors. *Molecules 23*, 762. doi: 10.3390/molecules23040762

Chapter 8

Plant Phytochemicals from the Chihuahuan Semi-desert with Possible Herbicidal Actions

Alisa Clementina Barroso-Ake, Roberto Arredondo-Valdés, Rodolfo Ramos-González, Elan Iñaki Laredo-Alcalá, Cristóbal Noé Aguilar-González, Juan Alberto Ascacio-Valdés, Mayela Govea, Anna Iliná, and Marco Antonio Tucuch-Peréz

CONTENTS

8.1 INTRODUCTION

Crop production is an essential component of agriculture, as it is the primary human activity that provides food security in the world. Effective crop production begins with managing phytosanitary problems. However, one of the farmers main problems is controlling and managing weeds. During the past 60 years, the excessive use of herbicides for the protection of plants in agriculture has caused a negative impact at different levels, causing secondary effects such as the presence of residues in food, high toxicity in humans, as well as in non-target organisms, and the emergence of resistance in different weed species. These events have helped reduce the inappropriate consumption of chemically synthesised herbicides in developing nations, giving rise to new safe, and innovative strategies (Rob et al., 2020; Seiber et al., 2014; Pvalea et al., 2019).

Various natural, plant-derived components have been accepted as an environmentally friendly alternative. However, farmers have first relied on highly toxic synthetic herbicides. The new alternatives in the control and management of weeds are based on bioherbicides, consisting of phytochemicals and compounds derived from plants. Various chemical components in plants have become an essential source for optimisation and research in discovering new products, from essential oils to complex mixtures. Botanical herbicides are biodegradable, less persistent,

DOI: 10.1201/9781003251255-8

and have low toxicity compared to chemicals (Ebadollahi et al., 2020; Magierowicz et al., 2020; Yang et al., 2020;).

Some plants contain various phytochemical compounds with antioxidant activity in the Chihuahuan semi-desert. This is due to the ability of these types of plants to grow in extreme weather conditions (Tucuch et al., 2020). The phytochemical compounds of plants from semi-arid regions are considered to develop potential products with secondary metabolites with a mechanism against pathogens and phytopathogenic microorganisms present in the agricultural and food industry and their everyday use in traditional medicine.

8.2 WEEDS

Plants that interfere with man's purposes in a particular place and time are considered weeds, either in an agricultural crop or a stage. Weeds have adaptive conditions that allow them to invade crops. These interferences are significantly associated with significant reductions in crop yields and decreases in the harvested product's quality (Gibson et al., 2008). They present multiple morphological, physiological, and behavioral characteristics, which allow them to be classified as broadleaf (dicots) and narrow-leaved (monocots) weeds. Broadleaf weeds have unique characteristics in the veins of leaves because they are network-shaped or reticulated, they have cotyledonary leaves with two seminal leaves in the seedlings, and the roots show vertical growth. Narrow-leaved weeds (monocotyledons) are divided into grasses and sedges; grasses have leaves in an alternate arrangement and parallel veins, a single seed leaf in their seedlings, and a fibrous root system, while sedges have triangular stems and leaves. They grow in the form of rosettes from the base of the stem and present inflorescence. The life cycle of weeds is classified as perennial, annual, or biennial. The cycle of perennials is more than two years, and under favorable conditions, they grow indefinitely. Their reproduction is by seed, often vegetatively, using tubers, stolons, bulbs, or rhizomes. The annuals complete their life cycle more diminutive than a year, in summer (May–September) and winter (October–April); biennials are not very common because they complete their life cycle in two years—in the first year, the plant forms the rosette and a deep taproot, and in the second year, they flower, mature, and die. Regarding their habitat, they can be aquatic or terrestrial. (Cronquist, 1981; Novelli and Cámpora, 2015; Cimmyt, 2021).

Weeds are one of the tremendous phytosanitary problems that cause farmers significant production and economic losses around the world, and it is widely known that these losses exceed those caused by any other agricultural problem such as insects and diseases (Abouziena et al., 2016). Its difficulty to manage and control is due to its genetic potential because weeds have resistance to adverse conditions, crossing mechanisms that guarantee survival, and their ability to spread over long distances (Bajwa, 2014).

It is estimated that direct losses in the agri-food sector caused by weeds are close to 10% of global production (FAO, 2016). In México, the economic damage due to the emergence of weeds or invasive species is on average 30% of crop yields. (Mexican Association of Weed Science, 2010).

Weeds are the main challenge for producers because they cause direct and indirect damage that affect crops in different ways, such as competition for nutrients, light, water, the release of toxic allelopathic substances to crops, creation of favorable habitats for the proliferation of phytopathogenic microorganisms and insects, and by making work on the farm difficult (López, 2003; FAO, 2016).

8.2.1 Weeds Associated with Agriculture

In the agronomic sector, weeds represent plants that grow out of place, discontinuing crop activity and causing significant losses; therefore, they do not have a fair economic value for farmers, they interfere with the production capacity and development of crops due to competition for physical space, light, water, and, nutrients, or due to the production of substances harmful to the crop (Fusagri, 1985; Pitty and Muñoz, 1991).

It indicates that weeds represent one of the significant agriculture problems globally because of their invasive action that favors competition with crops, simultaneously behaving as hosts of pests and diseases. Therefore, farmers must implement management and control models that minimise their interference with the target crop and, in this way, avoid and reduce the exorbitant increase in production costs. Under agricultural production conditions, man considers weeds to be undesirable plants; however, they constitute the most outstanding economic element of the entire pest complex, including plant pathogens, mites, insects, nematodes, and vertebrates (Labrada and Parker, 1996). The Food and Agricultural Organization (FAO) estimates that the annual losses caused by weeds in agriculture, mainly in developing countries, to be 125 million tons in the agri-food sector, an amount capable of feeding 250 million people. Producers in the field use more than 40% of their working day in weeding operations and even then they are prone to extraordinary losses because of the weeds competition with the crops.

These interruptions are related precisely to the decrease in the quality of the harvested food and the significant reductions in crop yields (Gibson et al., 2008) or to the contingent role of weeds as hosts of pests that are harmful to the crop (Norris and Kogan, 2005).

However, despite both global (Dill et al., 2008) and local (Casafe, 2012) increases in the management and control of weeds, which is mostly the application of herbicides because of their undeniable contribution to the management of productive systems, weeds continue to cause numerous problems related to escape processes (Scursoni et al., 2007) or the development of populations with resistance activity (Jasieniuk et al., 1996; Vidal et al., 2010; Ferraro and Ghersa, 2013). Furthermore, these deficiencies in control and management cause a feedback effect (Chapin et al., 1996) in which the increase in the application of chemical products is increasingly significant and necessary, thus worsening the effects of escape and resistance, which cyclically lead to further increases in the applied doses.

8.3 MANAGEMENT AND CONTROL OF WEEDS ASSOCIATED WITH AGRICULTURE

In agriculture, weed infestations and their behavior frequently change because of climate change, ecological change, and intensive management practices (Chauhan et al., 2006, 2014). As a guarantee, the existing management options must be modified to effectively control these changes. The most frequently used method of weed management developing countries is manual or mechanical, while in technologically developed regions, the dominant method is chemical management (Chauhan, 2012; Chauhan and Gill, 2014). That depends on the differentiation of management practices: The resources and the availability of labor (Zimdahl, 2013). Modern agriculture requires an updated and modified weed management regimen to solve the problems associated with traditional techniques (Bajwa, 2014). Weed management mechanisms based on ecology and traditional techniques can help aggravate problems of environmental pollution due to the high residuality of chemical products, resistance to herbicides, biological invasion, yield losses, and weed diversification (Chauhan, 2013; Chauhan and Johnson, 2010; Chauhan et al., 2010; Singh, 2012). Considering these potential problems and opportunities, unconventional and non-chemical weed management strategies can be of great use to farmers, such as crop nutrient management, improved tillage, weed seed predation, tolerant crops to herbicides, bioherbicide allelopathy, and thermal techniques.

One of the management strategies research has focused on is crop–weed interaction, which deeply studies crop yield losses caused by weed intervention and the population dynamics of weeds (Ghersa et al., 2000). Although there are descriptive approaches to the behavior of specific weed populations, the development of predictive models that cover the mechanical and functional process of managing weeds is still rudimentary (Grundy, 2003). The inclusion of descriptive measurement models of these processes would improve the estimates of the damage caused by weeds, bearing in mind that the moments and magnitude of the emergence and competition of the different weed populations are vital aspects to define their potential damage to crops (Vleeshouwers, 1997).

Population ecologists consider the abundance of weeds to analyse the practices that are carried out to manage them. This work aims to know the relative magnitude of the regulatory factors of the size of the weed population. Knowledge of how these factors interact, whether natural or manufactured, helps to evaluate alternative long-term and short-term weed-control practices and also helps to define the particular role of biological traits of individual weed species. For most crop-associated weed species, insufficient research has been done on their full-cycle dynamics under various management regimes. The most common has been that research efforts have been directed at particular stages of the weed cycle, depending on the influence of control technologies within agricultural management or based on an already accepted approach. If the goal is to achieve integrated weed management, including biological control, all phases of the life cycle of a weed species must be considered (Mortime, 1990). The probability of survival to maturity and seed production of weeds depends on the competitive ability of cultivable plants and the effectiveness of weed control practices. Mortality of established weeds can be substantial (>80%), depending on the accompanying crop (Lotz, 1991). Selective herbicides can cause high mortality levels, but the application rates and timing often require precise attention, particularly when applying post-emergence treatments. Variation in age/growth stage within individual plants of weed populations and climate may be factors that make the outcome of chemical control unpredictable. From the little evidence, crop plant competition and herbicides often act synergistically, causing weed mortality and reducing the number of surviving plants (Powles, 1990).

8.4 MECHANISMS AND MODE OF ACTION OF HERBICIDES

Herbicide is a chemical that interrupts or inhibits growth and development. Herbicides are used extensively in agriculture, industry, and urban areas because, if used correctly, they provide efficient weed control at a low cost (Peterson et al., 2001). However, if they are not applied correctly, herbicides can cause damage to cultivated plants, the environment, and even the people who apply them.

There is often confusion when referring to the name of herbicide. An herbicide label contains three names: The chemical name, the common name, and the trade name. For example, the herbicide sold under the trade name of Gesaprim has the common name of atrazine, which is its active ingredient, and the chemical name is 6-chloro-N-ethyl-N'-(1-methyl ethyl)-1, 3,5, triazine-2,4-diamine. The agrochemical company uses the trade name to promote the sale of its trademark and is commonly the best-known name of an herbicide. The common name is the generic name given to the active ingredient. Appropriate authorities approve it, such as the Weed Science Society of America (WSSA) and the International Organization for Standardization (IOS). The chemical name describes the chemical composition of the compound—herbicide (Caseley, 1996).

Chemical weed control is carried out through herbicides and is one of the main tools in modern agriculture. However, using herbicides requires technical knowledge for the correct choice and efficient and timely application of these products (Anderson, 1996).

Weeds, both annual and perennial, have specific anatomical and morphological characteristics that determine the quality and the length of time an herbicide must be retained and absorbed. Annuals are effectively controlled with a contact herbicide. On the other hand, systemic herbicides are very advantageous against established perennial weeds.

The most helpful way of classifying herbicides is according to their mode of action (Duke and Dayan, 2001; Schmidt, 2005). The mode of action is the sequence of events that occurs from the herbicide's absorption to the plant's death. Herbicides with the same mode of action have the same uptake and transport behavior and produce similar symptoms on treated plants (Gusolus and Curran, 1996). In addition, the classification of herbicides according to their mode of action makes it possible to predict, in a general way, their weed control spectrum, application time, crop selectivity, and persistence in the soil (Ashton and Crafts, 1981). Finally, this type of classification allows for the design of the most efficient chemical weed-control programs and avoids the possible adverse effects of the use of herbicides, such as residuality in the soil, the change of weed species, and the development of weed biotypes resistant to herbicides (Regehr and Morishita, 1989). The mode of action refers to the events that cause herbicides and the mechanism of action to the site or specific biochemical process that is affected and can be reviewed in Table 8.1 (Gunsolus and Curran, 1996).

TABLE 8.1 DIFFERENT MECHANISMS OF ACTION BY HERBICIDES

Type of Action	Mode of Action	Chemical Family
Growth Regulators	Altered hormonal balance act on protein synthesis, cell elongation, and respiration	Phenoxycarboxylic, benzoic, pyridine carboxylic, and quinoline carboxylic
Seedling Growth Inhibitors	Inhibition of mitosis division in cells	Dinitroanilines, chloroacetamides, ticarbamates, and benzoic acids
Photosynthesis Inhibitors	Interruption of electron flow in photosystem II, induction of destruction of carotenoids and chlorophyll	Amides, Benzothiazolins, Phenylureas, Triazinones, Triazolinones, Nitriles, Uracils
Pigment Synthesis Inhibitors	Carotenoid inhibition, induction of chlorophyll destruction	Benzoylpyrazoles, triazoles, pyridazinones, isoxazoles, isoxazolidinones, and triketones
Lipid Synthesis Inhibitors	Inhibition of lipid synthesis by the enzyme acetyl coenzyme-a carboxylase (ACCase)	Cyclohexanediones and aryloxyphenoxypropionates

8.5 PHYTOCHEMICALS AS BIOHERBICIDES

The phytochemicals present in plants, with biological activity, are the main interests in the development of compounds with potential for bioherbicides because the phytochemical molecules present in plant extracts contain low toxicity concentrations and are effective and friendly to the environment. In this context, México is a country recognised for its incredible biodiversity of endemic plants; in the semi-desert regions, there is a great variety of plants that contain a set of chemical compounds of secondary metabolites, as an environmental defense mechanism, due to the capacity the plants have to grow under extreme weather conditions. The efficacy of these compounds depends mainly on the organisms involved and the environment in which they develop. The chemical compounds found in these plants consist of a complex mixture of esters, ketones, alcohols, terpenes, aldehydes, polyphenols, carbohydrates (Wong-Paz et al., 2015).

As discussed above, natural products from plants offer an infinite source of chemical structures, as they have different modes of action because the chemical compounds they contain come from organisms that have different chemical defenses, as they depend on an environment in which skills are developed to cope with a wide variety of ecological situations. Therefore, they are ideal clues for discovering new natural products for weed control and management in this context. However, some drawbacks can be found in the development of new bioherbicides because the structures of natural products are usually complicated, containing one or more stereocenters, which can lead to complicated and expensive production on a large scale. Large-scale elaborations are necessary to carry out bioassays at greenhouse and field levels, followed by the design and commercialization of the final agrochemical product. A simplification of their structures in search of the 'active part' of the molecule usually leads to a loss of the phytotoxic effect (Duke et al., 2000).

Despite these drawbacks, some commercial herbicides using natural products are available on the market. As mentioned above, an investigation familiar with the high efficacy of phytotoxins is that of phytopathogenic microorganisms, as natural products have been used and verified, coming from natural products for the confrontation of phytotoxins against them, obtaining favorable results for the farmers. However, natural compounds can affect weeds and have rarely received attention for obtaining new products, except for glufosinate, a synthetic version of phosphinothricin, a degradation product of bialaphos biosynthesised by *Streptomyces hygroscopicus* and *Streptomyces viridochromogenes*. The nucleotide analog hydantoin, a natural compound of *S. hygroscopicus*, is one of the most efficient natural products, as several studies of structure-activity relationships have been reported for it and its analogs with patents related to its herbicidal effect (Nakajima et al., 1991; Agasimundin et al., 1998; Hanessian et al., 1999). On the side of plant chemicals and their derivatives, some have been developed to the stage of commercial herbicide. So is the case of *cinmethylin*, a natural derivative of monoterpene 1,4-cineole, benzyl group that reduces its volatility (Grayson et al., 1987). In triketones, a group of natural herbicides includes the *sulcotrione*, among others, which are structurally related to *leptospermone*, belonging to the genus *Callistemon*. Triketones act through a mode of action not previously used for any herbicide (Lee et al., 1997).

The systematic synthesis of plant chemical compounds, in which phytotoxic effects and physicochemical properties can be observed for the search of new bioherbicide products, is currently one of the most promising sources of alternative compounds for the control and management of weeds. Furthermore, the information on the mode of action of natural phytotoxins leads to the conclusion that they usually have modes of action different from those of traditional chemically synthesised herbicides (Duke et al., 1997). Therefore, experimental techniques have been developed to obtain herbicides based on natural products—the high throughput screening (H.T.S.) method being one of the first remarkable methods used on several synthesised and tested compounds.

Experimental work on the structure-activity relationship for the development of herbicides from natural products has been carried out on compounds from microorganisms (Duke et al., 1996) and plants (Macías et al., 2001). Several structural modifications were found to cause an increase in phytotoxicity through the modulation of lipophilicity on sesquiterpene lactones (Macías et al., 2005) and benzoxazine (Macías et al., 2006). In addition, several works generated knowledge on biocommunication phenomena in benzoxazinone-producing plants and helpful information for developing models of bioherbicides composed of benzoxazinone. These investigations offer interest from the practical aspects of the knowledge of new natural chemical substances, as this will extend to the minimum use of uncontrolled amounts of chemical products, which is a desirable alternative for agriculture, meeting the requirements of environmental and quality regulations of the consumer and the food market.

Kaab and Rebey (2019) studied the herbicidal activity of two semi-desert Asteraceae plants (*Artemisia herba*-alba and *Cynara cardunculus*), to which the plants inhibited seed germination and weed seedling growth. The Asteraceae family has been found to harbor the most prominent biocidal substances for agriculture and is a good source for isolating and purifying allelopathic secondary metabolites (Bessada et al., 2015; Watanabe et al., 2014). In addition, the crude methanolic extract of *C. cardunculus* had the best post-emergence herbicidal activity under greenhouse conditions. *C. cardunculus* leaves are known to be a good source of polyphenols (Omezzine et al., 2011; Falleh et al., 2008) and could inhibit the development of other invasive weeds such as barnyard grass *(Echinochloa crus-Galli)* and *Brachiaria* sp. (Rial et al., 2014). Many authors have shown that the phytotoxic effect of plant extracts was related to the presence of compounds. Methanol seems to be the best solvent to extract phenols because of its good solubility (Ben Hadj Ali et al., 2014).

8.6 CONCLUSION

Weeds are a phytosanitary problem; if there is no adequate control, they can cause significant economic losses in the agri-food sector. Most studies have focused on chemical management and control; however, this has brought multiple problems affecting the health of farmers. Therefore, studies of semi-desert plants have been carried out to obtain phytochemicals for the development of new alternatives for the management of phytosanitary problems, in this context, adequate control of weeds, which requires scientific knowledge in the area because the effectiveness of the search for chemical compounds, for economic, environmental, and social sustainability, will largely depend on it.

8.7 ACKNOWLEDGMENTS

The National Science and Technology Council (**CONACYT**) is thanked for supporting the development of the document by financing the 'Nano and microencapsulated bioherbicides loaded with plant extracts from the Chihuahuan semi-desert for the control of plant development' project (**320692**).

REFERENCES LIST

Abouziena, H.F., & Haggag, W.M. (2016). Weed control in clean agriculture: a review. *Planta Daninha, 34*, 377–392.

Agasimundin, Y.S., Mumper, M.W., & Hosmane, R.S. (1998). Inhibitors of glycogen phosphorylase b: synthesis, biochemical screening, and molecular modeling studies of novel analogs of hydantocidin. *Bioorg Med Chem, 6*, 911–923.

Anderson, W.P. (1996). *Weed Science: Principles* (3rd ed., 338 p). St. Paul: West Publishing Co.

Ashton, F.M. and Craft, A. S. (1981). *Mode of Action of Herbicides*. New York: Wiley.

Asociación Mexicana de la Ciencia de la Maleza (2010). [fecha de Consulta 6 de marzo de 2022]. Disponible en: https://somecima.com/memorias/

Bajwa, A.A. (2014). Sustainable weed management in conservation agriculture. *Crop Protection, 65*, 105–113.

Bajwa, A.A., Anjum, S.A., Nafees, W., Tanveer, M., & Saeed, H.S. (2014). Impact of fertilizer use on weed management in conservation agriculture-a review. *Pakistan Journal of Agricultural Research, 27*(1), 68–78.

Ben El Hadj Ali, I., Bahri, R., Chaouachi, M., Boussai, M., & Harzallah-Skhiri, F., (2014). Phenolic content, antioxidant and allelopathic activities of various extracts of *Thymus numidicus poir. Organs. Industrial Crops and Products, 62*, 188–195.

Bessada, S. M., Barreira, J. C., Oliveira, M. B. P. (2015). Asteraceae species with most prominent bioactivity and their potential applications: A review. *Industrial Crops and Products, 76*, 604–615.

CASAFE, Cámara Argentina de Sanidad Agropecuaria y Fertilizantes. (2012). *Evolución del mercado de herbicidas en Argentina* online. www.casafe.org

Caseley, J.C. (1996). Herbicidas. In: Labrada, R., J.C. Caseley, & C. Parker (eds. *Manejo de malezas para países en desarrollo. Estudio FAO Producción y Protección Vegetal 120*. Roma, Italia: Organización de las Naciones Unidas para la Agricultura y la Alimentación. http://www.fao.org/docrep/T1147S/t1147s0e.htm#TopOfPage

Chapin, F.S., Torn, M.S., & Tateno, M. (1996). Principles of ecosystem sustainability. *American Naturalist, 148*, 1016–1037.

Chauhan, B.S. (2012). Weed ecology and weed management strategies for dry-seeded rice in Asia. *Weed Technol, 26*, 1–13.

Chauhan, B.S. (2013). Strategies to manage weedy rice in Asia. *Crop Prot, 48*, 51–56.

Chauhan, B.S., & Gill, G.S. (2014). Ecologically based weed management strategies. In Chauhan, B.S. & Mahajan, G (eds.), *Recent Advances in Weed Management* (pp. 1–11). New York: Springer.

Chauhan, B.S., & Johnson, D.E. (2010). The role of seed ecology in improving weed management strategies in the tropics. *Adv Agron, 105*, 221–262.

Chauhan, B.S., Gill, G.S., & Preston, C. (2006). Tillage system effects on weed ecology, herbicide activity and persistence: a review. *Aust J Exp Agric, 46*, 1557–1570.

Chauhan, B.S., Migo, T., Westerman, P.R., & Johnson, D.E. (2010). Postdispersal predation of weed seeds in rice fields. *Weed Res, 50*, 553–560.

CIMMYT. (2021). *La importancia de conocer las malezas. Consultado 11 de marzo de 2022*. Disponible en: https://repository.cimmyt.org/bitstream/handle/10883/20925/62340.pdf

Cronqüist, A. (1981). *An Integrated System of Classification of Flowering Plants*. New York: Columbia üniversity Press (CJ.S.A.).

Dill, G.M., CaJacob, C.A., & Padgette, S.R. (2008). Glyphosate resistant crops: adoption, use and future considerations. *Pest Management Science, 64*, 326–331.

Duke, S.O., & Dayan, F.E. (2001). Classification and mode of action of the herbicides. In: XXI. De Prado, R. & J.V. Jorrín (eds.), *Uso de Herbicidas en la Agricultura del Siglo* (pp. 31–44). España: Servicio de Publicaciones. Universidad de Córdoba.

Duke, S.O., Abbas, H.K., Amagasa, T., & Tanaka, T. (1996). Phytotoxins of microbial origin with potential for use as herbicides. *Crit Rev Appl Chem, 35*, 82–113.

Duke, S.O., Dayan, F.E., Hernández, A., Duke, M.V., & Abbas, H.K. (1997). Natural products as leads for new herbicide modes of action. In: Proceedings Brighton Crop Protection Conference—Weeds (pp. 579–586). Brighton, UK.

Duke, S.O., Dayan, F.E., Romagni, J.G., & Rimando, A.M. (2000). Natural products as sources of herbicides: current status and future trends. *Weed Res, 40,* 99–111.

Ebadollahi, A., Ziaee, M., & Palla, F. (2020). Essential oils extracted from different species of the Lamiaceae plant family as prospective bioagents against several detrimental pests. *Molecules, 25,* 2020.

Falleh, H., Ksouri, R., Chaieb, K., Karray-Bouraoui, N., Trabelsi, N., & Boulaaba, M. (2008). Phenolic composition of *Cynara cardunculus L.* organs, and their biological activities. *Comptes rendus Biologies, 331,* 372–379.

FAO. (2016). *Manejo de malezas para países en desarrollo. Consultado 25 de febrero de 2022.* Disponible en: http://www.fao.org/docrep/t1147s/t1147s00.htm#Contents

Ferraro, D.O., & Ghersa, C.M. (2013). Fuzzy assessment of herbicide resistance risk: Glyphosate-resistant johnsongrass, *Sorghum halepense (L.) Pers.*, in Argentina's croplands. *Crop Protection, 51,* 32–39.

FUSAGRI. (1985). Control de malezas serie petróleo y agricultura. *Fundación servicio para el agricultor. FUSAGRI, 2*(8), 9–26.

Ghersa, C.M., Benech-Arnold, R.L., Satorre, E.H., & Martýnez-Ghersa, M.A. (2000). Advances in weed management strategies. *Field Crops Research, 67,* 95–104.

Gibson, D.J., Millar, K., Delong, M., Connolly, J., Kirwan, L., Wood, A.J., & Young, B.J. (2008). The weed community affects yield and quality of soybean (*Glycine max L.*). *Journal of the Science of Food and Agriculture, 88,* 371–381.

Grayson, B.T., Williams, K.S., & Freehauf, P.A. (1987). The physical and chemical properties of the herbicide cinmethylin (SD 95481). *Pest Sci, 21,* 143–153.

Grundy, A.C. (2003). Predicting weed emergence: A review of approaches and future challenges. *Weed Research, 43,* 1–11.

Gunsolus, J.L., & Curran, W.S. (1996). Herbicide mode of action and injury symptoms. *North Central Extension Publication, 377,* 14 p.

Hanessian, S., Lu, P.P., Sanceau, J.Y., Chemla, P., Gohda, K., Fonne-Pfister, R., Prade, L., & Cowan-Jacob, S.W. (1999). An enzyme-bound bisubstrate hybrid inhibitor of adenylosuccinate synthetase. Angew. Chem. *In.t Ed. 38,* 3160–3162.

Jasieniuk, M., BruleBabel, A.L., & Morrison, N.I. (1996). The evolution and genetics of herbicide resistance in weeds. *Weed Science 44*: 176–193.

Kaab, S.B., Rebey, I.B., Hanafi, M., Hammi, K.M., Smaoui, A., Fauconnier, M.L., & Ksouri, R. (2020). Screening of Tunisian plant extracts for herbicidal activity and formulation of a bioherbicide based on Cynara cardunculus. *South African Journal of Botany, 128,* 67–76.

Labrada, R., & Parker, C. (1996).El control de malezas en el contexto del manejo integrado de plagas. En: Manejo de malezas para países en desarrollo. Estudio FAO. *Producción y Protección Vegetal, 120,* 3–9.

Lee, D.L., Prisbylla, M.P., & Cromarthie, T.H. (1997). The discovery and structural requirements of inhibitors of p-hydroxyphenylpyruvate dioxygenase. *Weed Sci, 45,* 601–609.

López-Ramírez, R. (2003). *Control Químico de la Maleza en el Cultivo de la Calabacita (Cucurbita pepo L.). Universidad Autónoma de Chapingo.* Tesis de Licenciatura. 48 pp.

Lotz, L.A.P., Groenveld, R.M.W., & Habekott, V.O.B. (1991). Reduction of growth and reproduction of Cyperus esculentus by specific crops. *Weed Research, 31,* 153–160.

Macías, F.A., Marín, D., Oliveros-Bastidas, A., & Molinillo, J.M.G. (2006). Optimization of benzoxazinones as natural herbicide models by lipophillicity enhancement. *J. Agric. Food. Chem. 54,* 9357–9365.

Macías, F.A., Molinillo, J.M.G., Galindo, J.C.G., Varela, R.M., Simonet, A.M., & Castellano, D. (2001). The use of allelopathic studies in the search for natural herbicides. *J. Crop. Prod., 4*:237–255

Macías, F.A., Velasco, R.F., Castellano, D., & Galindo, J.C.G. (2005). Application of Hansch's model to guaianolide ester derivatives: a quantitative structure-activity relationship study. *J. Agric. Food. Chem., 53*:3530–3539

Magierowicz, K., Górska-Dabrik, E., & Golan, K. (2020). Effects of plant extracts and essential oils on the behavior of *Acrobasis advenella* (Zinck.) caterpillars and females. *J. Plant Dis. Prot., 127,* 63–71.

Mortimer, A.M. (1990). The biology of weeds. In: Hance, R.J. & Holly, K. (Eds.), *Weed Control Handbook: Principles* (8va ed., pp. 1–42). Blackwell Scientific Publications.

Nakajima, M., Itoi, K., & Takamatsu, Y. (1991). Hydantocidin: a new compound with herbicidal activity. *J. Antibiot., 44,* 293–300.

Norris, R.F., & Kogan, M. (2005). Ecology of interactions between weeds and arthropods. *Annual. Rev. Entomology, 50*, 479–503.

Novelli, D., & Cámpora, M.C. (2015). Malezas, la expresión de un sistema. *RIA. Revista de Investigaciones Agropecuarias, 41*(3), 241–247. [fecha de Consulta 6 de Marzo de 2022]. ISSN: 0325-8718. Disponible en: https://www.redalyc.org/articulo.oa?id=86443147005

Omezzine, F., Ladhari, A., Rinez, A., & Haouala, R. (2011). Potent herbicidal activity of *Inula crithmoides L. Scientia Horticulturae, 130*, 853–861.

Pavela, R., Maggi, F., Iannareli, R., & Benelli, G. (2019). Plant extracts for developing mosquito larvicides: From laboratory to the field, with insights on the modes of action. *Acta Trop., 193*, 236–271.

Peterson, D.E., Thompson, C.R., Regehr, D.L., & Al-Khatib, K. (2001). Herbicide mode of action. *Kansas State University, C-715*, 24 p.

Pitty, A., & Muñoz, L. (1991). *Guía práctica para el manejo de malezas. El Zamorano* (223 p). Tegucigalpa: Escuela Agrícola Panamericana.

Powles, S., & Howat, P.(1990). Herbicide resistant weeds in Australia. *Weed Technology, 4*, 178–185.

Regehr, D. L., & Morishita, D. W. 1989. Questions and Answers on Managing Herbicide-Resistant Weeds. Manhattan, KS: Kansas State University Extension Rep. MF-926. 8 p.

Rial, C., Novaes, P., Varela, R., Molinillo, J.A., & Macias, F., 2014. Phytotoxicity of Cardoon (*Cynara cardunculus*) allelochemicals on standard target species and weeds. *Agricultural and Food Chemistry*. https://doi.org/10.1021/jf501976h.

Rob, M.M., Hossen, K., Iwasaki, A., Suenaga, K., & Kato-Nogushi, H. (2020). Phytotoxic activity and identification of phytotoxic substances from *Schumanniathus dichotomous. Plants, 9*, 102.

Schmidt, R.R. (2005). Clasificación de los herbicidas según su modo de acción. Comité de acción contra la resistencia a herbicidas (HRAC). www.plantprotection.org/HRAC/Spanish_classification.htm.

Scursoni, J.A., Forcella, F., & Gunsolus, J. (2007). Weed escapes and delayed weed emergence in glyphosate-resistant soybean. *Crop Protection, 26*, 212–218.

Seiber, J.N., Coats, J., Duke, S.O., & Gross, A.D. (2014). Biopesticides: State of the art and future opportunities. *J. Agric. Food Chem., 62*, 11613.

Singh, S., Punia, S.S., Singh, A., & Brar, A.P. (2012). Weed control efficacy of trifluralin in cotton in N-W. India. *Har. J. Agron., 28*, 1–10.

Tucuch-Pérez, M.A., Arredondo-Valdés, R., & Hernández-Castillo, F.D. (2020). Antifungal activity of phytochemical compounds of extracts from Mexican semi-desert plants against Fusarium oxysporum from tomato by microdilution in plate method. *Nova Scientia, 12*(25). https://doi.org/10.21640/ns.v12i25.2345

Vidal, R.A., Rainero, H.P., Kalsing, A., & Trezzi, M.M. (2010). Prospección de las combinaciones de herbicidas para prevenir malezas tolerantes y resistentes al glifosato. *Planta Daninha, 28*, 159–165.

Vleeshouwers, L.M. (1997). Modeling the effect of temperature, soil penetration resistance, burial depth and seed weight on pre-emergence growth of weeds. *Annals of Botany, 79*, 553–563.

Watanabe, Y., Novaes, P., Varela, R.M., Molinillo, J.M.G., Kato-Noguchi, H., & Macías, F.A. (2014). Phytotoxic potencial of *Onopordum acanthium* L.(Asteraceae). *Chemistry & Biodiversit, 11*, 1247–1255.

Wong-Paz, J.E., Contreras-Esquivel, J.C., Rodríguez-Herrera,R., Carrillo-Inungaray, M.L., López, L.I., Nevárez-Moorillón, G.V., & Aguilar, C.N. (2015). Total phenolic content, in vitro antioxidant activity and chemical composition of plant extracts from semiarid Mexican region. *Asian Pacific Journal of Tropical Medicine, 8*(2), 104–111.

Yang, G.-Z., Zhang, J., Peng, J.-W., Zhang, Z.-J., Zhao, W.-B., Wang, R.-X., Ma, K.-Y., Li, J.-C., Liu, Y.-Q., & Zhao, Z.-M. (2020). Discovery of luotonin A analogs as potent fungicides and insecticides: Design, synthesis and biological evaluation inspired by natural alkaloid. *Eur. J. Med. Chem., 194*, 112253.

Zimdahl, R.L. (2013). *Fundamentals of Weed Science* (pp. 295–344). Waltham: Academic Press.

Chapter 9

Chemical and Bioactive Compounds from Mexican Desertic Medicinal Plants

Julio Cesar López-Romero, Heriberto Torres-Moreno, Arely del Rocio Ireta-Paredes, Ana Veronica Charles-Rodríguez, and María Liliana Flores-López

CONTENTS

9.1 INTRODUCTION

Deserts provide many benefits that can meet the demands of the local inhabitants and other surrounding communities. These benefits include water, food supply, medicines, and raw materials (Bidak et al., 2015). Deserts have great importance because of the large area that they occupy in the world and their specific characteristics as well as the diversity of interesting species they present (Serrano-Gallardo et al., 2017). Several studies have been conducted to understand the chemical and bioactive components of desert plants that are important for their application in folk medicine, from which humanity has benefited since ancient times. The therapeutic usage of indigenous plant products for ethnomedical and nutritional objectives has attracted the interest of researchers, focusing mainly on the search for bioactive compounds to which these effects can be attributed (Radha et al., 2021). The outstanding metabolic machinery of desert plants allows for the presence of metabolites of great chemical diversity, being their extracts, either as standardized natural products or as pure compounds, that are unlimited sources for the development of new drugs (Mustafa et al., 2017).

Phytochemicals are non-nutritive chemicals, which are responsible for the medicinal properties of plants, and according to the functions they play in their metabolism, they can be divided into two groups: Primary and secondary metabolites, considered essential and non-essential for the basic metabolic processes, respectively. Examples of primary metabolites are carbohydrates, amino acids, proteins, and chlorophylls (Mustafa et al., 2017). On the other hand, secondary metabolites, known as bioactive compounds, are commonly related to the protection

DOI: 10.1201/9781003251255-9

mechanism of plants against phytopathogenic agents and extreme environmental conditions. In general, the bioactive compounds can be classified as phenolic compounds, terpenes, and nitrogen-containing compounds, and their functionalities and properties will depend on their nature (Guía-García et al., 2022). It could be considered that nature developed complex biosynthetic pathways to produce phytochemicals during the evolution of plants as a chemical defense against microorganisms (viruses, bacteria, protozoa) and herbivores (insects, worms, mammals) (Efferth & Oesch, 2021). In recent years, 10,000 phytochemicals have been isolated from plants that are used in traditional medicine systems worldwide. In this context, natural products give a significant boost to drug discovery. In particular, the therapeutic areas of infectious diseases and oncology have benefited from numerous classes of bioactive compounds derived from natural sources (Qiu, 2014).

Increasingly, studies are being conducted to identify the diversity and abundance of bioactive compounds in wild plants in drylands and the influence of environmental factors on their chemical variations (Monreal-García et al., 2019). Recently, in a study conducted in the coastal desert area of Egypt, 234 native plant species were identified, of which 73% were associated with medicinal uses (Bidak et al., 2015). Al-Saleem et al. (2018) analyzed seven plant species growing in the Saudi Arabian desert – *Calendula tripterocarpa, Centarea sinaica, Centaurea pseudosinaica, Koelpinia linearis, Plectranthus arabicus, P. asirensis*, and *Tripleurospermum auriculatum* – finding in all investigated plants that lipids and phenolic compounds represent the highest chemical percentage, and different doses of the alcohol extract, up to 5000 mg/kg, did not produce any toxicity. Another study carried out with *Carya Illinoinensis, Selaginella lepidophylla, Euphorbia antisyphilitica*, and *Jatropha dioica* Sessé plants growing in the semi-desert zone of Coahuila and Durango, Mexico, the phytochemical analysis revealed the presence of flavonoids, lactons, quinons, triterpens, and sterols (Serrano-Gallardo et al., 2017). Quintanilla-Licea et al. (2016) reported the extraordinary therapeutic potential of two plants from the Mexican desert: *Pachycereus marginatus* and *Ibervillea sonorae*, traditionally used for cancer treatment, as they exhibited good cytotoxic activity.

It is important to identify native plants in desert regions of the world, as well as to preserve Indigenous peoples' knowledge about them. In Mexico, due to its geographical location, more than 60% of the territory is located between 20° N and 40° N, corresponding to the belt in which the deserts of the world are distributed, which is the reason why a great part of the country is arid. Mexico is covered by arid and semi-arid areas, which represent 75% of the national territory (Abd El-Ghani et al., 2017). The Mexican desert is characterized by extreme environmental conditions throughout the year, with temperatures exceeding 50°C in summer and below zero in winter, as well as high levels of ultraviolet (UV) radiation and drought (Conagua, 2022). In these areas, the native population identifies various plants for their medicinal applications, among which, *Acacia farnesiana* (huizache), *Agave lechuguilla* Torr. (lechuguilla), *Ambrosia confertiflora* DC (estafiate), *Asclepias subulate* Decne (Rush milkweed), *Bursera microphylla* (elephant tree), *Flourensia cernua* (hojasen), *Fouquieria splendens* (ocotillo), *Lippia graveolens* (Mexican oregano), *Lophocereus schottii* (senita cactus), *Prosopis glandulosa* var. torreyana (honey mesquite), and *Stegnosperma halifolium* (snake herb), among others, stand out. These plants have been reported for their antimicrobial (Hernández-García et al., 2018), antidiabetic (Esquivel-Gutiérrez et al., 2021), anticancer (Jiménez-Estrada et al., 2018; Rascón-Valenzuela et al., 2015; Rascón-Valenzuela et al., 2016; Orozco-Barocio et al., 2022), and anthelmintic (Zarza-Albarran et al., 2020) properties. This chapter reviews some native plants of the northern desert regions of Mexico

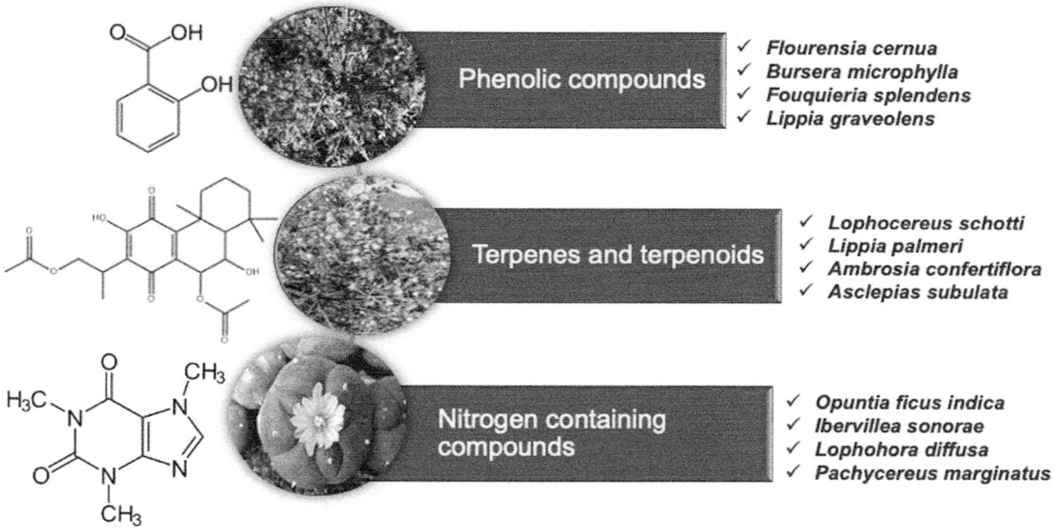

Figure 9.1 Classification of bioactive compounds by chemical groups and some examples of plants from Mexico's desert and semi-desert areas.

presented by chemical groups (Figure 9.1), identifying the medicinal uses that the native population preserves. It also presents perspectives on extraction techniques and protection systems to maintain chemical and bioactive properties of their phytochemicals.

9.2 PHENOLIC COMPOUNDS

Phenolic compounds are a diverse group of secondary metabolites produced by plants, characterized by having one or more phenol groups within their structure linked to aromatic or aliphatic groups (Alara et al., 2021). This group of compounds is synthesized through the shikimic acid and acetate-melonate pathway, and it has essential functions in plants to protect against different types of stress such as lesions, insect, and pathogen attacks, as well as protection against UV low/high radiation, temperatures, and drought (Khoddami et al., 2013; López-Romero et al., 2022). Within the phenolic compounds, there are various groups such as flavonoids, phenolic acids, anthocyanins, tannins, lignans, and lignins. These compounds have been associated with a wide variety of biological activities such as antioxidant, antimicrobial, antifungal, antiviral, antiproliferative, antitumor, anti-inflammatory, diabetes, among others (Soto-Hernández et al., 2017; Tanase et al., 2019; Jucá et al. al., 2020).

In particular, the extreme conditions in the Mexican desert promote the synthesis of phenolic compounds in many of the plants within this ecosystem, supporting the synthesis of phenolic compounds (Naikoo et al., 2019; Sharma et al., 2019). In this context, several investigations have been carried out focusing on identifying phenolic compounds in plants of the Mexican desert (Table 9.1).

TABLE 9.1 IDENTIFIED PHENOLIC COMPOUNDS IN SOME PLANTS FROM MEXICO'S DESERT AND SEMI-DESERT AREAS

Plant	Extraction Type	Identified Compounds	Reference
Fouquieria splendens	Stem/Flower	Cyanidin-3-glycoside, Pelargonidin-3-glycoside, Delphinidin-3-glycoside	Scogin (1977)
Fouquieria splendens	Leaves/Methanol	Isoquercitrin, Rutin, Scopolin, Caffeic acid, Ellagic acid	Scogin (1978)
Fouquieria splendens	Leaves/Stem/ Acetone	Apigenin, Apigenin 4'-methyl ether, Luteolin, Luteolin-3'-methyl ether, Kaempferol, Kaempferol-3-methyl, Kaempferol-3,4'-dimethyl ether, Quercetin, Quercetin 3-methyl, Quercetin 3,3'-dimethyl ether	Wollenweber and Yatskievych (1994)
Fouquieria splendens	Inflorescences/ Leaves/Ethanol (80%)	Quercetin-3-*O*-glycoside, Rutin, Myricetin derivative, Kaempferol derivative, Ellagic acid, Ellagic acid derivative	Monreal-García et al. (2019)
Fouquieria splendens	Leaves/Methanol	Kaempferol-3-β-glucoside, Quercetin, Catechol, Epigallocatechin gallate, Apigenin, Quercetin-3-β-glucoside, Kaempferol, Caffeic acid, Cinnamic acid, Chlorogenic acid, Ellagic acid, Gallic acid	López-Romero et al. (2022)
Bursera mycrophylla	Not specified	Dihydroclusin 9-acetate, 7-O-podophyllotoxinyl butyrate, Burseranin, Picropolygamain, Hemiariensin	Gigliarelli et al. (2018)
Bursera mycrophylla	Leaves/Fruits/ Ethanol	Kaempferol, Kaempferol glucoside, Catechin, Quercitin, Quercitin glucoside, Quercitin galloyl glucoside, Rutin, Quinic acid, Ellagic acid.	Vidal-Gutiérrez et al. (2020b)
Bursera mycrophylla	Leaves/Fruits/ Steam/Ethanol	Ariensin, Burseranin, Burseran, Dihydroclusin diacetate Picropoligamain, Hinokinin, 5′-desmethoxy yatein, 3,4-dimeth oxy-3′,4′-methylenedioxylignano-9,9′-epoxylignan-9′-ol, 7′, 8′-dehydropodophyllotoxin, 7′,8′-dehydroacetylpodophyllotoxin, β-peltatin A-methyl ether	Torres-Moreno et al. (2022)
Prosopis glandulosa	Leaves	Kaempferol, Kaempferol-3-O-rutinoside, Kaempferol-3-O-glucoside, Naringenin, Naringenin-chalcone, Quercetin, Rutin, Caffeic acid, Chlorogenic acid, *p*-Coumaric acid, Ferulic acid, Gallic acid, Hydroxybenzoic acid, Vanillic acid	González-Mendoza et al. (2018)
Prosopis glandulosa	Stem/Methanol/ Ethanol/Acetone	Esculetin. Quercetin, Rutin, Gallic acid	Assanga et al. (2020)

(*Continued*)

TABLE 9.1 (CONTINUED) IDENTIFIED PHENOLIC COMPOUNDS IN SOME PLANTS FROM MEXICO'S DESERT AND SEMI-DESERT AREAS

Plant	Extraction Type	Identified Compounds	Reference
Lophocereus schottii	Stem/Ethanol	Apigenin 7,4′-dimethyl ether, Epicatechin, Cyanidin, Cyanidin 3-O-glucoside, 2′,7-Dihydroxy-4′,5'-dimethoxyisoflavone, Gallocatechin, Kaempferol, Kaempferol-3-neohesperidoside, Kaempferol xiloside, Kaempferol 3-O-glucoside, Kaempferol 3-O-xylosyl-glucoside, Myricetin, Quercetin rhamnoside, Quercetin xyloside, Coumaroyl malic acid	Orozco-Barocio et al. (2022)
Flourensia cernua	Leaves/Ethanol	Ellagic acid, Catechin, Quercitin, Methyl gallate, Cinnamic acid, Hydroxycinnamic acid	Ruiz-Martinez et al. (2011)
Flourensia cernua	Leaves/Ethanol	Pyrogallol	Wong-Paz et al. (2015a)
Flourensia cernua	Stem/leaves/ Water-Ethanol	Chlorogenic acid, Coumarin acid, Quercetin	Wong-Paz et al. (2015b)
Flourensia cernua	Leaves/Water	Luteolin 7-O-rutinoside, Apigenin galactoside arabinoside	De León-Zapata et al. (2016)
Flourensia cernua	Leaves/Water	Quercetin-3-o-glucoside-7-o-glucoside, Luteolin-7-o-rutinoside, Apigenin-6-arabinoside-7-glucoside, Dimethyl epigallocatechin gallate	Medina-Morales et al. (2017)
Flourensia cernua	Leaves/Water	Rosmarinic acid, Floretin, Xanthohumol, Scopoletin, Luteolin-7-O-rutinoside, 6-C-glucosyl-8-C-arabinosyl apigenin, Isoschaftoside, *p* – Cumaric acid hexose	Alvarez-Pérez et al. (2020)
Acacia farnesiana	Pods/Chloroform, hexane, acetone, methanol, methanol/ water, and water	Caffeoylmalic acid, Digalloyl glucose, Gallic acid, Galloyl glucosa, Hydroxytyrosol acetate, Methyl gallate, Quinic acid	Delgadillo Puga et al. (2018)
Acacia farnesiana	Fruits/Hexane, chloroform, and methanol	Digallic acid, Ellagic acid, Methyl gallate, Ethyl gallate, Coumaric acid, Methyl digallate, Ellagic acid, Eriodictyol, Naringin, Prunin, Naringenin galloylglucopyranoside, (2*S*)-Naringenin 7-*O*-β-galloylglucopyranoside, Kaempferol, Naringenin, Acacetin, Naringenin 7-*O*-β-(4",6"-digalloylglucopyranoside), Chrysoeriol, Taxifolin	Hernández-García et al. (2019)

(*Continued*)

TABLE 9.1 (CONTINUED) IDENTIFIED PHENOLIC COMPOUNDS IN SOME PLANTS FROM MEXICO'S DESERT AND SEMI-DESERT AREAS

Plant	Extraction Type	Identified Compounds	Reference
Acacia farnesiana	Pods/Water-methanol	Ethyl gallate, Gallic acid, Methyl gallate, Naringin, Naringenin, Naringenin 7-O-*β*-51 (4", *6"-digalloylglucoside*), Naringenin 7-O-*β*- (*6"-galloylglucoside*),	Olmedo-Juarez et al. (2020)
Acacia farnesiana	Pods/Water-ethanol	Naringin, Naringenin, Naringenin-7-*O*-(6″-galloylglucoside), Naringenin 7-*O*-(4″, 6″-digalloylglucoside), Gallic acid, Methyl gallate, Ethyl gallate	Zarza-Albarrán et al. (2020)
Acacia farnesiana	Pods/Leaves/ Water	Apigenin 6,8-di-C-glucoside, Cyanidin 3-O-(6"-malonyl-3"-gluc osyl-glucoside), 3-Hydroxyphloretin 2'-O-xylosyl-glucoside, Isorhamnetin 3-O-glucoside 7-O-rhamnoside, Kaempferol 3-O-(6"-acetyl-galactoside) 7-O-rhamnoside, Patuletin 3-O-g lucosyl-(1->6)-[apiosyl(1->2)]-glucoside, Quercetin 3-O-rutinoside, 3-p-Coumaroylquinic acid, Gallic acid 4-O-glucoside, Gallic acid 3-O-gallate	Zapata-Campos et al. (2020)
Lippia graveolens	Aerial part/ Hexane-acetone	Naringenin, Rosmarinic acid	Martínez-Rocha et al. (2008)
Lippia graveolens	Aerial part/Chloroform	Quercetin *O*-hexoside, Scutellarein 7-*O*-hexoside, phloretin-2′-*O*-glucoside, Trihydroxy-methoxyflavone derivative, 6-O-Methylscutellarein	Leyva-López et al. (2016)
Lippia graveolens	Aerial part/ Water-ethanol	Apigenin 6,8-di-*C*-glucoside, Methyl 4,6-*O*-di-*O*-galloyl-β-D-glucopyranoside, 5,7,8-trihydroxycoumarin-5-β-glucopyrano side, (2*R*)- and (2*S*)-3′,4′,5,6-tetrahydroxyflavanone 7-*O*-β-glucopyranoside, (2*R*)- and (2*S*)-3′,4′,5,6-tetrahydroxyflavanone 7-*O*-β-glucopyranoside, (2*S*)-3′,4′,5,8-tetrahydroxyflavanone 7-*O*-β-glucopyranoside, 6-Hydroxyluteolin-7-*O*-hexoside, Taxifolin, Phlorizin, Eriodictyol, 2′′-O-(3′′′,4′′′-dimethoxybenzoyl) orientin, Ikarisoside F, Quercetin, Phloretin, Naringenin, Hispidulin, Cirsimaritin, Pinocembrin, Galangin, Genkwanin	Cortés-Chitala et al. (2021)

(Continued)

TABLE 9.1 (CONTINUED) IDENTIFIED PHENOLIC COMPOUNDS IN SOME PLANTS FROM MEXICO'S DESERT AND SEMI-DESERT AREAS

Plant	Extraction Type	Identified Compounds	Reference
Lippia graveolens	Aerial part/Ethanol	Quercetin-3-O-glucoside, Rutin, Taxifolin, Neohesperidin, Floridzin, Eriodictyol, Luteolin, Quercetin, Naringenin, Apigenin, Acacetin Quinic acid, Protocatechuic acid, 4-hydroxybenzoic acid, Caffeic acid, Coumaric acid, 2-Hydroxybenzoic acid	Frías-Zepeda & Rosales-Castro, 2021
Lippia graveolens	Aerial part/Methanol	Sakuranetin, Cirsimaritin, Pinocembrin, Naringenin	Quintanilla-Licea et al. (2020)
Lippia graveolens	Aerial part/Methanol	Kaempferol-glucoside, Pentahydroxyfavone derivative, Trihydroxyfavone hexoside, Taxifolin, Phloridzin, Dihydrokaempferol, Tetrahydroxyfavone derivative 1, Kaempferol, 3,6,2′,3′-tetrahydroxyfavone, Eriodictyol, Quercetin, Apigenin, Naringenin, Kaempferide, Tetrahydroxyfavone derivative 2, Cirsimaritin, Pinocembrin, Cafeic acid, Coumaric acid	Picos-Salas et al. (2021)
Agave lechugilla	Not specified/Ethanol	afzelequin 4β-8 quercetin, Kaempferol, Quercetin	Anguiano-Sevilla et al. (2018)
Agave lechugilla	Guishe/Ethanol	Apigenin, Apigenin 7-O-glycoside, Apigenin 7-O-rutinoside, Catechin, Cyanidin, Cyanidin 3-O-glycoside, Cyanidin O-diglycoside, Delphinidin, Delphinidin 3-O-glycoside, Flavanone, Hesperidin, Isorhamnetin, Isorhamnetin-glycoside, Isorhamnetin 3-O-rutinoside, Isorhamnetin diglycoside 1, Isorhamnetin diglycoside 2, Isorhamnetin triglycoside 1, Isorhamnetin triglycoside 2, Kaempferol, Kaempoferol 3-O-glycoside, Kaempferol 3-O-rutinoside, Kaempoferol diglycoside, Kaempoferol triglycoside, Myricetin, Myricetin 3-O-glycoside, Myricetin diglycoside, Naringenin, Naringenin O-rutinoside, Quercetin, Quercetin-3-O-xyloside, Quercetin 3-O-glycoside 1, Quercetin 3-O-glycoside 2, Quercetin 3-O-rutinoside, Quercetin 3-O-diglycoside 1, Quercetin 3-O-diglycoside 2	Morreeuw et al. (2021a)

(*Continued*)

TABLE 9.1 (CONTINUED) IDENTIFIED PHENOLIC COMPOUNDS IN SOME PLANTS FROM MEXICO'S DESERT AND SEMI-DESERT AREAS

Plant	Extraction Type	Identified Compounds	Reference
Agave lechugilla	Leaves/Etanol-water	Apigenin, Apigenin-7-O-glucoside, Apigenin-7-O-rutinoside, Catechin, Epicatechin gallate, Cyanidin, Cyanidin-3-O-glucoside, Cyanidin-3-O-rutinoside, Cyanidin-3,5-O-diglucoside, Cyanidin methyl diglycoside I, Cyanidin methyl glycoside II, Delphinidin, Delphinidin-3-O-glucoside, Delphinidin methyl glycoside, Gentiodelphin, Procyanidin, Procyanidin tetramer, Hesperidin, Hesperidin methylchalcone, Isorhamnetin, Isorhamnetin-glucoside, Isorhamnetin-glycoside, Isorhamnetin-glucosyl-pentoside, Isorhamnetin3-O-rutinoside, Isorhamnetin-hexosyl-hexosyl-pentoside, Isorhamnetin-3-gentiotrioside, Isorhamnetin triglycoside, Kaempferol, Kaempoferol-3-O-glucoside, Kaempferol-3-O-rhamnoside, Kaempferol rhamnose-malic acid, Kaempferol-hexose-malic acid, Afzelin O-gallate, Kaempferol-3-O-rutinoside, Kaempferol -3-O-(6-p-coumaroyl)-glucoside, Kaempferol-3-rhamnoside-7 -xylose-rhamnoside, Kaempoferol-3,7,4'-triglucoside, Myricetin, Dihydromyricetin, Myricetin-3-O-rhamnoside, Myricetrin O-gallate, Myricetin-hexoside-pentoside, Naringenin, Naringenin-7-O-rutinoside, Quercetin, Dihydroquercetin, Quercetin-3-O-xyloside, Quercetin-3-O-rhamnoside, Quercetin-3-O-glycoside, Quercetin-3-O-rutinoside, Quercetin-3-O-dirhamnoside, Quercetin-3-O-diglycoside, Quercetin-O-di-p-cou maroyl-rhamnopyranoside	Morreeuw et al. (2021b)
Agave lechugilla	Bagasse/Ethanol	Flavanone, Isorhamnetin, Catechin, Cyanidin, Delphinidin, Hesperidin, Quercetin, Apigenin, Kaempferol, Naringenin	Morreeuw et al. (2021c)

In a first phase, the presence of phenolic compounds has been identified through phytochemical screening in some plants, such as *A. rzedowskiana, A. impressa, A. ornithobroma, A. schidigera, A. angustifolia, A. lechuguilla, Annona muricata, A. farnesiana, Bromelina pinguin, B. microphylla, B. hindsiana, Bucida buceras, C. illinoinensis, F. cernua, Gutierrezia microcephala, J. cinerea, J. cordata, J. dioica, Leucophyllum frutescens, Larrea tridentata, L. graveolens, Morinda citrifolia, Opuntia ficus-indica, Olneya tesota, Phoradendron bollanum, Phoradendron californicum, P. glandulosa, P. laevigata, Parkinsonia aculeata, Quercus* sp., *Struthanthus quercicola, S. lepidophylla,* and *Yucca filifera,* which were found distributed in the deserts of Chihuahua, Nuevo Leon, Coahuila, San Luis Potosí, Puebla, Durango, Sonora, Sinaloa, and Tepic (Pío-León et al., 2009; Mendez et al., 2012; Ahumada-Santos et al., 2013; Menchaca et al., 2013; Iloki-Assanga et al., 2015; Salido et al., 2016; Sánchez et al., 2016; Serrano-Gallardo et al., 2017; Assanga et al., 2020; Tucuch-Pérez et al., 2020; García-García et al., 2021; Vega-Ruiz et al., 2021; Alcántara-Quintana et al., 2022).

Different studies have focused on the analysis of these bioactive compounds in the Mexican desert plants, with the goal of determining the profile and concentration of phenolic compounds through different spectroscopic and spectrometric techniques. An example of this is *F. splendens,* which has been characterized to establish its phenolic compounds profile. The studies showed that the extracts of leaves and stems of this plant had phenolic compounds (57.1 to 479.64 mg gallic acid equivalent (GAE)/g), flavonoids (6.81 to 29.68 mg quercetin equivalent (QE)/g), tannins (1.37 to 2.82 mg epicatechin equivalent (EE)/g), and anthocyanins (0.27 to 1.46 mg cyanidin-3-glucoside equivalent (C3gE)/g). Likewise, chromatography studies showed the presence of flavonoids, such as apigenin, apigenin 4'-methyl ether, catechol, epigallocatechin gallate, kaempferol, kaempferol-3-β-glucoside, kaempferol-3-methyl ether, kaempferol-3, 4´-dimethyl ether, isoquercitrin, luteolin, luteolin-3´-methyl ether, myrecetin, quercetin, quercetin-3-β-glucoside, quercetin 3-methyl ether, quercetin 3,3´-dimethyl ether, rutin, and scopolin, where apigenin, myrecetin, rutin, quercetin-3-o-glycoside, and kaempferol-3,4´-dimethyl ether were the majority compounds. Regarding phenolic acids, caffeic acid, chlorogenic acid, cinnamic acid, ellagic acid, and gallic acid were identified, the latter being the most abundant. Regarding anthocyanins, the presence of cyanidin-3-glycoside, pelargonidin-3-glycoside, and delphinidin-3-glycoside was demonstrated. Studies focusing on analyzing biological potential showed that this plant has a high antioxidant potential and the ability to inhibit the proliferation of cancer cell lines, such as cervical cancer (Scogin, 1977; Scogin, 1978; Wollenweber and Yatskievych, 1994; Monreal-García et al., 2019; López-Romero et al., 2022). Another species analyzed with a notable phenolic compound profile is *B. mycrophylla*. Extracts obtained from different parts of this plant (leaves, fruits, and resins) have presented a concentration of total phenols between 104.3 to 243.4 mg GAE/g. In addition, the presence of some flavonoids and phenolic acids was identified: Kaempferol, kaempferol glucoside, catechin, quercitin, quercitin glucoside, quercitin galloyl glucoside, rutin, quinic acid, and ellagic acid. Other phenolic compounds identified were lignans, including hinokinin, picropoligamain, hemiariensin, burseranin, burseran, ariensin, dihydroclusin diacetate, 5′-desmethoxy yatein, 3,4-dimethoxy-3′,4′-methylenedioxylignano-9,9′-epoxylignan-9′-ol, 7′,8′-Dehydropodophyllotoxin, β-peltatin A-methyl ether, 7′, 8′-dehydroacetylpodophyllotoxin, 7-O-podophyllotoxinyl butyrate, (8R,8′R)- dihydroclusin 9-acetate, dihydroclusin 9-acetate, and dihydroclusin 9′-acetate. The compounds identified in these studies were associated with high antioxidant and anti-inflammatory potential. An antiproliferative effect was also demonstrated against cervical, lung, and colon cancer cell lines, as well as low cytotoxicity in non-cancer cell lines (Gigliarelli et al., 2018; Vidal-Gutiérrez et al., 2020b; Torres-Moreno et al., 2022). *P. glandulosa* is a widely distributed plant in arid and semi-arid areas of Mexico. Various studies have shown that extracts from stems and leaves of this plant are a source of phenolic compounds such as total phenols (1.15 to 242.02 mg GAE/g), total flavonoids (75 to 190 μg/mg), flavanones and flavanols (48.66 to 119.88 mg CE/g), and condensed tannins (≈100 mg CE/g). In addition, HPLC (high performance liquid chromatography) analysis has allowed the determination of some flavonoids (esculetin, kaempferol, kaempferol-3-O-rutinoside, kaempferol-3-O-glucoside, naringenin, naringenin-chalcone, quercetin, and rutin, the former being the most abundant) and phenolic acids (caffeic acid, p-coumaric acid, chlorogenic acid, cerulic acid, gallic acid, hydroxybenzoic acid, and vanillic acid). *P. glandulosa* presents a notable antioxidant potential, which is associated with the high composition of phenolic compounds (Michel-López et al., 2014; González-Mendoza et al., 2018; Assanga et al., 2020). *L. schottii*, an endemic plant from the Sonoran Desert, is recognized as a great source of phenolic compounds (stem extracts), as about 80% of the phytochemical compounds belonging to flavonoids. In addition, it has been studied that the concentration of phenolic compounds and flavonoids is 73 mg GAE/g and 5 mg CE/g, respectively. Detailed profile studies showed the presence of flavonoids (apigenin 7,4′-dimethyl ether, epicatechin, cyanidin, cyanidin 3-O-glucoside, 2′,7-dihydroxy-4′,5'-dimethoxyisoflavone, gallocatechin, kaempferol, kaempferol-3-neohesperidoside, kaempferol xyloside, kaempferol 3-O-glucoside, kaempferol 3-O-xylosyl-glucoside, myricetin, quercetin rhamnoside, and quercetin xyloside) and phenolic acids (coumaroyl malic acid). These studies associated the effect of these compounds with the antioxidant and antiproliferative potential against murine lymphoma (Morales-Rubio et al., 2010; Orozco-Barocio et al., 2022).

F. cernua is a plant mainly distributed in the Coahuila and Chihuahua Deserts, for which phenolic compound profiles have been widely characterized. In stem and leaf extracts, total phenols (0.52 to 635 mg GAE/g) and total tannins (580 mg GAE/L) have been quantified. Among the last group, hydrolysable tannins (0.018–4.76 mg/g of polyphenols) and condensed tannins (0.0002–4.85 mg/g of polyphenols). Within the phenol profile, the following compounds have been identified: Apigenin galactoside arabinoside, apigenin-6-arabinoside-7-glucoside, catechin, 6-C-glucosyl-8-C-arabinosyl apigenin, dimethyl epigallocatechin gallate, isoschaftoside, floretin, luteolin-7-O-rutinoside, luteolin-7-O-rutinoside, methyl gallate, pyrogallol, quercetin, quercetin-3-O-glucoside-7-O-glucoside, quercetin-3-O-glucoside-7-O-glucoside, scopoletin, xanthohumol, coumarin acid, p–cumaric acid hexose, cinnamic acid, chlorogenic acid, ellagic acid, hydroxycinnamic acid, and rosmarinic acid. Previous information may explain the high antioxidant potential exhibited by this plant, demonstrating a high capacity to inhibit phytopathogenic fungi of importance to agroindustrial food (*Rhizopus stolonifer, Botrytis cinerea, Fusarium oxysporum,* and *Colletotrichum gloeosporioides*) (Ruiz-Martínez et al., 2011; Wong-Paz et al., 2015a; Wong-Paz et al., 2015b; De León-Zapata et al., 2016; Medina-Morales et al., 2017; Jasso de Rodríguez et al., 2019; Alvarez-Perez et al., 2020). Another plant distributed around the country, particularly in desert areas, is *A. farnesiana*. Studies have focused on the analysis of extracts from pods, fruits, and leaves, which have shown values of total phenols between 39.9 to 213 mg GAE/g, hydrolyzable tannins between 0.72 to 5.02 mg GAE/g, and condensed tannins between 9.52 to 45.28 mg CE/g. Analysis by chromatographic techniques have shown the presence of flavonoids, including acacetin, apigenin 6,8-di-C-glucoside, chrysoeriol, eriodictyol, cyanidin 3-O-(6"-malonyl-3"-glucosyl-glucoside), 3- hydroxyphloretin, isorhamnetin 3-O-glucoside 7-O-rhamnoside, kaempferol, kaempferol 3-O-(6"-acetyl-galactoside) 7-O-rhamnoside, naringin, naringenin, naringenin 7-O-β-(4",6"-digalloylglucopyranoside), naringenin galloylglucopyranoside, (2S)-naringenin 7-O-β-galloylglucopyranoside, naringenin 7-O-β-51 (4",6"-digalloylglucoside), naringenin 7-O-(4″,6″-digalloylglucoside), naringenin 7-O-β- (6"-galloylglucoside), prunin, patuletin 3-O-glucosyl-(1->6)-[apiosyl(1->2)]-glucoside, quercetin 3-O-rutinoside, taxifolin, and 2'-O-xylosyl-glucoside; and phenolic acids: caffeoylmalic acid, coumaric acid, 3-p-coumaroylquinic acid, digalloyl glucose, digallic acid, ellagic acid, ethyl gallate, gallic acid, gallic acid 4-O-glucoside, gallic acid 3-O-gallate, galloyl glucose, hydroxytyrosol acetate, methyl gallate, methyl digallate, and quinic acid. Recent research has shown an association between the extracts and some biological activities, such as antioxidant and anti-inflammatory. Also, the ability to inhibit the growth of the pathogenic microorganism *Mycobacterium tuberculosis* has been observed and a notable antiparasitic effect, showing the ability to affect the eggs and larvae of *Haemonchus contortus* (Delgadillo Puga et al., 2015; Delgadillo Puga et al., 2018; Hernández-García et al., 2019; Olmedo-Juárez et al., 2020; Zarza-Albarrán et al., 2020; Zapata-Campos et al., 2020).

L. graveolens is a plant that is distributed in various ecosystems in Mexico, mainly in arid and semi-arid areas. Studies of this plant have focused on the analysis of the aerial part, which has been shown to contain total phenols (4.28–4.54 mg GAE/mL; 4.95–104 mg GAE/g), and total flavonoids (0.43–97.5 mg QE/g). On the other hand, its phenolic compounds profile has been widely studied. In this sense, flavonoids have been identified, such as acacetin, apigenin, apigenin 6,8-di-C-glucoside, cirsimaritin, dihydrokaempferol, 2′ ′-O-(3′ ′ ′,4′ ′ ′-dimethoxybenzoyl), eriodictyol, floridzin, galangin, genkwanin, hispidulin, 7-O-hexoside, 6-hydroxyluteolin-7-O-hexoside, ikarisoside F, quercetin, quercetin O-hexoside, quercetin-3-O-glucoside kaempferide, kaempferol, kaempferol-glucoside, luteolin, methyl 4,6-O-di-O-galloyl-β-D-glucopyranoside, 6-O-methylscutellarein, naringenin, neohesperidin, phlorizin, phloridzin, phloretin, pinocembrin, phloretin-2′-O-glucoside, pentahydroxyfavone derivative, rutin, sakuranetin, scutellarein, taxifolin, trihydroxyfavone hexoside, tetrahydroxyfavone derivative 1, 3,6,2′,3′-tetrahydroxyfavone, trihydroxy-methoxyflavone derivative, tetrahydroxyfavone derivative 2, 5,7,8-trihydroxycoumarin-5-β-gl ucopyranoside, (2R)- and (2S)-3′,4′,5,6-tetrahydroxyflavanone 7-O-β-glucopyranoside, (2R)- and (2S)-3′,4′,5,6-tetrahydroxyflavanone 7-O-β-glucopyranoside, and (2S)-3′,4′,5,8-tetrahydroxyflavanone 7-O-β-glucopyranoside. Other identified phenolic compounds are phenolic acids, such as caffeic acid, coumaric acid, 4-hydroxybenzoic acid, 2-hydroxybenzoic acid, protocatechuic acid, quinic acid, and rosmarinic acid. The evaluated extracts of this plant showed a high antioxidant and anti-inflammatory effect, and the extracts exhibited α-amylase and α-glucosidase inhibitory effect. Additionally, isolated flavonoids showed the capacity to inhibit the growth of *Entamoeba histolytica* (Martínez-Rocha et al., 2008; Flores-Martínez et al., 2016; Leyva-López et al., 2016; Clarenc Aarland et al., 2020; Quintanilla-Licea et al., 2020; Cortés-Chitala et al., 2021; Frías-Zepeda and Rosales-Castro, 2021; Picos-Salas et al., 2021).

A different plant with a widely studied phenolic compound profile is *A. lechuguilla*, which has shown that the content of total phenols varies between 2.69 to 37.45 mg GAE/g with total flavonoids between 1.47 to 19.62 mg QE/g. The chemical characterization of leaves extracts, guishe, and baggase, has shown a diversity of phenolic compounds, highlighting the presence of flavonoids such as apigenin, apigenin-7-O-glucoside, apigenin-7-O-rutinoside, afzelequin 4β-8 quercetin, catechin, epicatechin gallate, cyanidin, cyanidin-3-O-glucoside, cyanidin-3-O-rutinoside, cyanidin-3,5-O-diglucoside, cyanidin methyl diglycoside I, cyanidin methyl glycoside II, delphinidin, delphinidin-3-O-glucoside, delphinidin methyl glycoside, flavanone, gentiodelphin, procyanidin, procyanidin tetramer, hesperidin, hesperidin methylchalcone, isorhamnetin, isorhamnetin-glucoside, isorhamnetin-glycoside, isorhamnetin-glucosyl-pentoside, isorhamnetin3-O-rutinoside, isorhamnetin- hexosyl-hexosyl-pentoside, isorhamnetin-3-gentiotrioside, isorhamnetin triglycoside, kaempferol, kaempoferol-3-O-glucoside, kaempferol-3-O-rhamnoside, kaempferol 3-O-rutinoside, kaempoferol diglycoside, kaempoferol triglycoside, kaempferol rhamnose-malic acid, kaempferol-hexose-malic acid, afzelin O-gallate, kaempferol-3-O-rutinoside, kaempferol-3-O-(6-p-coumaroyl)-glucoside, kaempferol-3-rhamnoside-7-xylose -rhamnoside,

Kaempoferol-3,7,4'-triglucoside, myricetin, dihydromyricetin, myricetin-3-O-rhamnoside, myricetrin O-gallate, myricetin-hexoside-pentoside, naringenin, naringenin-7-O-rutinoside, quercetin, dihydroquercetin, quercetin-3-O-xyloside, quercetin-3-O-rhamnoside, quercetin-3-O-glycoside, quercetin-3-O-rutinoside, quercetin-3-O-dirhamnoside, quercetin-3-O-diglycoside, and quercetin-O-di-p-coumaroyl-rhamnopyranoside. Within the characterization, the main compounds in the evaluated extracts are flavanone and isorhamnetin. Thus, an association between the extracts and antioxidant activity has been observed, as well as a promising anti-proliferative activity against different lines of lung adenocarcinoma, being the mechanism of action associated with the induction of apoptosis (Anguiano-Sevilla et al., 2018; Morreeuw et al., 2021a–c).

9.3 TERPENES

Terpenes are a class of bioactive compounds formed by isoprene units, a five-carbon building block (Perveen, 2018; Masyita et al., 2022). Terpenes are classified according to isoprene units in monoterpenes (two units), sesquiterpenes (three units), diterpenes (four units), triterpenes (six units) and tetraterpenes (eight units) (Aldred et al., 2009). In addition, according to their structure characteristics, terpenes can be subdivided in acyclic (linear) – an example is the monoterpene *β*-myrcene – and cyclic (ring-shaped), such as *p*-cymene (Buckle, 2015). Meanwhile, terpenoids are a class of secondary metabolites derived from terpenes with different functional groups and oxidized methyl groups (Perveen, 2018; Masyita et al., 2022). Terpenes and terpenoids play a fundamental role in basic plant functions such as growth and development. In addition, they regulate specialized functions such as interaction of the plant with the environment, resistance/tolerance of the plant to environmental stress, and defense against predators or pathogens. Furthermore, the pharmacological properties (antiviral, antidiabetic, anticancer, antimicrobial, anti-inflammatory, and antioxidant, among others) of these types of metabolites are extensively recognized; therefore, they have been considered good prospects for the development of new drugs (Cox-Georgian et al., 2019). Diverse investigations have established that the plants of the Mexican desert are an important source of terpenes and terpenoids.

For its multiple pharmacological properties, the chemical composition of the famous cactus *L. schotii* has been investigated (Bravo Hollis & Sánchez Mejorada, 1978; Orozco-Barocio et al., 2013), leading to isolation and characterization of diverse terpenoids and triterpenoids, such as Lophenol, lupeol, locereol, lathosterol, schottenol, spinasterol, $^{\Delta 7}$-stigmasterol, 24-methylenelophenol, 5α-campest-7-ene-3β-ol, and 5α-cholesta-7,14-dien-3β-ol (Djerassi et al., 1958; Campbell and Kircher, 1980). A study carried out by Orozco-Barocio et al. (2022) identified in the ethanol extract from stems the sterols sitosterol glucoside, poriferasta-8,22,25-trieno, and schottenol ferulate, as well as the terpenoid phytofluene; these compounds have been associated with the antitumoral effect of the plant.

F. splendens is other medicinal plant from the Mexican desert that has attracted attention for its multiple biological properties (López-Romero, 2022). Chemical studies in *F. splendens* have allowed the isolation and characterization of the saponin ocotillol (Warnhoff and Halls, 1965), which has been demonstrated to have antitumor activity (Wang et al., 2013). Due to the great interest for large-scale production of ocotillol, Morales-Cepeda et al. (2021) established an expensive method to isolate ocotillol from ocotillo logs using a batch reactor. On the other hand, the terpenoids fouquierol and isofouquierol have been isolated from ocotillo flowers (Nevárez-Prado et al., 2021). A study carried out by Monreal-García et al. (2019) showed that the flowers of *F. splendens* subsp. *splendens* had the highest concentration of total carotenoids in comparison to *F. splendens* subsp. *campanulate*, which is related to the effect of environmental conditions on the chemical composition of ocotillo subspecies.

Due to its multiple medicinal properties and its agro-industrial uses, the terpene composition of the *F. cernua* has been investigated (Jasso de Rodríguez et al., 2022). Through chromatography-gas spectrometry, diverse mono- and sesquiterpenoids compounds have been identified in *F. cernua* leaves: -myrcene, 3-carene, limonene, 1,8-cineole, borneol, *cis*-jasmone, β-caryophyllene, caryophyllene oxide, and globulol (Estell et al., 1994). Estell et al. (2016) demonstrated that the UV-light restriction modifies the concentration of santolina triene, tricyclene, α-thujene, α-pinene, camphene, and other terpene compounds in *F. cernua* leaves. Also, it was demonstrated that the profile and the concentration of these compounds varies according to the position of the leaves, the age of the plant (Estell et al., 2013), and its growth stage (Fredrickson et al., 2007).

In the search of new anticancer agents, plants from the Mexican desert such as *A. subulate, S. halimifolium* (Benth, 1844), and *Jacquinia macrocarpa Cav.* spp. *pungens (A. Gray) B. Ståhl* have been investigated. Bioguided fractionation of *A. subulata* methanolic extract led to the isolation and characterization of the cardenolide glycosides 12,16-dihydroxycalotropin, calotropin, corotoxigenin 3-*O*-glucopyranoside, and desglucouzarin, which effectively reduced the cancer cell proliferation via apoptotic cell death (Rascón-Valenzuela et al., 2016; Rascón-Valenzuela et al., 2015). Recently, Meneses-Sagrero et al. (2017) isolated and characterized spinasterol from methanol extract of *S. malifolium* stems, a triterpene with antiproliferative effect on cancer cells. Vidal-Gutiérrez et al. (2020a) isolated and characterized, for first time, primulasaponin from the methanol extract of the fruit shells of *J. macrocarpa*. In addition, in this investigation, the antiproliferative effect of the compound against cancer cells was demonstrated.

Because of the outstanding potential of *A. confertiflora* to inhibit the growth of *Mycobacterium tuberculosis* (Robles-Zepeda et al., 2013), a study carried out by Coronado-Aceves et al. (2016) aimed to isolate the responsible compounds of the mycobactericidal activity of the plant, which led to purified and characterized sesquiterpene lactones reynosin, santamarine, and 1,10-epoxyparthenolide from dichloromethane extract from aerial parts. *In vitro* assays showed that these compounds have great activity against clinical strains of *M. tuberculosis*, particularly reynosin and santamarine.

To enhance the use of *L. palmeri* S. Wats (Verbenaceae) as a food flavoring and preservative, its antimicrobial properties and the composition of its essential oils have been investigated. Ortega-Nieblas et al. (2011) reported that two chemotypes, *p*-cymene/thymol and carvacrol chemotypes, from the state of Sonora, Mexico, present different profiles of essential oils. Among the most abundant compounds that were reported for both chemotypes are *p*-cymene, thymol, carvacrol, and γ-terpinene, in addition to another series of compounds reported for both chemotypes (Table 9.2).

9.4 NITROGEN-CONTAINING COMPOUNDS

Nitrogen-containing medicinal compounds have been used since ancient times, as they are constituents of many drugs (e.g., morphine, quinine, and taxol) (Puri et al., 2018). Betalains are the main group of nitrogen-containing compounds; they are the yellow and violet pigments that substitute anthocyanins in plants belonging to the family Caryophyllales (Achatocarpaceae, Aizoaceae, Amaranthaceae, Basellaceae, Cactaceae, Chenopodiaceae, Didiereaceae, Halophytaceae, Hectorellaceae, Nyctaginaceae, Phytolaccaceae, Portulacaceae, and Stegnospermataceae) (Gandía-Herrero, 2013). The most common and widely known source of betalains are those belonging to the Amaranthaceae (namely, *Beta vulgaris* L. and *Amaranthus* sp.) and Cactaceae families (namely, *Opuntia* sp. and *Hylocereus* sp.) (Gengatharan, 2015).

TABLE 9.2 IDENTIFIED TERPENES IN SOME PLANTS FROM MEXICO DESERT AND SEMI-DESERT AREAS

Plant	Extraction Type	Compound	Reference
Lophocereus schotti	Aerial part/Chloroform-methanol (2:1)	Lophenol, Locereol, Lathosterol, Schottenol, Spinasterol, 24-Methylenelophenol, 5α-campest-7-ene-3β-ol 5α-cholesta-8,14-dien-3β-ol	Campbell and Kircher (1980)
	Aerial part/*n*-Octyl alcohol	Lophenol, Schottenol, Lupeol, $^{\Delta 7}$-stigmasterol	Djerassi et al. (1958)
	Stems/Ethanol	Sitosterol glucoside Schottenol ferulate Phytofluene Phytofluene	Orozco-Barocio et al. (2022)
Fouquieria splendens	Bark/Ether	Ocotillol	Warnhoff and Halls (1965)
	Logs	Ocotillol	Morales-Cepeda et al. (2021)
	Flowers/Ether	Fouquierol Isofouquierol	Nevárez-Prado et al. (2021)
Flourensia cernua	Leaves/Ether	Myrcene, 3-carene, Limonene, 1,8-cineole Borneol, *cis*-jasmone, β-caryophyllene, Caryophyllene oxide, Globulol	Estell et al. (1994)
	Leaves/Ethanol	Santolina triene, Tricyclene, α-Thujene, α-Pinene, Camphene, Sabinene, β-Pinene, Myrcene, Mesitylene, Yomogi alcohol, 3-Carene, α-Terpinene, *p*-Cymene, Limonene, 1,8-cineole, (Z)-β-Ocimene, (E)-β-Ocimene, Artemisia alcohol, Terpinolene,trans-sabinene hydrate, *Cis-p*-Menth-2-en-1-ol, α Campholenal, *trans*-Pinocarveol, Isoborneol Borneol, Terpin-4-ol, *m*-Cymen-8-ol, α-Terpineol, Myrtenal, Myrtenol, cis-Chrysanthenyl acetate, Bornyl acetate, Carvacrol, α-Cubebene, Eugenol, Cyclosativene, α-Copaene,	Estell et al. (2016)

(*Continued*)

TABLE 9.2 (CONTINUED) IDENTIFIED TERPENES IN SOME PLANTS FROM MEXICO DESERT AND SEMI-DESERT AREAS

Plant	Extraction Type	Compound	Reference
		β-Bourbonene, β-Cubebene, (Z)-Jasmone, (E)-Caryophyllene, α-Humulene, Allo-Aromadendrene, Drima-7,9(11)-diene, γ-Muurolene, Germacrene D, β-Selinene, epi-Cubebol, Bicyclogermacrene, α-Muurolene, γ-Cadinene, *cis*-Calamenene, Δ-Cadinene, Cadina-1,4-diene, Elemol, Longicamphenylone, Ledol, Germacrene D-4-ol, Spathulenol, Caryophyllene oxide, β-Oplopenone, 1-epi-Cubenol, Hinesol, (Z)-methyl jasmonate, β-Eudesmol, Selin-11-en-4-α-ol, Bulnesol, (Z)-Methyl epi-jasmonate α-Bisabolol, Eudesma-4(15),7-dien-1-β-ol, Oplopanone, Xanthorrhizol, β-Acoradienol, Nootkatone, Cryptomeridiol, Flourensiadiol, α-Bisabolene epoxide	

(*Continued*)

TABLE 9.2 (CONTINUED) IDENTIFIED TERPENES IN SOME PLANTS FROM MEXICO DESERT AND SEMI-DESERT AREAS

Plant	Extraction Type	Compound	Reference
Asclepias subulata	Aerial parts/Methanol	12,16-Dihydroxycalotropin, Calotropin, Corotoxigenin, 3-*O*-glucopyranoside, Desglucouzarin	Rascón-Valenzuela et al. (2016); Rascón-Valenzuela et al. (2015)
Stegnosperma malifolium	Stems/Methanol	Spinasterol	Meneses- Sagrero et al. (2017)
Ambrosia confertiflora	Aerial parts/Dichlorometane	Reynosin Santamarine 1,10-Epoxyparthenolide	Coronado-Aceves et al. (2016)
Lippia palmeri	Leaves/Essential oils	p–Cymene, Thymol, Carvacrol, γ–Terpinene, iso–Aromandrene, p–Thymol, Longipinene epoxide, α–Eudesmol, Limonene–6–ol, Cariofileno, Thymol acetate, α–Bisabolene, Myrcene, α–Caryophyllene, Linalool, Terpinene–4–ol	Ortega-Nieblas et al. (2011)

The term *betalain* was introduced to describe these pigments that are immonium derivates from betalamic acid (Wohlpart, 1968). Betalamic acid is the chromophore common to all betalain pigments (Strack et al., 2003). The betalamic acid addition residue determines the pigment classification as betacyanin or betaxanthin (Azeredo et al., 2009). The condensation of this structure with amines or their derivatives generates yellow betaxanthins (Latin: *beta* = beet; Greek: *xanthos* = yellow), whereas the condensation of betalamic acid with cyclo-dopa [cyclo-3-(3,4-dihydroxyphenylalanine)] or its glucosyl derivatives gives rise to violet betacyanins (Latin: *beta* = beet; Greek: *kyaneos* = blue) (Khan and Giridhar, 2015). The resulting glycosides of betacyanins can be linked to acylation groups, leading to several structures (Sigurdson et al., 2017). Four structural classes of betacyanins have been reported: Betanin, gomphrenin, amaranthine, and bougainvillein. These structures differ in the attachment of glucosyl groups to oxygen atoms at the o-position on the cyclo-dopa radical (Belhadj Slimen et al., 2017). To date, about 70 betalains have been identified in nature, including about 50 betacyanins and 20 betaxanthins. They are found in fruits and seeds such as beetroot (*B. vulgaris*), amaranth (*Amaranthus* spp.), pears (*Opuntia* spp.), and dragon fruit (*Stenocereus* spp.), among others (Cai et al., 2005; Stintzing and Carle, 2007). They are water-soluble nitrogenous pigments located in the vacuoles of plant tissues responsible for the red-violet (base form, betacyanins, where the betalamic acid appears condensed with *cyclo*-dihydroxyphenylanine or *cyclo*-DOPA) and yellow-orange (acid form, betaxanthin derivates from betalamic acid) colors, which display stability over the pH range of 3–7 (Stintzing et al., 2006).

Although betalain's biosynthesis and regulatory pathways are currently only partially understood, and their origins are unknown, both betalain and anthocyanin compounds are functionally interchangeable, resulting in complex evolutionary relationships (Li, et al, 2019). Betalains are secondary metabolites derived from the amino acid L-tyrosine (Tzin, 2010). The pathway involved in betalain biosynthesis begins with the hydroxylation of L-tyrosine to 3, -dihydroxyphenylalanine or L-DOPA (Dawson, 2009) through the monophenolase activity of the enzyme tyrosinase (or polyphenoloxidase) (Grandia-Herrero and García-Carmona, 2013). Betalains play an important role in human health because of their biological and pharmacological properties, such as antioxidant, anticancer, antilipidemic, antibacterial, etc. (Gengatharan et al., 2015; Vulić et al., 2013; Gandía-Herrero et al., 2016; Khan, 2016; Belhadj et al., 2017; Miguel, 2018; Yong et al., 2018).

Some studies are concerned with the betanin isolation, involving comprehensive mechanisms to extract the pure compounds from plant sources. Among betanin purification studies, chromatographic methods, including HPLC using reverse phase columns, are found to be the most effective (Strack et al., 2003). However, studies about the stability of these molecules or their antioxidant capacity after purification and storage are limited (Tonon et al., 2010). One of the main limitations that prevent the potential use of betanins has been their instability. Therefore, in recent years, several studies have sought to stabilize them and increase their commercial applications (Castro-Enríquez, 2020).

The *Cacteaceae* family is a betalain plant source with high potential. Prickly pear (*Opuntia*) and pear (*Cereus*, *Hylocereus*, and *Selenicerus*) are the most common betalain sources for food coloring (Mizrahi et al., 1997; Stinzing et al., 2006). *Opuntia* is characterized by its tolerance to different soils, temperatures, and humidity. Mexico has great genetic variation, with a wide variety of fruit pulp tones (red, white, and yellow) and with a wide harvest period including early (May), intermediate (August), early ripening (August), and late (November) (Nobel & De la Barrera 2003). Betalains represent a good source of pigments, vitamins, minerals, amino acids, and sugars in the mucilage fraction (Frati et al., 1992), while the cactus fruit is widely used in the food industry (Mobhammer et al., 2005).

Alkaloids are a group of nitrogen-containing compounds that have been used medicinally since ancient times. Alkaloids possess a wide range of pharmacological and therapeutic effects (Al-Safi, 2021). They are mainly represented by nitrogenous organic compounds in plants, fungi, bacteria, and organisms (Dang, 2012), which are an important group with a variety of biological activities (Uzor, 2020). Alkaloids can be classified by their source combined with their chemical structure (Qiu, 2014). They represent one of the most important types of natural products because of their large number and structural diversity and complexity (Yang, 2021). The presence of nitrogen in their structure is the peculiar chemical feature of alkaloids. However, because of the huge structural diversity, alkaloid classification is extremely challenging. More recent classifications are based on carbon skeletons and/or their biochemical precursor (Chiocchio, 2021). There are various classifications for alkaloids, but the most popular classification divides whole alkaloids into three parts. The first are called true alkaloids, which are derived from amino acid and heterocyclic compounds with nitrogen (e.g., tropanes, plyamines, piperidines, quinolizidineas, indolizidines, isoquinolines, tetrahydroisoquinolin, phenylethylamines, indoles, and β-carbolines). Second **are** protoalkaloids; these types of alkaloids have nitrogen, and they are derived from amino acids, and on this basis, they are subsequently divided into phenylethylamino alkaloids, pyrrolizidine alkaloids, and terpenoid indole alkaloids (Roy, 2017). And third are pseudo alkaloids – those that do not originate from amino acid aromatic alkaloids, ephedra alkaloids, purine alkaloids, or sesquiterpene alkaloids, such as isoprenoid alkaloids including mono- (from geraniol), di- (from geranylgeranyl-PP), and triterpene (from cholesterol) derivatives. The latter are called steroidal alkaloids (Aniszewski, 2015).

Alkaloids are widely distributed in the plant kingdom and exist mainly in higher plants, such as species of the family Ranunculaceae, Leguminosae, Papaveraceae, Menispermaceae, and Loganiaceae (Wang, 2009). They play an essential role in human medicine and in the natural life of an organism's defense. Alkaloids account for about 20% of the known secondary plant metabolites (Kaur et al., 2015). In plants, alkaloids protect them from predators and regulate their growth (Doak et al., 2014). Therapeutically, they are especially known to be anesthetic, cardioprotective, and anti-inflammatory. Some alkaloids used in clinical settings include morphine, strychnine, quinine, ephedrine, and nicotine (Kurek, 2019). Recently, interest has returned to biologically active natural products, spurred on by very active developments in the field of traditional medicine and their potential in drug discovery (Atanasov et al., 2021). However, alkaloids are not presented as lead compounds to explore the commercialization and licensing of new drugs (Amirkia and Heinrich, 2015).

There are many reports of plants distributed around the world that show the presence of alkaloids with pharmacological effects, including ajmaline (antiarrhytmic), colchicine (anti-irritant), emetine (antitoxin, emetic), ergot alkaloids (vasoconstrictor, hallucinogenic, inotropic, glaucine (antitussive), morphine (analgesic) nicotine (nicotinic ecetylcholine receptor agoinst), phytostigmine (acethylcholinesterase inhibitor), quinidine (antiarrhytmic), quinine (antipyretic, fever), resperpine (lower blood pressure), tobocuraine (muscle relaxant), vinblastine, vincristine (antitumor), and many other healing effects (Cushnie et al., 2014; Qiu et al, 2014). Recently, a sleep-inducing effect was reported for *Lophophora diffusa* (peyote), a plant used in rituals and ceremonies in Mexico, which was associated to the presence of isoanhalamine, isoanhaldine, isoanhalonidine, and isopellotine (Chan et al., 2021). It is important to point out that this plant is regulated for its exploitation and consumption; therefore, its research is also limited. The

TABLE 9.3 IDENTIFIED NITROGEN-CONTAINING COMPOUNDS PRESENT IN SOME PLANTS FROM MEXICO DESERT AND SEMI-DESERT AREAS AND THEIR ACTIVITIES

Plant	Identified Compounds	Activity	Reference
Opuntia ficus indica	Betanin	Antiproliferative activity in K562 cells	Sreekanth (2007)
Opuntia robusta Wendi (Camuesa)	Betanin, isobetanin, betanidin, and isobetanidin	Source of pigments	Castellanos-Santiago (2008)
Opuntia spp.	Betalain	Anticancer (Chronic myeloid leukemia cell line K562); and antiproliferative (Human colon cancer cell line HT29).	Sreekanth (2007); Serra (2013)
Opuntia matudae	Betalain	Antibacteria (*E. coli* O157:H7)	Hayek (2012)
Pachycereus marginatus (DC) Britton & Rose	Isoquinoline alkaloids	Anticancer lymphoma L5178Y-R	Shetty et al. (2012)
Ibervillea sonorae (S. Watson) Greene	Cucurbitacines	Hypoglycemic activity	Andrade-Cetto and Heinrich (2005) Zapata-Bustos et al. (2014)
Lophophora diffusa (peyote)	Isoanhalamine Isoanhaldine Isoanhalonidine Isopellotine	Sleep-inducing effect	Chan et al. (2021)

Mexican desert could represent an important source of alkaloids for medicinal uses; however, studies are still required in this regard. Examples of different activities of nitrogen-containing compounds present in some desert plants are shown in Table 9.3.

9.5 PERSPECTIVES ON EXTRACTION AND BIOACTIVITY PROTECTION

The interesting medicinal properties of plants from desert areas, particularly in Mexico, have led to the generation of various lines of research focused on the design of extraction techniques in addition to seeking alternatives to increase and protect the properties of their bioactive compounds.

To obtain pure compounds or plant extracts (a mixture of bioactive compounds), techniques classified as conventional or non-conventional are used. Conventional extraction techniques, such as hydrodistillation and Soxhlet extraction, have been successful in recovering essential oils from aromatic plants (Avila-Sosa et al., 2011; Wright et al., 2013).

Non-conventional techniques (e.g., microwave and ultrasound) usually guarantee higher recovery yields, in addition to better bioactivity, as they use less aggressive conditions and shorter extraction times (Guía-García et al., 2022). Non-conventional techniques can be useful to isolate phenolic compounds for which more specific procedures are required. For example, Guía-García et al. (2021) reported higher values of ellagic acid from *R. microphylla* fruits only when ohmic heating extraction was used compared to conventional agitation. Ultrasound-assisted extraction has also been reported to be effective for extracting phenolic compounds from the Mexican desert plants *J. dioica*, *F. cernua*, *Turnera diffusa*, and *Eucalyptus camaldulensis*, which showed good antioxidant activity (Wong-Paz et al., 2015a). Recently, supercritical fluid extraction (SFE) has become one of the most advanced techniques for obtaining bioactive compounds and plant extracts. It is based on the fact that the solvents used for the extraction retain the properties of gas and liquid at the same time (supercritical state); in this way, the solvent diffuses as a gas and dissolves the content as a liquid. CO_2 is the most common solvent for SFE, as it is cheap, environmentally friendly, and generally recognized as safe (GRAS), but other solvents are also investigated to extract compounds that cannot be recovered using CO_2 alone (Al-Maqtari et al., 2021). To date, there are few studies using SFE to extract bioactive compounds from Mexican desert plants, mainly because its implementation requires high investment costs. However, recent works have shown the potential of this technique to efficiently extract oleoresins from *L. graveolens* HBK, which exhibited outstanding antimicrobial activity against two multidrug-resistance strains (*Enterococcus faecalis* and *Staphylococcus aureus*) (Calva-Cruz, et al., 2022).

On the other hand, the susceptibility to degradation of bioactive compounds is widely known, especially under certain conditions of temperature, humidity, and light, among others. In this context, the encapsulation in micro- or nanosystems of pure compounds and plant extracts can protect of their bioactivity while allowing a controlled release, leading to a prolonged effect (Kavetsou et al., 2019). During the encapsulation process, the target compounds (core) are trapped within a wall material (e.g., polysaccharides, lipids, and proteins), forming capsules or fibers, thus improving their physical and chemical stability (Zhang et al., 2016). The shape, size, and functionality will depend on the technique and conditions (e.g., wall materials and concentration) used. Some of the simplest and most economical techniques for encapsulation of bioactive compounds are coacervation, solvent displacement, and layer-by-layer (LbL). While techniques such as spray-drying, freeze-drying, and electrospraying can be more expensive, they are very effective for the formation of micro- and nanostructures (Guía-García et al., 2022). Even though encapsulation has been successfully applied for the protection of bioactive compounds from several natural sources (Fabra et al., 2016; Rodrigues et al., 2020), there are few works focused on plants of the Mexican desert. One of these is the recent work reported by Jasso de Rodríguez et al. (2019), in which plant extracts of *F. cernua*, *F. microphylla*, and *F. retinophylla* were encapsulated using coacervation technique and alginate as wall material. The produced capsules had spherical-shape and were classified as microcapsules (size range of 2.1–68.8 μm), which showed improved stability during release in an *in vitro* gastrointestinal digestion system, as well as strong antioxidant activity compared to unencapsulated extracts.

Given the great potential of Mexican desert plants, novel extraction and encapsulation techniques should be investigated to achieve higher yields, functionality, and broader bioactivity of their bioactive compounds and extracts. This represents an area of opportunity for the development of new drugs with prolonged effects (Figure 9.2).

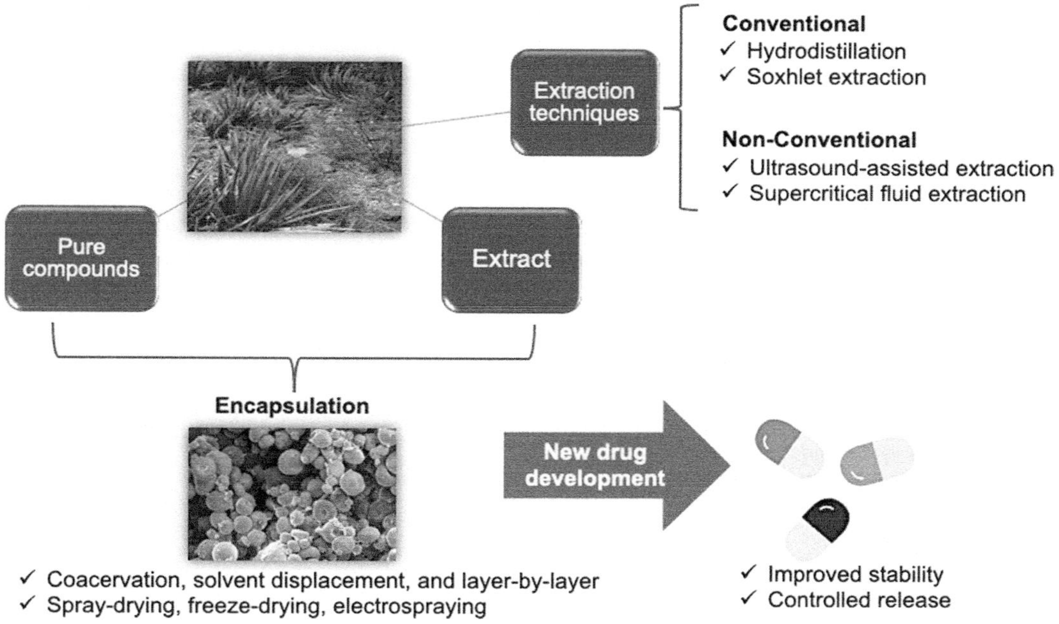

Figure 9.2 Extraction and encapsulation of pure compounds and extracts from Mexico's desert and semi-desert areas for the development of new drugs.

9.6 CONCLUSIONS

Plants from Mexico's semi-desert and desert areas are an excellent source of chemical compounds, such as phenolic compounds, triterpenes, and alkaloids, and their = presence and production are associated with the extreme conditions of growth and development. Extracts or pure compounds isolated from these plants have several biological activities. Thus, these plants could represent a feasible alternative for the development of new pharmacological therapies throughout the compound's extraction or semi-purified fractions; however, it is necessary to carry out toxicity studies to guarantee their possible application.

REFERENCES

Abd El-Ghani, M. M., Huerta-Martínez, F. M., Hongyan, L., & Qureshi, R. (2017). Plant responses to hyperarid desert environments. *Plant Responses to Hyperarid Desert Environments, Rzedowski 1959*, 1–598. https://doi.org/10.1007/978-3-319-59135-3

Ahumada-Santos, Y. P., Montes-Avila, J., de Jesús Uribe-Beltrán, M., Díaz-Camacho, S. P., López-Angulo, G., Vega-Aviña, R., ... & Delgado-Vargas, F. (2013). Chemical characterization, antioxidant and antibacterial activities of six Agave species from Sinaloa, Mexico. *Industrial Crops and Products*, *49*, 143–149.

Al-Maqtari, Q. A., Al-Ansi, W., Mahdi, A. A., Al-Gheethi, A. A. S., Mushtaq, B. S., Al-Adeeb, A., Wei, M., & Yao, W. (2021). Supercritical fluid extraction of four aromatic herbs and assessment of the volatile compositions, bioactive compounds, antibacterial, and anti-biofilm activity. *Environmental Science and Pollution Research*, *28*, 25479–25492. https://doi.org/10.1007/s11356-021-12346-6

Al-Saleem, M. S., Awaad, A. S., Alothman, M. R., & Alqasoumi, S. I. (2018). Phytochemical standardization and biological activities of certain desert plants growing in Saudi Arabia. *Saudi Pharmaceutical Journal, 26*(2), 198–204. https://doi.org/10.1016/j.jsps.2017.12.011

Al-Snafi, A. E., Al-Saedy, H. A., Talab, T. A., Majid, W. J., El-Saber Batiha, G., & Jafari-Sales, A. (2021). The bioactive ingredients and therapeutic effects of Marrubium vulgare – A review. *International Journal of Biological and Pharmaceutical Sciences Archive, 1*(2), 9–21.

Alara, O. R., Abdurahman, N. H., & Ukaegbu, C. I. (2021). Extraction of phenolic compounds: A review. *Current Research in Food Science, 4*, 200-214.

Alcántara-Quintana, L. E., Arjona-Ruiz, C., de Loera, D., Gamboa-León, R., & Terán-Figueroa, Y. (2022). In vitro inhibitory and proliferative cellular effects of different extracts of Struthanthus quercicola: A preliminary study. *Evidence-Based Complementary and Alternative Medicine, 2022*, 1–9.

Aldred, E. M., Buck, C., & Vall, K. (2009). Chapter 22 – Terpenes. In E. M. Aldred, C. Buck & K. Vall (Eds.), *Pharmacology a handbook for complementary healthcare professionals* (pp. 167–174). Churchill Livingstone. https://doi.org/10.1016/B978-0-443-06898-0.00022-0

Alvarez-Perez, O. B., Ventura-Sobrevilla, J. M., Ascacio-Valdés, J. A., Rojas, R., Verma, D. K., & Aguilar, C. N. (2020). Valorization of Flourensia cernua DC as source of antioxidants and antifungal bioactives. *Industrial Crops and Products, 152*, 112422.

Amirkia, V., & Heinrich, M. (2015). Natural products and drug discovery: A survey of stakeholders in industry and academia. *Front. Pharmacol., 6*, 237.

Andrade-Cetto, A. and Heinrich, M. (2005). Mexican plants with hypoglycaemic effect used in the treatment of diabetes. *Journal of Ethnopharmacology, 99*, 325-348. https://doi.org/10.1016/j.jep.2005.04.019

Anguiano-Sevilla, L. A., Lugo-Cervantes, E., Ordaz-Pichardo, C., Rosas-Trigueros, J. L., & Jaramillo-Flores, M. E. (2018). Apoptosis induction of agave lechuguilla torrey extract on human lung adenocarcinoma cells (SK-LU-1). *International Journal of Molecular Sciences, 19*(12), 3765.

Aniszewski, T. (2015). *Alkaloids: Chemistry, Biology, Ecology, and Applications* (2nd ed.). Elsevier.

Assanga, S. B. I., Luján, L. M. L., Ruiz, J. C. G., McCarty, M. F., Cota-Arce, J. M., Espinoza, C. L. L., ... & Ángulo, D. F. (2020). Comparative analysis of phenolic content and antioxidant power between parasitic Phoradendron californicum (toji) and their hosts from Sonoran Desert. *Results in Chemistry, 2*, 100079.

Atanasov, A. G.; Zotchev, S. B.; Dirsch, V. M.; Supuran, C. T. Natural products in drug discovery: Advances and opportunities. *Nat. Rev. Drug Discov.* 2021, *28*, 1–17.

Avila-Sosa, R., Gastélum-Reynoso, G., García-Juárez, M., de la Cruz Meneses-Sánchez, M., Navarro-Cruz, A. R., & Dávila-Márquez, R. M. (2011). Evaluation of different Mexican plant extracts to control anthracnose. *Food and Bioprocess Technology, 4*(4), 655–659. https://doi.org/10.1007/s11947-009-0318-4

Azeredo, H. M. (2009). Betalains: properties, sources, applications, and stability–a review. *International Journal of Food Science & Technology, 44*(12), 2365-2376.

Belhadj Slimen, I., Najar, T., & Abderrabba, M. (2017). Chemical and antioxidant properties of betalains. *Journal of Agricultural and Food Chemistry, 65*(4), 675-689.

Bidak, L. M., Kamal, S. A., Waseem, M., & Halmy, A. (2015). Goods and services provided by native plants in desert ecosystems: Examples from the northwestern coastal desert of Egypt. *Global Ecology and Conservation, 3*, 433–447. https://doi.org/10.1016/j.gecco.2015.02.001

Bravo Hollis, H., Sánchez Mejorada, H. (1978). *Las cactáceas de México*. Universidad Nacional Autónoma de México.

Buckle, J. (2015). Chapter 3 - Basic plant taxonomy. In. J. Buckle, *Clinical Aromatherapy* (3rd ed., pp. 37–72). https://doi.org/10.1016/B978-0-7020-5440-2.00003-6

Cai, Y.Z., Sun, M., & Corke, H. (2005). Characterization and application of betalain pigments from plants of the Amaranthaceae. *Trends Food Sci Technol 16*, 370–376.

Calva-Cruz, O. D. J., Badillo-Larios, N. S., De León-Rodríguez, A., Espitia-Rangel, E., González-García, R., Turrubiartes-Martinez, E. A., Castro-Gallardo, A., & Barba de la Rosa, A. P. (2022). Lippia graveolens HBK oleoresins, extracted by supercritical fluids, showed bactericidal activity against multidrug resistance Enterococcus faecalis and Staphylococcus aureus strains. *Drug Development and Industrial Pharmacy, 47*(10), 1546–1555. https://doi.org/10.1080/03639045.2021.2008417

Campbell, C. E., & Kircher, H. W. (1980). Senita cactus: A plant with interrupted sterol biosynthetic pathways. *Phytochemistry, 19*(12), 2777-2779. https://doi.org/10.1016/S0031-9422(00)83969-2

Castellanos-Santiago, E., & Yahia, E. M. (2008). Identification and quantification of betalains from the fruits of 10 Mexican prickly pear cultivars by high-performance liquid chromatography and electrospray ionization mass spectrometry. *Journal of Agricultural and Food Chemistry, 56*(14), 5758-5764.

Castro-Enríquez, D. D., Montaño-Leyva, B., Toro-Sánchez, D., Juaréz-Onofre, J. E., Carvajal-Millan, E., Burruel-Ibarra, S. E., ... & Rodríguez-Félix, F. (2020). Stabilization of betalains by encapsulation: A review. *Journal of Food Science and Technology, 57*(5), 1587-1600.

Chan, C. B., Poulie, C. B., Wismann, S. S., Soelberg, J., & Kristensen, J. L. (2021). The alkaloids from lophophora diffusa and other "false peyotes". *Journal of Natural Products, 84*(8), 2398–2407. https://doi.org/10.1021/acs.jnatprod.1c00381

Chiocchio, I., Mandrone, M., Tomasi, P., Marincich, L., & Poli, F. (2021). Plant secondary metabolites: An opportunity for circular economy. *Molecules, 26*(2), 1–31.

Clarenc Aarland, R., Castellanos-Hernández, O. A., Rodríguez-Sahagún, A., & Acevedo-Hernández, G. J. (2020). Effect of saline stress on the morphology and phytochemistry of in vitro grown mexican oregano (Lippia graveolens Kunth). *Biotecnia, 22*(3), 131-137.

CONAGUA (Comisión Nacional del Agua, México) (2022). Estaciones meterologicas automáticas. Available in Estaciones Meteorológicas Automáticas (EMA's) (conagua.gob.mx).

Coronado-Aceves, E. W., Velázquez, C., Robles-Zepeda, R. E., Jiménez-Estrada, M., Hernández-Martínez, J., Gálvez-Ruiz, J. C., & Garibay-Escobar, A. (2016). Reynosin and santamarine: Two sesquiterpene lactones from Ambrosia confertiflora with bactericidal activity against clinical strains of Mycobacterium tuberculosis. *Pharmaceutical Biology, 54*(11), 2623–2628. https://doi.org/10.3109/13880209.2016.1173067

Cortés-Chitala, M. D. C., Flores-Martínez, H., Orozco-Ávila, I., León-Campos, C., Suárez-Jacobo, Á., Estarrón-Espinosa, M., & López-Muraira, I. (2021). Identification and quantification of phenolic compounds from Mexican oregano (Lippia graveolens HBK) hydroethanolic extracts and evaluation of its antioxidant capacity. *Molecules, 26*(3), 702.

Cox-Georgian, D., Ramadoss, N., Dona, C., & Basu, C. (2019). Therapeutic and medicinal uses of terpenes. *Medicinal Plants: From Farm to Pharmacy*, 333-359. https://doi.org/10.1007/978-3-030-31269-5_15

Cushnie, T.P., Cushnie, B., & Lamb, A.J. (2014). Alkaloids: An overview of their antibacterial, antibiotic-enhancing and antivirulence activities. *Int J Antimicrob Agents, 44*(5), 377–386.

Dang, T. T. T., Onoyovwi, A., Farrow, S. C., & Facchini, P. J. (2012). Biochemical genomics for gene discovery in benzylisoquinoline alkaloid biosynthesis in opium poppy and related species. *Methods in Enzymology, 515*, 231–266.

Dawson, T. L. (2009). Biosynthesis and synthesis of natural colours. *Color Technol, 125*, 61–73. https://doi.org/10.1111/j.1478-4408.2009.00177.x

De León-Zapata, M. A., Pastrana-Castro, L., Rua-Rodríguez, M. L., Alvarez-Pérez, O.B., Rodríguez-Herrera, R., & Aguilar, C. N. (2016). Experimental protocol for the recovery and evaluation of bioactive compounds of tarbush against postharvest fruit fungi. *Food Chemistry, 198*, 62–67.

Delgadillo Puga, C., Cuchillo Hilario, M., Espinosa Mendoza, J. G., Medina Campos, O., Molina Jijón, E., Díaz Martínez, M., ... & Pedraza Chaverri, J. (2015). Antioxidant activity and protection against oxidative-induced damage of Acacia shaffneri and Acacia farnesiana pods extracts: In vitro and in vivo assays. *BMC Complementary and Alternative Medicine, 15*(1), 1–8.

Delgadillo Puga, C., Cuchillo-Hilario, M., Navarro Ocaña, A., Medina-Campos, O. N., Nieto Camacho, A., Ramírez Apan, T., ... & Pedraza-Chaverri, J. (2018). Phenolic compounds in organic and aqueous extracts from Acacia farnesiana pods analyzed by ULPS-ESI-Q-oa/TOF-MS. In vitro antioxidant activity and anti-inflammatory response in CD-1 mice. *Molecules, 23*(9), 2386.

Djerassi, C., Krakower, G., Lemin, A., Liu, L. H., Mills, J., Villotti, R. (1958). The neutral constituents of the cactus Lophocereus schottii. The structure of Lophenol--4α-Methyl-Δ7-cholesten-3β-ol: A link in sterol biogenesis 1-3. *Journal of the American Chemical Society, 80*(23), 6284–6292. https://doi.org/10.1021/ja01556a031

Doak, B. C., Over, B., Giordanetto, F., & Kihlberg, J. (2014). Oral druggable space beyond the rule of 5: Insights from drugs and clinical candidates. *Chem. Biol., 21*, 1115–1142.

Efferth, T., Oesch, F. (2021). Repurposing of plant alkaloids for cancer therapy: Pharmacology and toxicology. *Seminar in Cancer Biology, 68*, 143–163. https://doi.org/10.1016/j.semcancer.2019.12.010

Esquivel-Gutiérrez, E. R., Manzo-Avalos, S., Peña-Montes, D. J., Saavedra-Molina, A., Morreeuw, Z. P., & Reyes, A. G. (2021). Hypolipidemic and antioxidant effects of guishe extract from Agave lechuguilla, a Mexican plant with biotechnological potential, on streptozotocin-induced diabetic male rats. *Plants, 10*(11), 2492.

Estell, R. E., Fredrickson, E. L., & James, D. K. (2016). Effect of light intensity and wavelength on concentration of plant secondary metabolites in the leaves of Flourensia cernua. *Biochemical Systematics and Ecology, 65*, 108–114. https://doi.org/10.1016/j.bse.2016.02.019

Estell, R. E., Havstad, K. M., Fredrickson, E. L., & Gardea-Torresdey, J. L. (1994). Secondary chemistry of the leaf surface of Flourensia cernua. *Biochemical Systematics and Ecology, 22*(1), 73–77. https://doi.org/10.1016/0305-1978(94)90116-3

Estell, R. E., James, D. K., Fredrickson, E. L., & Anderson, D. M. (2013). Within-plant distribution of volatile compounds on the leaf surface of Flourensia cernua. *Biochemical Systematics and Ecology, 48*, 144–150. https://doi.org/10.1016/j.bse.2012.11.020

Fabra, M. J., Flores-López, M. L., Cerqueira, M. A., Jasso de Rodríguez, D., Lagaron, J. M., & Vicente, A. A. (2016). Layer-by-Layer technique to developing functional nanolaminate films with antifungal activity. *Food and Bioprocess Technology, 9*(3), 471–480. https://doi.org/10.1007/s11947-015-1646-1

Flores-Martínez, H., León-Campos, C., Estarrón-Espinosa, M., & Orozco-Ávila, I. (2016). Optimización del proceso de extracción de sustancias antioxidantes a partir del orégano mexicano (Lippia graveolens HBK) utilizando la metodología de superficie de respuesta (MSR). *Revista Mexicana de Ingeniería Química, 15*(3), 773–785.

Frati Munari, A. C., Vera Lastra, O., & Ariza Andraca, C. R. (1992). Evaluation of nopal capsules in diabetes mellitus. *Gac. Med. Mex. 128*, 431–436.

Fredrickson, E. L., Estell, R. E., & Remmenga, M. D. (2007). Volatile compounds on the leaf surface of intact and regrowth tarbush (Flourensia cernua DC) Canopies. *Journal of Chemical Ecology, 33*, 1867–1875. https://doi.org/10.1007/s10886-007-9360-8

Frías-Zepeda, M. E., & Rosales-Castro, M. (2021). Effect of extraction conditions on the concentration of phenolic compounds in Mexican oregano (Lippia graveolens Kunth) residues. *Revista Chapingo Serie Ciencias Forestales, 27*(3), 367–381.

Gandía-Herrero, F., Escribano, J., & García-Carmona, F. (2016). Biological activities of plant pigments betalains. *Crit Rev Food Sci Nutr, 56*, 937–945.

Gandía-Herrero, F, & García-Carmona, F. (2013). Biosynthesis ofbetalains: Yellow and violet plant pigments. *Trends Plant Sci 18*, 334–343. https://doi.org/10.1016/j.tplants.2013.01.003

García-García, J. D., Anguiano-Cabello, J. C., Arredondo-Valdés, R., Candido del Toro, C. A., Martínez-Hernández, J. L., Segura-Ceniceros, E. P., ... & Ilyina, A. (2021). Phytochemical characterization of Phoradendron bollanum and Viscum album subs. austriacum as Mexican mistletoe plants with antimicrobial activity. *Plants, 10*(7), 1299.

Gengatharan, A., Dykes, G.A., & Choo, W.S. (2015). Betalains: natural plant pigments with potential application in functional foods. *LWT Food Sci Technol, 64*, 645–649. https://doi.org/10.1016/j.lwt.2015.06.0526.

Gigliarelli, G., Zadra, C., Cossignani, L., Robles Zepeda, R. E., Rascón-Valenzuela, L. A., Velázquez-Contreras, C. A., & Marcotullio, M. C. (2018). Two new lignans from the resin of Bursera microphylla A. gray and their cytotoxic activity. *Natural Product Research, 32*(22), 2646-2651.

González-Mendoza, D., Troncoso-Rojas, R., Gonzalez-Soto, T., Grimaldo-Juarez, O., Cecena-Duran, C., Duran-Hernandez, D., & Gutierrez-Miceli, F. (2018). Changes in the phenylalanine ammonia lyase activity, total phenolic compounds, and flavonoids in Prosopis glandulosa treated with cadmium and copper. *Anais da Academia Brasileira de Ciências, 90*, 1465–1472.

Guía-García, J. L., Charles-Rodríguez, A. V, López-Romero, J. C., Torres-Moreno, H., Genisheva, Z., Robledo-Olivo, A., Reyes-Valdés, M. H., Ramírez-Godina, F., García-Osuna, H. T., & Flores-López, M. L. (2021). Phenolic composition and biological properties of Rhus microphylla and Myrtillocactus geometrizans fruit extracts. *Plants, 10*, 1–15. https://doi.org/10.3390/plants10102010

Guía-García, J. L., Charles-Rodríguez, A. V, Reyes-Valdés, M. H., Ramírez-Godina, F., Robledo-Olivo, A., García-Osuna, H. T., Cerqueira, M. A., & Flores-López, M. L. (2022). Micro and nanoencapsulation of bioactive compounds for agri-food applications: A review. *Industrial Crops & Products, 186*(June), 1–13. https://doi.org/10.1016/j.indcrop.2022.115198

Hayek, S. A., & Ibrahim, S. A. (2012). Antimicrobial activity of xoconostle pears (Opuntia matudae) against Escherichia coli O157: H7 in laboratory medium. *International Journal of Microbiology, 2012*, 1–6.

Hernández-García, E., García, A., Avalos-Alanís, F. G., Rivas-Galindo, V. M., Rodríguez-, J., Alcántara-Rosales, V. M., Camacho-Corona, M. R., Avalos-Alanís, F. G., Rivas-Galindo, V. M., Alcántara-Rosales, V. M., & Delgadillo-Puga, C. (2018). Chemical composition of Acacia farnesiana (L) Willd fruits and activity against Mycobacterium tuberculosis and dysentery bacteria. *Journal of Ethnopharmacology.* https://doi.org/10.1016/j.jep.2018.10.031

Hernández-García, E., García, A., Garza-González, E., Avalos-Alanís, F. G., Rivas-Galindo, V. M., Rodríguez-Rodríguez, J., ... & del Rayo Camacho-Corona, M. (2019). Chemical composition of Acacia farnesiana (L) wild fruits and its activity against Mycobacterium tuberculosis and dysentery bacteria. *Journal of Ethnopharmacology, 230*, 74-80.

Iloki-Assanga, S. B., Lewis-Luján, L. M., Lara-Espinoza, C. L., Gil-Salido, A. A., Fernandez-Angulo, D., Rubio-Pino, J. L., & Haines, D. D. (2015). Solvent effects on phytochemical constituent profiles and antioxidant activities, using four different extraction formulations for analysis of Bucida buceras L. and Phoradendron californicum. *BMC Research Notes, 8*(1), 1–14.

Irais, C., Javier, M., & Miguel-angel, T.-L. (2017). *Antimicrobial activity and toxicity of plants from northern Mexico*. April 2020.

Jasso de Rodríguez, D., Puente-Romero, G. N., Díaz-Jiménez, L., Rodríguez-García, R., Ramírez-Rodríguez, H., Villarreal-Quintanilla, J. A., Flores-López, M. L., Carrillo-Lomelí, D. A., & Genisheva, Z. A. (2019). In vitro gastrointestinal digestion of microencapsulated extracts of Flourensia cernua, F. microphylla, and F. retinophylla. *Industrial Crops and Products, 138*(March), 111444. https://doi.org/10.1016/j.indcrop.2019.06.007

Jasso de Rodríguez, D., Victorino-Jasso, M. C., Rocha-Guzmán, M. E., Moreno-Jiménez, M. R., Díaz-Jiménez, L., Rodríguez-García, R., Villareal-Quintanilla, J. Á., Peña-Ramos, F. M., Carrillo-Lomelí, D. A., Genisheva, Z. A., Flores-López, M. L. (2022). Flourensia retinophylla: An outstanding plant from northern Mexico with antibacterial activity. *Industrial Crops and Products, 85*. https://doi.org/10.1016/j.indcrop.2022.115120.

Jiménez-Estrada, M., Velázquez-Contreras, C., Garibay-Escobar, A., Sierras-Canchola, D., Lapizco-Vázquez, R., Ortiz-sandoval, C., Burgos-Hernández, A., & Robles-Zepeda, R. E. (2013). In vitro antioxidant and antiproliferative activities of plants of the ethnopharmacopeia from northwest of Mexico. *BMC Complementary and Alternative Medicine, 13*, 1–8. http://www.biomedcentral.com/1472-6882/13/12

Jucá, M. M., Cysne Filho, F. M. S., de Almeida, J. C., Mesquita, D. D. S., Barriga, J. R. D. M., Dias, K. C. F., ... & Vasconcelos, S. M. M. (2020). Flavonoids: biological activities and therapeutic potential. *Natural Product Research, 34*(5), 692-705.

Kaur, R. A. J. B. I. R., & Arora, S. A. R. O. J. (2015). Alkaloids-important therapeutic secondary metabolites of plant origin. *J Crit Rev, 2*(3), 1-8.

Kavetsou, E., Koutsoukos, S., Daferera, D., Polissiou, M. G., Karagiannis, D., Perdikis, D. C., & Detsi, A. (2019). Encapsulation of mentha pulegium essential oil in yeast cell microcarriers: An approach to environmentally friendly pesticides [research-article]. *Journal of Agricultural and Food Chemistry, 67*(17), 4746–4753. https://doi.org/10.1021/acs.jafc.8b05149

Khan MI (2016). Plant betalains: Safety, antioxidant activity, clinical efficacy, and bioavailability. *Compr Rev Food Sci Food Saf, 15*, 316–330.

Khan, M. I., & Giridhar, P. (2015). Plant betalains: Chemistry and biochemistry. *Phytochemistry, 117*, 267-295.

Khoddami, A., Wilkes, M. A., & Roberts, T. H. (2013). Techniques for analysis of plant phenolic compounds. *Molecules, 18*(2), 2328-2375.

Kumar, M., Puri, S., Pundir, A., Bangar, S. P., & Changan, S. (2021). Composition of selected medicinal plants for therapeutic uses from cold desert of Western Himalaya. *Plants, 1429*(10).

Kurek, J. (Ed.) (2019). Introductory chapter: Alkaloids—Their importance in nature and for human life. In *Alkaloids—Their Importance in Nature and Human Life*. IntechOpen.

Leyva-López, N., Nair, V., Bang, W. Y., Cisneros-Zevallos, L., & Heredia, J. B. (2016). Protective role of terpenes and polyphenols from three species of Oregano (Lippia graveolens, Lippia palmeri and Hedeoma patens) on the suppression of lipopolysaccharide-induced inflammation in RAW 264.7 macrophage cells. *Journal of Ethnopharmacology, 187*, 302–312.

Li, G., Meng, X., Zhu, M., & Li, Z. (2019). Research progress of betalain in response to adverse stresses and evolutionary relationship compared with anthocyanin. *Molecules, 24*(17), 1–14.

López-Romero, J. C., Torres-Moreno, H., Rodríguez-Martínez, K. L., Ramírez-Audelo, V., Vidal-Gutiérrez, M., Hernández, J., ... & González-Aguilar, G. A. (2022). Fouquieria splendens: A source of phenolic compounds with antioxidant and antiproliferative potential. *European Journal of Integrative Medicine, 49*, 102084.

Martínez-Rocha, A., Puga, R., Hernández-Sandoval, L., Loarca-Piña, G., & Mendoza, S. (2008). Antioxidant and antimutagenic activities of Mexican oregano (Lippia graveolens Kunth). *Plant Foods for Human Nutrition, 63*(1), 1–5.

Masyita, A., Mustika Sari, R., Dwi Astuti, A., Yasir, B., Rahma Rumata, N., Emran, T. B., Nainu, F., Simal-Gandara, J. (2022). Terpenes and terpenoids as main bioactive compounds of essential oils, their roles in human health and potential application as natural food preservatives. *Food Chem: X, 13*, 100217. https://doi.org/10.1016/j.fochx.2022.100217

Medina-Morales, M. A., López-Trujillo, J., Gómez-Narváez, L., Mellado, M., García-Martínez, E., Ascacio-Valdés, J. A., ... & Aguilera-Carbó, A. (2017). Effect of growth conditions on β-glucosidase production using Flourensia cernua leaves in a solid-state fungal bioprocess. *3 Biotech, 7*(5), 1–6.

Menchaca, M. D. C. V., Morales, C. R., Star, J. V., Or, A., Morales, M. E. R., Nuntilde, M. A., & Gallardo, L. B. S. (2013). Antimicrobial activity of five plants from Northern Mexico on medically important bacteria. *African Journal of Microbiology Research, 7*(43), 5011-5017.

Mendez, M., Rodríguez, R., Ruiz, J., Morales-Adame, D., Castillo, F., Hernández-Castillo, F. D., & Aguilar, C. N. (2012). Antibacterial activity of plant extracts obtained with alternative organics solvents against food-borne pathogen bacteria. *Industrial Crops and Products, 37*(1), 445-450.

Meneses-Sagrero, S. E., Navarro-Navarro, M., Ruiz-Bustos, E., Del-Toro-Sánchez, C. L., Jiménez-Estrada, M., & Robles-Zepeda, R. E. (2017). Antiproliferative activity of spinasterol isolated of Stegnosperma halimifolium (Benth, 1844). *Saudi Pharmaceutical Journal, 25*(8), 1137–1143. https://doi.org/10.1016/j.jsps.2017.07.001

Michel-López, C. Y., González-Mendoza, D., Ruiz-Sánchez, E., & Zamora-Bustillos, R. (2014). Modifications of photochemical efficiency, cellular viability and total phenolic content of Prosopis glandulosa leaves exposed to copper. *Chemistry and Ecology, 30*(3), 227–232.

Miguel, M. (2018). Betalains in some species of the Amaranthaceae family: A review. *Antioxidants, 7*, 53.

Mizrahi, Y., Nerd, A. & Nobel, P. S. (1997). Cacti as crops. *Horticultural Reviews, 18*, 291–320.

Mobhammer, M. R., Stintzing, F. C. & Carle, R. (2005). Colour studies on fruit juice blends from Opuntia and Hylocereus cacti and betalain-containing model solutions derived therefrom. *Food Research International, 38*, 975–981.

Monreal-García, H. M., Almaraz-Abarca, N., Ávila-Reyes, J. A., Torres-Ricario, R., González-Elizondo, M. S., Herrera-Arrieta, Y., & Gutiérrez-Velázquez, M. V. (2019). Phytochemical variation among populations of Fouquieria splendens (Fouquieriaceae). *Botanical Sciences, 97*(3), 398-412.

Morales-Cepeda, A. B., Macclesh Del Pino-Pérez, L. A., Marmolejo, M., Rivera-Armenta, J. L., Peraza-Vázquez, H. (2021). Isolation of ocotillol/ocotillone from Fouquieria splendens (Ocote) using a batch reactor. *Preparative Biochemistry and Biotechnology, 52*(5): 540-548. https://doi.org/10.1080/10826068.2021.1972425

Morales-Rubio, E., Treviño-Neávez, J. F., & Viveros-Valdez, E. (2010). Free radical scavenging activities of Lophocereus schottii (Engelmann). *International Journal of Natural and Engineering Sciences, 4*(1), 65-68.

Morreeuw, Z. P., Castillo-Quiroz, D., Ríos-González, L. J., Martínez-Rincón, R., Estrada, N., Melchor-Martínez, E. M., ... & Reyes, A. G. (2021a). High throughput profiling of flavonoid abundance in Agave lechuguilla residue-valorizing under explored mexican plant. *Plants, 10*(4), 695.

Morreeuw, Z. P., Escobedo-Fregoso, C., Ríos-González, L. J., Castillo-Quiroz, D., & Reyes, A. G. (2021b). Transcriptome-based metabolic profiling of flavonoids in Agave lechuguilla waste biomass. *Plant Science, 305*, 110748.

Morreeuw, Z. P., Ríos-González, L. J., Salinas-Salazar, C., Melchor-Martínez, E. M., Ascacio-Valdés, J. A., Parra-Saldívar, R., ... & Reyes, A. G. (2021c). Early optimization stages of Agave lechuguilla bagasse processing toward biorefinement: Drying procedure and enzymatic hydrolysis for flavonoid extraction. *Molecules, 26*(23), 7292.

Mustafa, G., Arif, R., Atta, A., Sharif, S., & Jamil, A. (2017). Contents list available at RAZI publishing bioactive compounds from medicinal plants and their importance in drug discovery in Pakistan. *Matrix Science Pharma (MSP), 1*(1), 17–26.

Naikoo, M. I., Dar, M. I., Raghib, F., Jaleel, H., Ahmad, B., Raina, A., Khan, F. A., & Naushin, F. (2019). Chapter 9 – Role and regulation of plants phenolics in abiotic stress tolerance: An overview. In M. I. R. Khan, P. S. Reddy, A. Ferrante, & N. A. Khan (Eds.), *Plant signaling molecules* (pp. 157–168). Woodhead Publishing. https://doi.org/https://doi.org/10.1016/B978-0-12-816451-8.00009-5

Nevárez-Prado, L., Rocha-Gutiérrez, B. A., Neder-Suárez, D., Cordova-Lozoya, M. T., Ayala Soto, J. G., Salazar Balderrama, M. I., Ruiz Anchondo J. R., Hernández-Ochoa, L. R. (2021). El género Fouquieria: descripción y revisión de aspectos etnobotánicos, fitoquímicos y biotecnológicos. *Tecnociencia Chihuahua, 15*, 186–220. I: https://doi.org/10.54167/tecnociencia.v15i3.840

Nobel, P. S., & De la Barrera, E. (2003). Tolerances and acclimation to low and high temperatures for cladodes, fruits and roots of a widely cultivated cactus, Opuntia ficus-indica. *New Phytologist, 157*(2), 271–279.

Olmedo-Juárez, A., Zarza-Albarran, M. A., Rojo-Rubio, R., Zamilpa, A., Gonzalez-Cortazar, M., Mondragón-Ancelmo, J., ... & Mendoza-de Gives, P. (2020). Acacia farnesiana pods (plant: Fabaceae) possesses antiparasitic compounds against Haemonchus contortus in female lambs. *Experimental Parasitology, 218*, 107980.

Orozco-Barocio, A., Paniagua-Domínguez, B. L., Benítez-Saldaña, P. A., Flores-Torales, E., Velázquez-Magaña, S., Nava, H. J. A. (2013). Cytotoxic effect of the ethanolic extract of Lophocereus schottii: a Mexican medicinal plant. *African Journal of Traditional, Complementary, and Alternative Medicines, 10*(3), 397-404.

Orozco-Barocio, A., Robles-Rodríguez, B. S., del Rayo Camacho-Corona, M., Méndez-López, L. F., Godínez-Rubí, M., Peregrina-Sandoval, J., ... Ortuno-Sahagun, D. (2022). In vitro anticancer activity of the polar fraction from the Lophocereus schottii ethanolic extract. *Frontiers in Pharmacology, 13*, 1–11.

Ortega-Nieblas, M., Robles-Burgueño, M., Acedo-Félix, E., González-León, A., Morales-Trejo, A., & Vázquez-Moreno, L. (2011). Chemical composition and antimicrobial activity of oregano (Lippia palmeri S. Wats) essential oil. *Revista Fitotecnia Mexicana, 34*(1), 11–17.

Perveen, S. (2018). Introductory chapter: Terpenes and terpenoids. In S. Perveen & A. Al-Taweel (Eds.), *Terpenes and Terpenoids*. IntechOpen. https://doi.org/10.5772/intechopen.79683

Picos-Salas, M. A., Gutiérrez-Grijalva, E. P., Valdez-Torres, B., Angulo-Escalante, M. A., López-Martínez, L. X., Delgado-Vargas, F., & Heredia, J. B. (2021). Supercritical CO2 extraction of oregano (Lippia graveolens) phenolic compounds with antioxidant, α-amylase and α-glucosidase inhibitory capacity. *Journal of Food Measurement and Characterization, 15*(4), 3480-3490.

Pío-León, J. F., López-Angulo, G., Paredes-López, O., Uribe-Beltrán, M. D. J., Díaz-Camacho, S. P., & Delgado-Vargas, F. (2009). Physicochemical, nutritional and antibacterial characteristics of the fruit of Bromelia pinguin L. *Plant Foods for Human Nutrition, 64*(3), 181–187.

Puri, S. K., Habbu, P. V., Kulkarni, P. V., & Kulkarni, V. H. (2018). Nitrogen containing secondary metabolites from endophytes of medicinal plants and their biological/pharmacological activities-A review. *Systematic Reviews in Pharmacy, 9*(1), 22–30. https://doi.org/10.5530/srp.2018.1.5

Qiu S., Sun H., Zhang A. H., Xu H. Y., Yan G. L, Han Y., & Wang X. J. (2014). Natural alkaloids: basic aspects, biological roles, and future perspectives. *Chin J Nat Med, 12*(6), 401–406.

Quintanilla-Licea, R., Gomez-Flores, R., Samaniego-Escamilla, M. Á., Hernández-Martínez, H. C., Tamez-Guerra, P., & Morado-Castillo, R. (2016). Cytotoxic effect of methanol extracts and partitions of two Mexican desert plants against the murine lymphoma L5178Y-R. *American Journal of Plant Sciences, 7*(11), 1521.

Quintanilla-Licea, R., Vargas-Villarreal, J., Verde-Star, M. J., Rivas-Galindo, V. M., & Torres-Hernández, Á. D. (2020). Antiprotozoal activity against Entamoeba histolytica of flavonoids isolated from Lippia graveolens kunth. *Molecules*, *25*(11), 2464.

Radha, M. K., Puri, S., Pundir, A., Bangar, S. P., Changan, S., ... Mekhemar, M. (2021). Evaluation of nutritional, phytochemical and mineral composition of selected medicinal plants for therapeutic uses from cold desert of Western Himalaya. *Plants*, *10*(7), 1–16.

Rascón-Valenzuela, L. A., Velázquez, C., Garibay-Escobar, A., Medina-Juárez, L. A., Vilegas, W., Robles-Zepeda, R. E. (2015). Antiproliferative activity of cardenolide glycosides from Asclepias subulata. *Journal of Ethnopharmacology*, *171*, 280–286. 10.1016/j.jep.2015.05.057

Rascón-Valenzuela, L. A., Velázquez, C., Garibay-Escobar, A., Vilegas, W., Medina-Juárez, L. A., Gámez-Meza, N., & Robles-Zepeda, R. E. (2016). Apoptotic activities of cardenolide glycosides from Asclepias subulata. *Journal of Ethnopharmacology*, *193*, 303–311. https://doi.org/10.1016/j.jep.2016.08.022

Robles-Zepeda, R. E., Coronado-Aceves, E. W., Velázquez-Contreras, C. A., Ruiz-Bustos, E., Navarro-Navarro, M., & Garibay-Escobar, A. (2013). In vitro anti-mycobacterial activity of nine medicinal plants used by ethnic groups in Sonora, Mexico. *BMC Complementary and Alternative Medicine*, *13*, 1–6. https://doi.org/10.1186/1472-6882-13-329

Rodrigues, R. M., Ramos, P. E., Cerqueira, M. F., Teixeira, J. A., Vicente, A. A., Pastrana, L. M., Pereira, R. N., & Cerqueira, M. A. (2020). Electrosprayed whey protein-based nanocapsules for β-carotene encapsulation. *Food Chemistry*, *314*(November 2019), 126157. https://doi.org/10.1016/j.foodchem.2019.126157

Roy, A. (2017). A review on the alkaloids an important therapeutic compound from plants. *International Journal of Plant Biotechnology*, *3*(2), 1–9.

Ruiz-Martínez, J., Ascacio, J. A., Rodríguez, R., Morales, D., & Aguilar, C. N. (2011). Phytochemical screening of extracts from some Mexican plants used in traditional medicine. *Journal of Medicinal Plants Research*, *5*(13), 2791-2797.

Salido, A. A. G., Assanga, S. B. I., Luján, L. M. L., Ángulo, D. F., Espinoza, C. L. L., & ALA, S. A. (2016). Composition of secondary metabolites in Mexican plant extracts and their antiproliferative activity towards cancer cell lines. *International Journal of Science*, *5*, 63–77.

Sánchez, E., Morales, C. R., Castillo, S., Leos-Rivas, C., García-Becerra, L., & Martínez, D. M. O. (2016). Antibacterial and antibiofilm activity of methanolic plant extracts against nosocomial microorganisms. *Evidence-based Complementary and Alternative Medicine: eCAM*, *2016*, 1–8.

Scogin, R. (1977). Anthocyanins of the Fouquieriaceae. *Biochemical Systematics and Ecology*, *5*(4), 265-267.

Scogin, R. (1978). Leaf phenolics of the Fouquieriaceae. *Biochemical Systematics and Ecology*, *6*(4), 297-298.

Serra, A. T., Poejo, J., Matias, A. A., Bronze, M. R., & Duarte, C. M. (2013). Evaluation of Opuntia spp. derived products as antiproliferative agents in human colon cancer cell line (HT29). *Food Research International*, *54*(1), 892-901.

Serrano-Gallardo, L. B., Castillo-Maldonado I., Borjón-Ríos C. G., Rivera-Guillén M. A., Morán-Martínez J., Téllez- López M. A., García-Salcedo J. J., Pedroza-Escobar D. & Vega-Menchaca M. del C. (2017). Antimicrobial activity and toxicity of plants from northern Mexico. *Indian Journal of Traditional Knowledge*, *16*(2), 203-207.

Sharma, A., Shahzad, B., Rehman, A., Bhardwaj, R., Landi, M., & Zheng, B. (2019). Response of phenylpropanoid pathway and the role of polyphenols in plants under abiotic stress. *Molecules*, *24*(13), 2452.

Shetty, A. A., Rana, M. K. and Preetham, S. P. (2012). Cactus: A medicinal food. *Journal of Food Science and Technology*, *49*, 530-536. https://doi.org/10.1007/s13197-011-0462-5

Sigurdson, G. T., Tang, P., & Giusti, M. M. (2017). Natural colorants: Food colorants from natural sources. *Annual Review of Food Science and Technology*, *8*, 261-280.

Soto-Hernández, M., Tenango, M. P., & García-Mateos, R. (Eds.) (2017). *Phenolic compounds: biological activity*. BoD–Books on Demand.

Sreekanth, D., Arunasree, M., Roy, K. R., Reddy, T. C., Reddy, G. V., & Reddanna, P. (2007). Betanin a betacyanin pigment purified from fruits of Opuntia ficus-indica induces apoptosis in human chronic myeloid leukemia Cell line-K562. *Phytomedicine*, *14*, 739–746.

Stintzing FC, Carle R (2007). Betalains–emerging prospects for food scientists. *Trends Food Sci Technol*, *18*, 514–525.

Stintzing, F. C., Kugler, F., Carle, R. & Conrad, J. (2006). First 13C- NMR assignments of betaxanthins. *Helvetica Chimica Acta, 89,* 1008–1016.

Strack, D., Vogt, T. & Schliemann, W. (2003). Recent advances in betalain research. *Phytochemistry, 62,* 247–269.

Tanase, C., Coşarcă, S., & Muntean, D. L. (2019). A critical review of phenolic compounds extracted from the bark of woody vascular plants and their potential biological activity. *Molecules, 24*(6), 1182.

Tonon, R. V., Brabet, C., & Hubinger, M. D. (2010). Anthocyanin stability and antioxidant activity of spray-dried açai (Euterpe oleracea Mart.) juice produced with different carrier agents. *Food Research International, 43*(3), 907–914.

Torres-Moreno, H., López-Romero, J. C., Vidal-Gutiérrez, M., Rodríguez-Martínez, K. L., Robles-Zepeda, R. E., Vilegas, W., & Velarde-Rodríguez, G. M. (2022). Seasonality impact on the anti-inflammatory, antiproliferative potential and the lignan composition of Bursera microphylla. *Industrial Crops and Products, 184,* 115095.

Tucuch-Pérez, M. A., Arredondo-Valdés, R., & Hernández-Castillo, F. D. (2020). Antifungal activity of phytochemical compounds of extracts from Mexican semi-desert plants against Fusarium oxysporum from tomato by microdilution in plate method. *Nova Scientia, 12*(25), 1–19.

Tzin, V., & Galili, G. (2010). New insights into the shikimate and aromatic amino acids biosynthesis pathways in plants. *Molecular Plant, 3*(6), 956–972.

Uzor, P. F. (2020). Alkaloids from plants with antimalarial activity: A review of recent studies. *Evidence-Based Complementary and Alternative Medicine, 2020,* 1–17.

Vega-Ruiz, Y. C., Hayano-Kanashiro, C., Gámez-Meza, N., & Medina-Juárez, L. A. (2021). Determination of chemical constituents and antioxidant activities of leaves and stems from Jatropha cinerea (Ortega) müll. arg and jatropha cordata (ortega) müll. arg. *Plants, 10*(2), 212.

Vidal-Gutiérrez, M., Torres Moreno, H., Jiménez Estrada, M., Escobar, G., Velazquez Contreras, C. A., Robles Zepeda, R. E. (2020a). Cytotoxic activity of primulasaponin isolated from Jacquinia macrocarpa Cav. spp. pungens (A. Gray) B. Ståhl. *Pharmacology Online, 1,* 121–1218.

Vidal-Gutiérrez, M., Robles-Zepeda, R. E., Vilegas, W., Gonzalez-Aguilar, G. A., Torres-Moreno, H., & López-Romero, J. C. (2020b). Phenolic composition and antioxidant activity of Bursera microphylla A. Gray. *Industrial Crops and Products, 152,* 112412.

Vulić, J., Ćanadanović-Brunet, J., Ćetković, G., Dilas, S., & Mandić, A. (2010). Beet root pomace: A good source of antioxidant phytochemicals. In *Review of faculty of engineering: Analecta technica Szegedinensia, 2–3,* 278–280.

Wang, Z. T., & Liang, G. Y. (2009). *Zhong yao hua xue.* Shanghai Scientific and Technical.

Wang, H., Yu, P., Bai, J., Zhang, J., Kong, L., Zhang, F., Du, G., Pei, S., Zhang, L., Jiang, Y., Tian, J., Fu, F. (2013). Ocotillol enhanced the antitumor acitivity of doxorubicin via p53-dependent apoptosis. *Evidence-Based Complementary and Alternative Medicine, 8.* https://doi.org/10.1155/2013/468537

Warnhoff, E. W., & Halls, C. M. M. (1965). Desert plant constituents: II. Ocotillol: an intermediate in the oxidation of hydroxy isoöctenyl side chains. *Canadian Journal of Chemistry, 43*(12), 3311–3321. https://doi.org/10.1139/v65-461

Wohlpart, A., & Mabry, T. J. (1968). On the light requirement for betalain biogenesis. *Plant Physiology, 43*(3), 457.

Wollenweber, E., & Yatskievych, G. (1994). External flavonoids of ocotillo (Fouquieria splendens). *Zeitschrift für Naturforschung C, 49*(9-10), 689-690.

Wong-Paz, J. E., Muñiz Márquez, D. B., Martínez Ávila, G. C. G., Belmares Cerda, R. E., & Aguilar, C. N. (2015a). Ultrasound-assisted extraction of polyphenols from native plants in the Mexican desert. *Ultrasonics Sonochemistry, 22,* 474–481. https://doi.org/10.1016/j.ultsonch.2014.06.001

Wong-Paz, J. E., Contreras-Esquivel, J. C., Rodríguez-Herrera, R., Carrillo-Inungaray, M. L., López, L. I., Nevárez-Moorillón, G. V., & Aguilar, C. N. (2015b). Total phenolic content, in vitro antioxidant activity and chemical composition of plant extracts from semiarid Mexican region. *Asian Pacific Journal of Tropical Medicine, 8*(2), 104-111.

Wright, C., Chhetri, B. K., & Setzer, W. N. (2013). Chemical composition and phytotoxicity of the essential oil of Encelia farinosa growing in the Sonoran Desert. *American Journal of Essential Oils and Natural Products, 1*(1), 18–22.

Yang, Y., Li, X., Zhang, C., Lv, L., Gao, B., & Li, M. (2021). Research progress on antibacterial activities and mechanisms of natural alkaloids: A review. *Antibiotics, 10*(3), 1–30.

Yong Y. Y., Dykes G., Lee S. M., Choo W. S. (2018). Effect of refrigerated storage on betacyanin composition, antibacterial activity of red pitahaya (Hylocereus polyrhizus) and cytotoxicity evaluation of betacyanin rich extract on normal human cell lines. *LWT, 91*, 491–497.

Zapata-Bustos, R., Alonso-Castro, Á. J., Gómez-Sánchez, M. and Salazar-Olivo, L. A. (2014). Ibervillea sonorae (Cucurbitaceae) induces the glucose uptake in human adipocytes by activating a PI3K-independent pathway. *Journal of Ethnopharmacology, 152*, 546–552. https://doi.org/10.1016/j.jep.2014.01.041

Zapata-Campos, C. C., García-Martínez, J. E., Chavira, J. S., Valdés, J. A. A., Morales, M. A. M., & Mellado, M. (2020). Chemical composition and nutritional value of leaves and pods of Leucaena leucocephala, Prosopis laevigata and Acacia farnesiana in a xerophilous shrubland. *Emirates Journal of Food and Agriculture, 32*, 723–730.

Zarza-Albarrán, M. A., Olmedo-Juárez, A., Rojo-Rubio, R., Mendoza-de Gives, P., González-Cortazar, M., Tapia-Maruri, D., ... & Zamilpa, A. (2020). Galloyl flavonoids from Acacia farnesiana pods possess potent anthelmintic activity against Haemonchus contortus eggs and infective larvae. *Journal of Ethnopharmacology, 249*, 112402.

Zhang, Z., Zhang, R., Zou, L., Chen, L., Ahmed, Y., Al Bishri, W., Balamash, K., & McClements, D. J. (2016). Encapsulation of curcumin in polysaccharide-based hydrogel beads: Impact of bead type on lipid digestion and curcumin bioaccessibility. *Food Hydrocolloids, 58*, 160–170. https://doi.org/10.1016/j.foodhyd.2016.02.036

Chapter 10

Edible Coating Based on Chia (*Salvia hispanica* L.) Functionalized with *Rhus microphylla* Fruit Extract to Improve the Cucumber (*Cucumis sativus* L.) Shelf Life

Ana Veronica Charles-Rodríguez, Maria Reyes de la Luz, Jorge L. Guía-García, Fidel Maximiano Peña-Ramos, Armando Robledo-Olivo, Antonio F. Hench-Cabrera, and María Liliana Flores-López

CONTENTS

10.1 INTRODUCTION

The agri-food industry suffers important product losses due to several biotic and abiotics factors, reaching the highest losses in fruit and vegetables during the postharvest stage (up to 40%) (Flores-López et al., 2016). Cucumber (*Cucumis sativus* L.) is one of the fruits with the highest margin of loss because of its high water content, and this promotes microbial attack with an average shelf life

DOI: 10.1201/9781003251255-10

between 5 to 7 days (Mohammadi et al., 2015). In Mexico, this crop occupies the fifth place in exports, representing an important market for the country, making development of new strategies to maintain its quality and extend its shelf life relevant (Trejo Valencia et al., 2018).

A novel alternative is the use of edible coatings, which act as a barrier against water loss, thus extending the postharvest quality and shelf life of the produce (Medeiros et al., 2012). In addition, edible coatings can act as vehicles of bioactive compounds or extracts from plants, leveraging their antioxidant and antimicrobial properties which could enhance the benefits of fruit and vegetables (Cai et al., 2020; Yin et al., 2019). Different materials can be used to produce coatings, but natural materials such as biopolymers (e.g., gums, alginates, mucilages.) are preferred (Hassan et al., 2018). The use of mucilages has become relevant because of their high fiber content; besides, these materials have a high water-retention capacity and solubility, which gives them viscous properties (Waghmare et al., 2021). In this regard, mucilage extracted from chia seeds has been used as edible material to enhance the shelf life of strawberries, reducing the enzymatic browning of the fruit (Mousavi et al., 2021). Moreover, the mucilage has showed remarkable properties when it is combined with essential oils, enhancing their antifungal properties against *Aspergillus* spp. and *Penicillium* spp. (Muñoz-Tébar et al., 2022).

On the other hand, the interesting activities of Mexican desert and semi-desert plants (e.g., antioxidant and antifungal) are the result of their tremendous metabolic machinery adapted to survive extreme climatic conditions (De León-Zapata et al., 2016). For example, the ethanol extracts of *Juglans* spp. and *Carya* sp. have been effective in inhibiting the growth of *Fusarium oxysporum* on tomato crops under greenhouse conditions (Jasso de Rodríguez et al., 2019). In addition, nanolaminate coatings based on chitosan and alginate were used as carriers of ethanol extracts of *Flourensia cernua*, proving to be an important barrier to fungal attack, as well as reducing the weight loss during postharvest storage of tomato fruits (20 °C for 15 days) (Salas-Méndez et al., 2019). The authors associated the results to the antioxidant and antimicrobial activities of the extracts of *F. cernua*. The functionalization of edible coatings for the development of postharvest technologies is an area of research receiving great interest, especially with the incorporation of bioactive compounds or extracts from Mexican desert and semi-desert plants. This chapter reports, for the first time, the effect of an edible coating based on chia mucilage containing *R. microphylla* fruit extract on the shelf life of cucumber fruits.

10.2 MATERIALS AND METHODS

10.2.1 Reagents

Chia seeds and cucumber fruits were purchased from a local market in Saltillo, Coahuila, Mexico. Peptone water and count plate agar were acquired from BD Difco (Spain). The 2,6-Dichlorophenolindophenol and ascorbic acid were obtained from Sigma-Aldrich (USA). All reagents used were analytical grade.

10.2.2 Mucilage Extraction

The mucilage extraction was performed following the method described by Charles-Rodríguez et al. (2020), with some modifications. Briefly, chia seeds were hydrated with distilled water (1:10, w/v) for 60 minutes (min) under continuous stirring at room temperature. The mixture was liquefied and centrifuged to separate the phases. The resulting mucilage was lyophilized (Labconco

Freezone 6.0, Expotech, Houston, TX, USA) at -55°C and 0.133 mbar for 24 h and stored in a dry place until use.

10.2.3 Shelf-life Assay in Cucumber Fruits

The fruits were washed with distilled water and dried in an oven. Then, the cucumbers were immersed in the coating solution (0.24% w/v lyophilized chia mucilage, 0.15% w/v sodium chloride, and 0.05% w/v glycerol) containing *R. microphylla* extract (C+R) (2500 mg/L) for 10 seconds (s) to apply the coating on the surface of the fruit. The coated cucumbers were placed in aluminum trays and dried at 30°C for 20 min in a convection oven (Biobase Biodustry Shandongo Co., Ltd., Jinan, China) to dry the coating. For control, uncoated fruits were used. Finally, the cucumbers were stored at 18°C for 15 days (d) at 85% relative humidity (RH), and the physicochemical and microbiological tests were performed at regular intervals (0, 3, 6, 9, and 15 d).

10.2.4 Weight Loss

The weight loss of cucumbers during storage was evaluated by the mass change. The results were expressed as percentage (%) and calculated using Equation 10.1.

$$\text{Weight loss}(\%) = \frac{W_0 - W_f}{W_0} x100 \tag{10.1}$$

where W_0 is the initial weight, and W_f is the respective weight of every 3 d.

10.2.5 Total Soluble Solids (TSS) and pH

The juice of cucumbers was obtained by weighing and liquefying 10 g of fruit pulp. For the determination of TSS, a juice droplet was placed on a refractometer (PROCKET-ATAGO, USA). The pH value was directly determined from juice using a potentiometer (HANNA, China).

10.2.6 Vitamin C

Vitamin C content was evaluated by the method of 2,6-dichlorophenolindophenol using an ascorbic acid standard curve (AOAC, 1990). The results were expressed as mg of vitamin per 100 g of fresh weight.

10.2.7 Color

The surface color was analyzed using a colorimeter (Minolta, USA). The luminosity (L^*), blue/red color (a^*), and yellow/blue tone (b^*) were evaluated. In addition, the hue angle (°H) was calculated according to Equation 10.2:

$$\text{Hue angle}(°\text{H}) = \tan^{-1}\frac{b^*}{a^*} \tag{10.2}$$

10.2.8 Microbiological Analysis

The microbiological assay followed the methodology described by Sogvar et al. (2016) with minor modifications. Briefly, 10 g of a sample were homogenized and diluted in 90 mL of sterile peptone water. Serial dilutions were poured on Petri dishes, then plate count agar was added, and the mixtures were homogenized. Subsequently, they were incubated at 30°C for 2 d for the determination of total mesophiles. The assay was performed twice, and the results were expressed as logarithm of colony forming units per gram (log CFU/g).

10.2.9 Statistical Analyses

All assays were made in triplicate, and the results were expressed as means ± standard deviations. One-way analyses of variance (ANOVA) were used to detect significant differences in each assay followed by Tukey's mean comparison test ($p < 0.05$) using Minitab software version 17.0 (State College, PA, USA).

10.3 RESULTS AND DISCUSSION

10.3.1 Weight Loss

Figure 10.1 shows the weight loss during the storage period. Uncoated cucumbers presented the highest ($p < 0.05$) water loss with a value of 5.1% at the end of the test, while coated cucumbers with C+R had a weight loss of 2.6% at 15 d of storage. The decrease in weight is associated to the loss of moisture and volatile compounds from the surface of fresh fruits, which cause some physiological changes (Brasil et al., 2012). The results obtained agree with those reported by Patel and Panigrahi (2019), in which cucumber fruits coated with 1.5 μM of starch and 2.5 μM with D-glucose had lower weight loss than the control during the storage period of 20 d at 4°C. The lower weight loss observed in fruits coated with C+R may be due to the barrier created by the chia mucilage and components of the *R. microphylla* fruit extract. Polysaccharide-based coatings have good barrier properties to O_2 and CO_2, but one disadvantage is the lower moisture

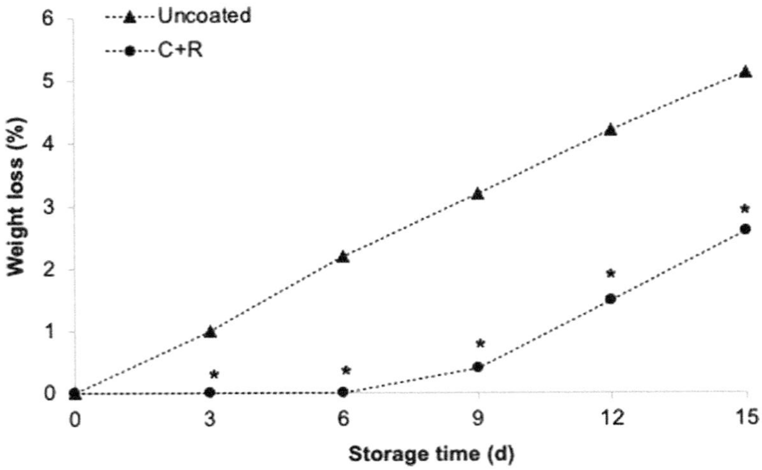

Figure 10.1 Weight loss (%) of uncoated (control) and coated (C+R) cucumber fruits. Significant differences between means (*$p < 0.05$) compared to uncoated cucumber fruits.

barrier properties because of the hydrophilic nature of polysaccharides (Flores-López et al., 2016). However, it seems that the synergy between chia mucilage and *R. microphylla* fruit extract can be more effective, as the intermolecular interactions between components are more stable (Charles-Rodríguez et al., 2020).

10.3.2 TSS and pH

The TSS values showed significant differences between treatments, as the fruits treated with C+R showed an increased value of 2.7%–3.1%, while the control showed values of 2.4%–2.9% (Figure 10.2a). The increase in TSS values observed during storage on the coated cucumbers has been reported for other fruits, such as goji fruit (Fan et al., 2019). This could be due by the hydrolysis of insoluble polysaccharides during the maturation process (Yang et al., 2019). A positive

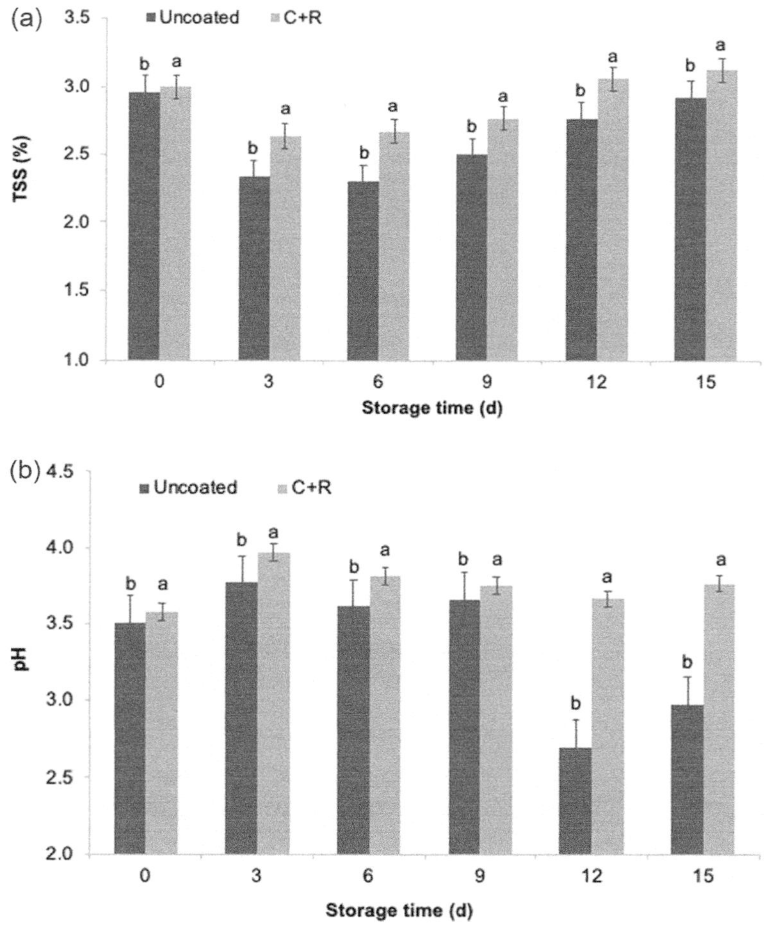

Figure 10.2 Total soluble sugars (a) and pH (b) values of uncoated (control) and coated (C+R) cucumber fruits. Different letters (a, b) on the same day indicate significant differences ($p < 0.05$) between treatments.

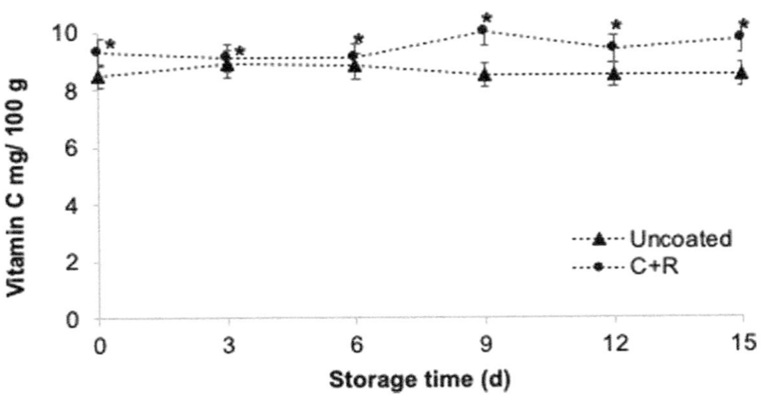

Figure 10.3 Vitamin C content throughout storage time of uncoated (control) and coated (C+R) cucumber fruits. Significant differences between means (*$p < 0.05$) compared to uncoated cucumber fruits.

effect was shown in the pH values (Figure 10.2b), in which the higher pH values in the coated cucumbers may be attributed to the formation of sugars from the organic acids and other complex molecules present in the fruit, being used as an energy reserve for the metabolic ripening process (Cenobio-Galindo et al., 2019).

10.3.3 Vitamin C

During the storage time, the cucumbers coated with C+R presented significant differences in the content of vitamin C. In general, the vitamin C values were kept in the range of 9.3–10.0 mg/100 g in coated cucumbers (Figure 10.3). A decrease in vitamin C was observed during storage because of the action of oxidase enzymes, such as phenol oxidase and ascorbic acid oxidase. Nevertheless, the use of extracts helps to reduce the activity of these enzymes by the extract's antioxidant capacities, which brings a minor decrease of vitamin C in the coated cucumbers (Guía-García et al., 2021; Oluwaseun et al., 2013).

10.3.4 Color

Color is an important attribute in the shelf life of fruits (Shahiri et al., 2013). The luminosity of the coated cucumber remained similar to the L^* value compared to the control; however, at 15 d of storage, the L^* value of the uncoated cucumber was significantly higher (Figure 10.4a). Advanced ripening processes in uncoated fruits during storage conditions and the accelerated respiration process can cause discoloration of cucumbers and accumulation of water on the surface (Manjunatha and Anurag, 2014). Furthermore, $°H$ values at 0, 3, and 6 d of storage showed ($p <$ 0.05) differences between treatments (Figure 10.4b). The $°H$ value represents the degree of green color of the cucumber peel (Cosme-Silva et al., 2017). In this study, the coated fruits presented values of 127°–131°, indicating a green-yellow color as reported by Jia et al. (2018). The ripening process significantly affects the fruit color by an accelerated respiration rate, which may cause a discoloration in cucumbers and a higher water accumulation on their surface (Manjunatha and Anurag, 2014). Green fruits suffer this process due to a decrease in their chlorophyll content

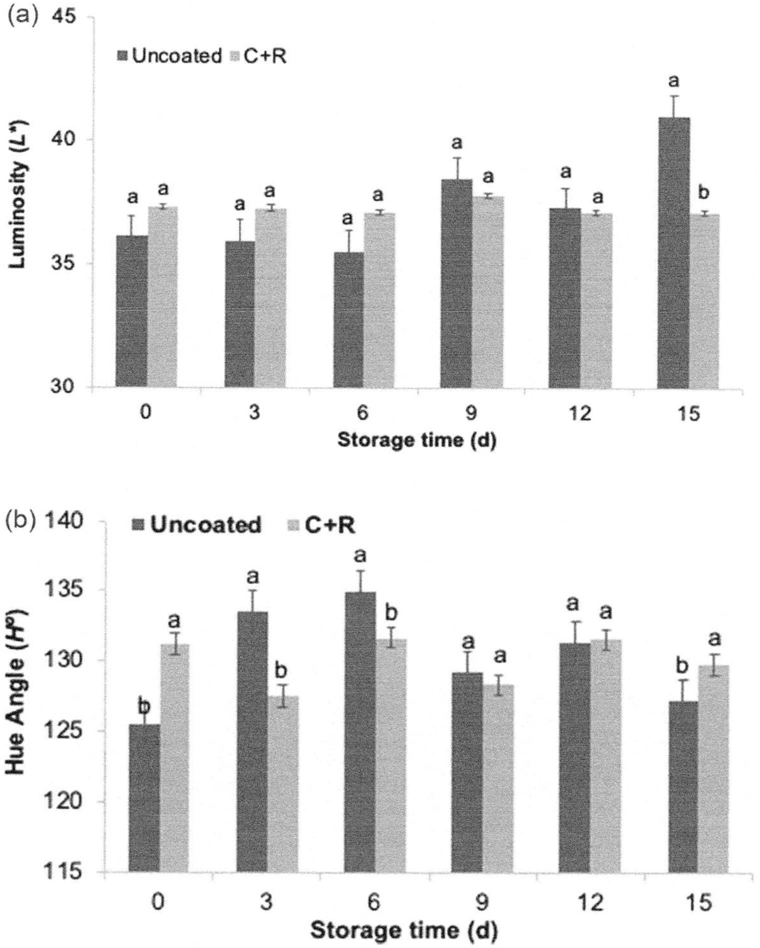

Figure 10.4 L^* = Luminosity (a) and H^o = huge angle (b) of uncoated (control) and coated (C+R) cucumber fruits. Different letters on the same day indicate significant differences ($p < 0.05$) between treatments.

by their erosion over time, promoting changes in the organoleptic properties (Moalemiyan and Ramaswamy, 2012).

10.3.5 Microbiological Analysis

The C+R treatment applied to cucumbers was effective in the growth inhibition of total mesophilic microorganisms. In control fruits, the total mesophilic count increased from 6.0 to 7.6 log CFU/g, and coated fruits were 5.3 to 6.6 log CFU/g at day 15 of storage (Figure 10.5). The application of functionalized edible coatings has demonstrated their capabilities to reduce the fungal decay and the growth of microorganisms in fruits (Chu et al., 2020; Correa-Pacheco et al., 2017; Guerreiro et al., 2016). The antimicrobial effect detected in the functionalized edible coating evaluated in this work can probably be related to the presence of phenolic compounds (e.g.,

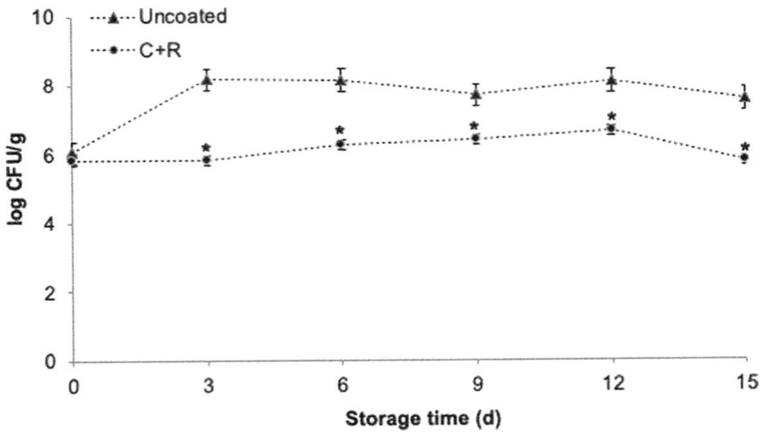

Figure 10.5 Counts of total mesophilic microorganisms throughout storage time of uncoated (control) and coated (C+R) cucumber fruits. Significant differences between means (*$p < 0.05$) compared to uncoated cucumber fruits.

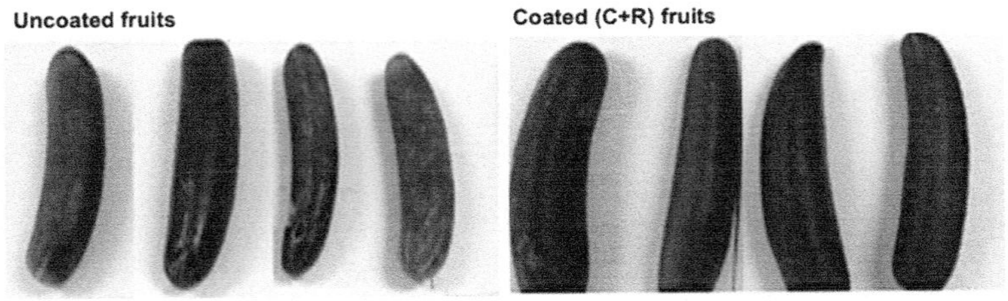

Figure 10.6 Uncoated (control) and coated (C+R) cucumber fruits on the day 15 of storage at 18 °C and 85% RH.

gallic acid, ferulic acid, ellagic acid, and epicatechin) previously reported in the *R. microphylla* fruit extracts (Charles-Rodríguez et al., 2020), which are involved in the plant protection against predators and microorganisms (Guía-García et al., 2021; Jasso de Rodríguez et al., 2019).

Under storage conditions at 18°C, the coated cucumbers showed a healthy appearance until 15 d, while the uncoated fruits presented green-yellow spots and fungal growth at day 9 of storage (Figure 10.6). The coated fruits maintained their dark green color during the test, mainly associated with the interaction of C+R on fruit surface and, consequently, a barrier to the respiration process (Capitani et al, 2016).

10.4 CONCLUSIONS

The edible coating based on chia mucilage functionalized with *R. microphylla* fruit extract showed to be an effective barrier to the weight loss and microbial attack in cucumber fruits. The C+R was able to retain the quality parameters of the cucumber fruits, as it prevented changes in color, TSS, pH, and vitamin C during the postharvest storage for 15 d at 18 °C and 85% RH.

This is the first scientific report on the functionality of *R. microphylla* fruit extract to extend the shelf life of cucumber fruits using edible coating chia mucilage as a vehicle. This technology is a promising postharvest tool to improve the postharvest quality of fruits and vegetables and can be considered as a feasible alternative on an industrial scale.

REFERENCES

AOAC. (1990). *Assoaciation of official analytical chemists* (A. of O. A. Chemists (ed.); Official M). Washington, DC: AOAC.

Brasil, I. M., Gomes, C., Puerta-Gomez, A., Castell-Perez, M. E., & Moreira, R. G. (2012). Polysaccharide-based multilayered antimicrobial edible coating enhances quality of fresh-cut papaya. *LWT - Food Science and Technology, 47*(1), 39–45. https://doi.org/10.1016/j.lwt.2012.01.005

Cai, C., Ma, R., Duan, M., Deng, Y., Liu, T., & Lu, D. (2020). Effect of starch film containing thyme essential oil microcapsules on physicochemical activity of mango. *LWT - Food Science and Technology, 131.* https://doi.org/10.1016/j.lwt.2020.109700

Capitani, M. I., Matus-Basto, A., Ruiz-Ruiz, J. C., Santiago-García, J. L., Betancur- Ancona, D. A., Nolasco, S. M., Tomás, M. C., & Segura-Campos, M. R. (2016). Characterization of biodegradable films based on Salvia hispanica L. protein and mucilage. *Food and Bioprocess Technology, 9*(8), 1276–1286. https://doi.org/10.1007/s11947-016-1717-y

Cenobio-Galindo, A. de J., Ocampo-López, J., Reyes-Munguía, A., Carrillo-Inungaray, M. L., Cawood, M., Medina-Pérez, G., Fernández-Luqueño, F., & Campos-Montiel, R. G. (2019). Influence of bioactive compounds incorporated in a nanoemulsion as coating on avocado fruits (Persea americana) during postharvest storage: Antioxidant activity, physicochemical changes and structural evaluation. *Antioxidants, 8.* https://doi.org/10.3390/antiox8100500

Charles-Rodríguez, A. V, Rivera-Solis, L. L., Martins, J. T., Genisheva, Z., Robledo-Olivo, A., González-Morales, S., López-Guarin, G., Martínez-Vázquez, D., Vicente, A., & Flores-López, M. L. (2020). Edible films based on black chia (Salvia hispanica L.) seed mucilage containing Rhus microphylla fruit phenolic extract. *Coatings 10,* 1–15.

Chu, Y., Gao, C. C., Liu, X., Zhang, N., Xu, T., Feng, X., Yang, Y., Shen, X., & Tang, X. (2020). Improvement of storage quality of strawberries by pullulan coatings incorporated with cinnamon essential oil nanoemulsion. *LWT - Food Science and Technology, 122,* 109054. https://doi.org/10.1016/j.lwt.2020.109054

Correa-Pacheco, Z. N., Bautista-Baños, S., Valle-Marquina, M. Á., & Hernández-López, M. (2017). The effect of nanostructured chitosan and chitosan-thyme essential oil coatings on Colletotrichum gloeosporioides growth in vitro and on cv Hass avocado and fruit quality. *Journal of Phytopathology, 165,* 297–305. https://doi.org/10.1111/jph.12562

Cosme-Silva, G. M., Silva, W. B., Medeiros, D. B., Salvador, A. R., Cordeiro, M. H. M., da Silva, N. M., Santana, D. B., & Mizobutsi, G. P. (2017). The chitosan affects severely the carbon metabolism in mango (Mangifera indica L. cv. Palmer) fruit during storage. *Food Chemistry, 237,* 372–378. https://doi.org/10.1016/j.foodchem.2017.05.123

De León-Zapata, M. A., Pastrana-Castro, L., Rua-Rodríguez, M. L., Alvarez-Pérez, O. B., Rodríguez-Herrera, R., & Aguilar, C. N. (2016). Experimental protocol for the recovery and evaluation of bioactive compounds of tarbush against postharvest fruit fungi. *Food Chemistry, 198,* 62–67. https://doi.org/10.1016/j.foodchem.2015.11.034

Fan, X. -J., Zhang, B., Yan, H., Feng, J. -T., Ma, Z. -Q., & Zhang, X. (2019). Effect of lotus leaf extract incorporated composite coating on the postharvest quality of fresh goji (Lycium barbarum L.) fruit. *Postharvest Biology and Technology, 148,* 132–140. https://doi.org/10.1016/J.POSTHARVBIO.2018.10.020

Flores-López, M. L., Cerqueira, M. A., de Rodríguez, D. J., & Vicente, A. A. (2016). Perspectives on utilization of edible coatings and nano-laminate coatings for extension of postharvest storage of fruits and vegetables. *Food Engineering Reviews, 8,* 292–305. https://doi.org/10.1007/s12393-015-9135-x

Guerreiro, A. C., Gago, C. M. L., Faleiro, M. L., Miguel, M. G. C., & Antunes, M. D. C. (2016). Edible coatings enriched with essential oils for extending the shelf-life of "Bravo de Esmolfe" fresh-cut apples. *International Journal of Food Science & Technology, 51,* 87–95. https://doi.org/10.1111/ijfs.12949

Guía-García, J. L., Charles-Rodríguez, A. V., López-Romero, J. C., Torres-Moreno, H., Genisheva, Z., Robledo-Olivo, A., Reyes-Valdés, H. M., Ramírez-Godina, F., García-Osuna, H. T., & Flores-López, M. L. (2021). Phenolic composition and biological properties of Rhus microphylla and Myrtillocactus geometrizans fruit extracts. *Plants, 10*, 2010. https://doi.org/10.3390/plants10102010

Hassan, B., Chatha, S. A. S., Hussain, A. I., Zia, K. M., & Akhtar, N. (2018). Recent advances on polysaccharides, lipids and protein based edible films and coatings: A review. *International Journal of Biological Macromolecules, 109*, 1095–1107. https://doi.org/10.1016/J.IJBIOMAC.2017.11.097

Jasso de Rodríguez, D., Gaytán-Sánchez, N. A., Rodríguez-García, R., Hernández-Castillo, F. D., Díaz-Jiménez, L., Villarreal-Quintanilla, J. A., Flores-López, M. L., Carrillo-Lomelí, D. A., & Peña-Ramos, F. M. (2019). Antifungal activity of Juglans spp. and Carya sp. ethanol extracts against Fusarium oxysporum on tomato under greenhouse conditions. *Industrial Crops and Products, 138*, 1–10. https://doi.org/10.1016/j.indcrop.2019.06.005

Jia, B., Zheng, Q., Zuo, J., Gao, L., Wang, Q., Guan, W., & Shi, J. (2018). Application of postharvest putrescine treatment to maintain the quality and increase the activity of antioxidative enzyme of cucumber. *Scientia Horticulturae, 239*(November 2017), 210–215. https://doi.org/10.1016/j.scienta.2018.05.043

Manjunatha, M., & Anurag, R. K. (2014). Effect of modified atmosphere packaging and storage conditions on quality characteristics of cucumber. *Journal of Food Science and Technology, 51*(11), 3470–3475. https://doi.org/10.1007/s13197-012-0840-7

Medeiros, B. G. de S., Pinheiro, A. C., Carneiro-Da-Cunha, M. G., & Vicente, A. A. (2012). Development and characterization of a nanomultilayer coating of pectin and chitosan - Evaluation of its gas barrier properties and application on "Tommy Atkins" mangoes. *Journal of Food Engineering, 110*, 457–464. https://doi.org/10.1016/j.jfoodeng.2011.12.021

Moalemiyan, M., & Ramaswamy, H. S. (2012). Quality retention and shelf-life extension in mediterranean cucumbers coated with a pectin-based film. *Journal of Food Research, 1*(3), 159. https://doi.org/10.5539/jfr.v1n3p159

Mohammadi, A., Hashemi, M., & Hosseini, S. M. (2015). Chitosan nanoparticles loaded with Cinnamomum zeylanicum essential oil enhance the shelf life of cucumber during cold storage. *Postharvest Biology and Technology, 110*, 203–213. https://doi.org/10.1016/J.POSTHARVBIO.2015.08.019

Mousavi, S. R., Rahmati-Joneidabad, M., & Noshad, M. (2021). Effect of chia seed mucilage/bacterial cellulose edible coating on bioactive compounds and antioxidant activity of strawberries during cold storage. *International Journal of Biological Macromolecules, 190*, 618–623. https://doi.org/10.1016/J.IJBIOMAC.2021.08.213

Muñoz-Tébar, N., Carmona, M., Ortiz De Elguea-Culebras, G., Molina, A., & Berruga, M. I. (2022). Chia seed mucilage edible films with origanum vulgare and satureja montana essential oils: Characterization and antifungal properties. *Membranes, 12*, 1–16. https://doi.org/10.3390/membranes12020213

Oluwaseun, A. C., Arowora Kayode, A., Bolajoko, F. O., Bunmi, A. J., & Olagbaju, A. R. (2013). Effect of edible coatings of carboxy methyl cellulose and corn starch on cucumber stored at ambient temperature. *Asian Journal of Agriculture and Biology, 1*, 133–140.

Patel, C., & Panigrahi, J. (2019). Starch glucose coating-induced postharvest shelf- life extension of cucumber. *Food Chemistry, 288*(February), 208–214. https://doi.org/10.1016/j.foodchem.2019.02.123

Salas-Méndez, E. de J., Vicente, A., Pinheiro, A. C., Ballesteros, L. F., Silva, P., Rodríguez-García, R., Hernández-Castillo, F. D., Díaz-Jiménez, M. de L. V., Flores-López, M. L., Villarreal-Quintanilla, J. Á., Peña-Ramos, F. M., Carrillo-Lomelí, D. A., & Jasso de Rodríguez, D. (2019). Application of edible nanolaminate coatings with antimicrobial extract of Flourensia cernua to extend the shelf-life of tomato (Solanum lycopersicum L.) fruit. *Postharvest Biology and Technology, 150*(June 2018), 19–27. https://doi.org/10.1016/j.postharvbio.2018.12.008

Shahiri Tabarestani, H., Sedaghat, N., & Alipour, A. (2013). Shelf life improvement and postharvest quality of cherry tomato (Solanum lycopersicum L.) fruit using basil mucilage edible coating and cumin essential oil. *International Journal of Agronomy and Plant Production, 4*(9), 2346–2353.

Sogvar, O. B., Koushesh Saba, M., & Emamifar, A. (2016). Aloe vera and ascorbic acid coatings maintain postharvest quality and reduce microbial load of strawberry fruit. *Postharvest Biology and Technology, 114*, 29–35. https://doi.org/10.1016/J.POSTHARVBIO.2015.11.019

Trejo Valencia, R., Sánchez Acosta, L., Fortis Hernández, M., Preciado Rangel, P., Gallegos Robles, M. Á., Antonio Cruz, R. del C., Vázquez Vázquez, C. (2018). Effect of seaweed aqueous extracts and compost on vegetative growth, yield, and nutraceutical quality of cucumber (Cucumis sativus L.) fruit. *Agronomy, 8*, 1–13. https://doi.org/10.3390/AGRONOMY8110264

Waghmare, R., Preethi, R., Moses, J. A., & Anandharamakrishnan, C. (2021). Mucilages: Sources, extraction methods, and characteristics for their use as encapsulation agents. *Critical Reviews in Food Science and Nutrition*, 1–22. https://doi.org/10.1080/10408398.2021.1873730

Yang, Z., Zou, X., Li, Z., Huang, X., Zhai, X., Zhang, W., Shi, J., & Tahir, H. E. (2019). Improved postharvest quality of cold stored blueberry by edible coating based on composite gum arabic/roselle extract. *Food and Bioprocess Technology, 12*, 1537–1547. https://doi.org/10.1007/s11947-019-02312-z

Yin, C., Huang, C., Wang, J., Liu, Y., Lu, P., & Huang, L. (2019). Effect of chitosan- and alginate-based coatings enriched with cinnamon essential oil microcapsules to improve the postharvest quality of mangoes. *Materials, 12*, 1–19.

Chapter 11

Larrea Tridentate

Bioactive Compounds, Biological Activities and Its Potential Use in Phytopharmaceuticals Improvement

Julio César López-Romero, Heriberto Torres-Moreno, Karen Lillian Rodríguez-Martínez, Alejandra del Carmen Suárez-García, Minerva Edith Beltrán-Martinez, and Jimena García-Dávila

CONTENTS

11.1 INTRODUCTION

Larrea tridentata, commonly known as 'gobernadora', 'arbusto de creosota', and 'hediondilla', is distributed throughout the deserts of North America, extending from the southwestern United States to central Mexico. In Mexico, this plant is distributed throughout desert areas in the states of San Luis Potosí, Coahuila, Durango, Sonora, Zacatecas, and Baja California Norte. In the United States, this plant can be found in the states of Arizona, California, Nevada, Texas, and New Mexico (Balderas et al., 2018).

This plant has been widely used in traditional medicine for different purposes. Most common applications include the use of infusion preparations and consumed as capsules (leaves or ground twigs) via oral consumption as natural treatments against varicella, tuberculosis,

DOI: 10.1201/9781003251255-11

sexually transmitted diseases, menstrual pain, and snake bites, among others (Arteaga et al., 2005; Gnabre et al., 2015).

Depending on the traditional use, different scientific studies have been carried out to validate its empirical properties and guarantee its safety. Relevant biological activities reported for this plant include antimicrobial, antifungal, antiparasitic, antiviral, antiproliferative, and antioxidant activity. These biological activities have been attributed to the presence of chemical compounds in the plant, such as phenolic compounds, terpenes, anthraquinones, alkaloids, and saponins (Soriano et al., 1999; Craigo et al., 2000; Vargas-Arispuro et al., 2005; Favela-Hernández et al., 2012; Garza, 2014; Aguirre-Joya et al., 2018).

Despite the relevant activities shown by *L. tridentata*, current toxicological studies that guarantee its safe application in humans are inconclusive and contradictory (Sheik et al., 1997; Portilla-de Buen et al., 2008). Previous scenarios have made the study of different drug development research levels possible, starting from the discovery phase, to the non-clinical phase, then finally clinical assays, which will reveal the toxicological effects of the drug candidates obtained from *L. tridentata* in order to guarantee the efficacy and safety of its products.

11.2 CHARACTERISTICS, DISTRIBUTION, AND MEDICINAL USES

Larrea tridentata (Sessé & Moc. ex DC.) Coville is a medicinal plant known by many common names, such as 'arbusto de creosota', 'hediondilla', 'guamis', 'gobernadora', 'sonora', 'tasajo' and 'jarilla'. In the Sonoran Desert in Mexico, the members of the Seri tribe also call it 'háaxat' or 'háajat'. The name of the genus, *Larrea*, is in honor of the scientist who first described it, J.A. Hernández de Larrea. The natives call the plant 'gobernadora' for its ability to inhibit the growth of surrounding plants and 'hediondilla' for its bad smell, particularly noticeable after it has rained (Gnabre et al., 2015; García-Castillo, 2012).

L. tridentata is a perennial bush of 1–3.5 m tall with flexible stems. Its leaves are dark green to yellowish green, 1 cm long by 5 mm wide, and covered with pubescents and a resinous material with a penetrating odor. *L. tridentata's* flowers are solitary with yellow petals. It presents globose fruits with five carpels, and its seeds are elongated and curved. Its flowering period is between February and April; however, it can flower with a minimum of 12 mm of precipitation (Figure 11.1) (Seinet, 2020).

L. tridentata is the most common bush in the deserts of North America, especially on dry pains and plateaus below 5,000 feet (1.576 m). It is not used to feed livestock (Seinet, 2020). It is distributed from the southwestern United States to Central Mexico, being abundant in the states of San Luis Potosí, Coahuila, Durango, Sonora, Zacatecas, and Baja California Norte. In the southwestern United States, this plant is distributed in the states of Arizona, California, Nevada, Texas, and New Mexico (Mata-Balderas et al., 2018).

The Pima and Maricopa Native American tribes of the southwestern United States use *L. tridentata* in the form of extracts or decoctions to empirically treat more than 50 diseases including chickenpox, tuberculosis, sexually transmitted diseases, period pain, snake bites (Arteaga et al., 2005; Gnabre et al., 2015), infections in the genitourinary and respiratory tracts, inflammation of the musculoskeletal system, skin damage, kidney problems, and diabetes, among others (Martins et al., 2013a). Traditional medicine also uses *L. tridentata* leaves for the treatment of cold, flu, skin sores, arthritis, sinusitis, gout, anemia, fungal infections, and cancer (Gnabre et al., 2015). Its antimicrobial and antiallergic properties as well as its effect on autoimmune diseases have also been widely recognized in traditional indigenous medicine (Arteaga et al.,

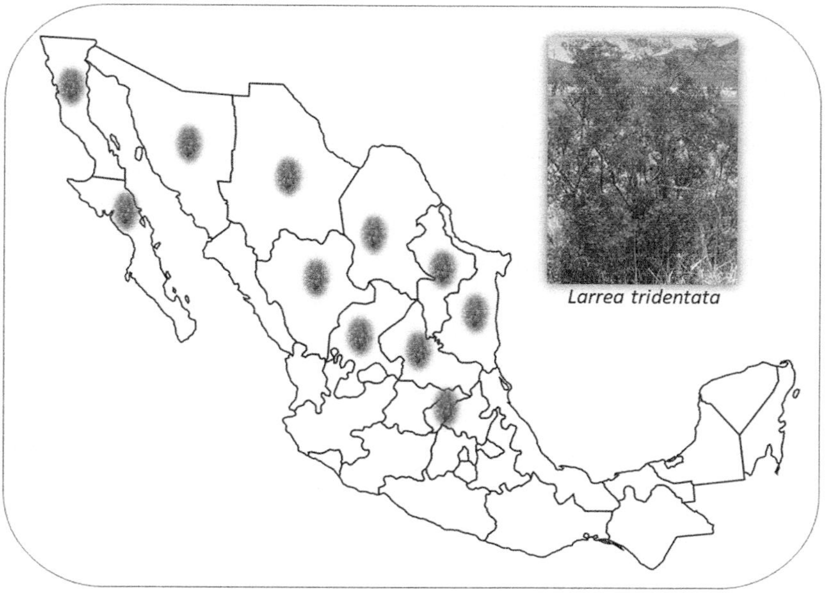

Figure 11.1 Characteristics and distribution of *Larrea tridentata*.

2005). Native healers in the southwest of North America use the aqueous extract of *L. tridentata*, popularly known as 'chaparral tea', for the treatment of gallbladder and kidney stones (Arteaga et al., 2005).

11.3 BIOACTIVE COMPOUNDS

Bioactive compounds are secondary metabolites produced by plants known as phytochemicals, which generally do not take part in the plant's growth and development (Bernhoft, 2010). These chemical compounds present a great structural variety and can be classified into three groups based on their structure: phenolic compounds, alkaloids, and terpenes/terpenoids, the latter being the most abundant in natural products (Azmir, 2013). Historically, these compounds have been associated with beneficial effects toward people's health via the consumption of natural products. In addition, several scientific studies have demonstrated that these compounds have a broad range of biological activities (Cör et al., 2018; Debnath et al., 2018; Li et al., 2014). The studies previously mentioned have served as the foundation of the development of current pharmacological therapies (Serrano et al., 2006).

11.3.1 Phenolic Compounds Identified in *L. tridentata*

Phenolic compounds are a diverse group of compounds that have aromatic rings and can present diverse complexity in their structure. Likewise, these structures may contain different groups as well as glycosides (Asim et al., 2018; Shahidi and Yeo, 2018). These compounds are mainly synthetized through proteins, tyrosine, and tryptophan, and currently, more than 8,000 types of these compounds have been identified (Chhikara et al., 2018). The most abundant group of these compounds are flavonoids, which are identified as flavonoles, flavones, flavanones, anthocyanins, chalcones, and isoflavones. The other group of phenolic compounds constitutes phenolic acids, which can be divided into hydroxycinnamic acids and hydroxybenzoic acids (Cianciosi et al., 2018). Additionally, other types of these compounds can be found synthetized by plants, which can be called tannins, lignans, and stilbenes (Aguilera, 2016). These compounds have shown a wide spectrum of biological activity such as antioxidant, antimicrobial, anti-inflammatory, antiviral, antiproliferative, and antifungal, among others (Li et al., 2014; Tungmunnithum et al., 2018).

One of the most commonly reported compounds in extracts obtained from leaves and above ground parts of *L. tridentata* are phenolic compounds. In this regard, different studies have demonstrated that the content of this type of compound varies between 68.55 to 487.13 ellagic acid equivalent/g of dry sample (Aarland et al., 2015; Martins et al., 2012; Martins et al., 2010b; Martins et al., 2011; Martins et al., 2013a; Sagaste et al., 2019; Skouta et al., 2018). Similarly, Hyder et al. (2002) analyzed the content of phenolic compounds from different constituents of *L. tridentata* (flowers, seeds, leaves, stems and roots). It was observed that the content of these compounds varied between ≈8 to ≈41 mg/g of dry sample, with a more prominent concentration in the stems, leaves, and roots, respectively. Likewise, another group of phenolic compounds quantified in *L. tridentata* are flavonoids, which have shown values that range between 0.88 to 24 mg quercetin equivalent/g of dry sample (Aarland et al., 2015; Martins et al., 2012; Martins et al., 2013a). Generally, the variations in phenolic compound and flavonoid content can be associated to methodological differences between studies, the location from which the plant was harvested, the harvesting season, the part of the plant used, the type of solvents used in the extraction, and the extraction method.

On the other hand, other studies have aimed to identify the phenolic compounds present in the plant (Figure 11.2). In this regard, one of the commonly found group compounds in extracts of *L. tridentata* are flavonoids, in which quercetin, kaempferol, catechin and epicatechin, gossypetin, herbacetin, quercetin-3-rhamno glucoside, quercetin-3-glucoside, quercetin-3'-methylether, and kaempferol-3-methylether have been identified (Bernhard, 1981; Chirikdjian, 1973; Martins et al., 2012; Martins et al., 2011; Martins et al., 2013a; Sagaste et al., 2019). In this sense, Skouta et al. (2018) obtained several factions of a hydroethanolic extract of *L. tridentata,* which were shown to contain different flavonoids, such as juglanin, liquiritin, 3',4',5,7-tetraacetoxyflavone and 3',4',5,7-tetramethylquercetin. Other investigations have focused on the isolation of these compounds, for example, Sakakibara et al. (1975) isolated two 8-trihydroflavonols, known as dimethoxytrihydroxyflavone and trihydroxytrimethoxyflavone (Abou-Gazar et al., 2004), and isolated and characterized six flavonoids through a methanolic extract: herbacetin 3,8,4'-trimethyl ether, 5,7,4'-trihydroxy-3,8,3'-trimethoxy flavone, apigenin, (+)-dihydroisorhamnetin, 5,7,4'-trihydroxy-3,8-dimethoxyflavone, and kaempferol-3-methyl ether. In the same context, Favela-Hernández et al. (2012) identified four new flavonoids in *L. tridentata*, which were named 5,4′-dihydroxy-3,7,8,3′-tetramethoxyflavone, 5,4′-dihydroxy-3,7,8-trimethoxyflavone, 5,4′-dihydroxy-7-methoxyflavone and 5,8,4′-trihydroxy-3,7-dimethoxyflavone. In a study carried out by Schmidt et al. (2012), six new phenolic compounds were also isolated and analyzed; they were named naringenin, 3'-O-methyltaxifolin, apigenin-7-methylether, apigenin-7,4'-dimethyl ether, kaempferol-3,7-dimethylether, and herbacetin-3,7-dimethylether. Similarly, other studies have demonstrated that extracts in this plant are the source of other flavonoid structures, for example, (S)-4', 5-dihydroxy-7- methoxyflavanone and 3-methoxy-6, 7, 4'-trihydroxyflavone (Bashyal et al., 2017; Lambert et al., 2005).

Another group of phenolic compounds identified in extracts of *L. tridentata* are phenolic acids (Figure 11.2). In this regard, different investigations have demonstrated the presence of chlorogenic acid, *p*-coumaric acid, ferulic acid, gallic acid, caffeic acid, and ellagic acid. In general, it

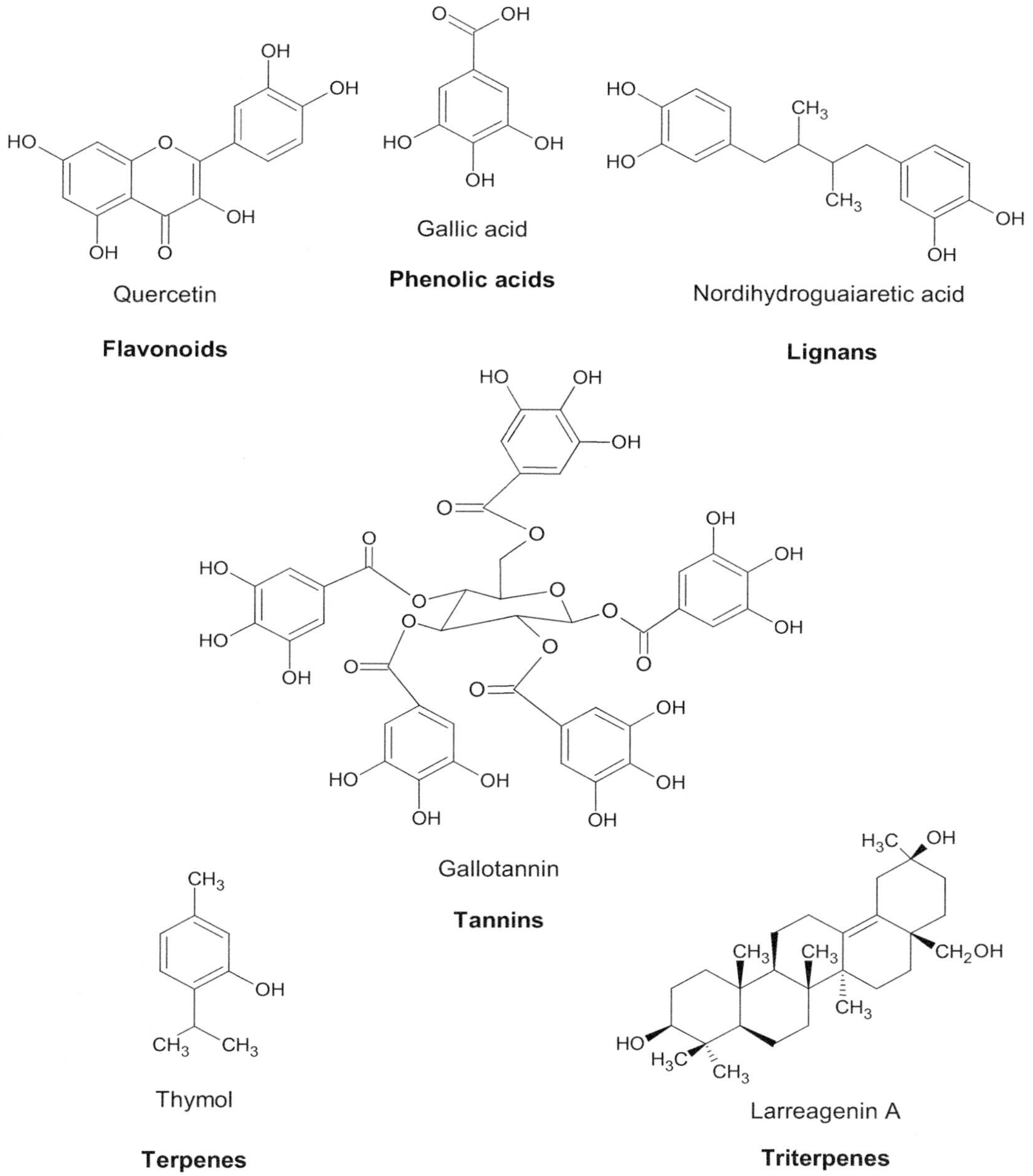

Figure 11.2 Groups of bioactive compounds identified in *Larrea tridentata*.

has been observed that the latter two phenolic acids were the most abundant in the extracts with 2.28 and 2.5 mg/g, respectively (Aarland et al., 2015; Aguilar et al., 2008; Sagaste et al., 2019; Ventura et al., 2008).

Studies have also reported the presence of tannins, another type of phenolic compounds, in the extract of *L. tridentata*. Hyder et al. (2002) determined the content of condensed tannins in flowers, seeds, leaves, stems, and roots extracts of *L. tridentata,* observing that the content varied between 0.1 to 1.8 mg/g of dry sample and that flower extracts presented the highest concentration. In the same context, Treviño-Cueto et al. (2007), Aguilar et al. (2008), and Ventura et al. (2008) determined the content of hydrolyzed tannins and condensed tannins in extracts from leaves of *L. tridentata.* It was observed that condensed tannins had a high concentration, which was 41%–68% higher than the concentration of hydrolyzed tannins.

The most abundant group of phenolic compounds in extracts of *L. tridentata* are lignans (Figure 11.2). Different studies have demonstrated that the mostly isolated and quantified lignan in *L. tridentata* is nordihydroguaiaretic acid (NDGA), which has shown variations in its concentration, presenting amounts that range from 2.12 to 66.5 mg/g of dry sample (Downum et al., 1988; Elakovich and Stevens, 1985; Hyder, 2002; Martins et al., 2010a; Martins et al., 2012; Martins et al., 2010b; Martins et al., 2011; Martins et al., 2013; Skouta et al., 2018; Vargas-Arispuro et al., 2005). Other studies have demonstrated that besides containing NDGA lignan, extracts of this plant also contain the compound with substitutions by the methyl group in their structure (Gnabre et al., 1996; Vargas-Arispuro et al., 2005). Likewise, other lignans have also been identified in this plant. For example, Konno et al. (1990) isolated six lignans from the extracts of leaves and stalks of *L. tridentata,* which were identified as lerreatricin, 3',3"-dimethoxylarreatricin, 4-epi-larreatricin, 3"-hydroxy-4-epi-larreatricin, larreatridenticin, and 3,4-dehy-drolarreatricin. Abou-Gazar et al. (2004) utilized a methanolic extract from the leaves of *L. tridentata*, isolating three lignans, which were named (7S,8S,7'S,8'S)-3,3',4'-trihydroxy-4-methoxy-7,7'-epoxylignan, meso-(rel 7S,8S,7'R,8'R)-3,4,3',4'-tetrahydroxy7,7'-epoxylignan, and (E)-4,4'-dihydroxy-7,7'-dioxol ign-8(8')-ene. In the same context, Lambert et al. (2005), isolated six lignans from a diethyl ether toluene extracts from flowers: 3,4'-dihydroxy-3',4-dimethoxy-6,7'-cyclolignan, 3'-demethoxy-isoguaiacin, three butane-type di-O-methylated lignans, and one butane-type tri-O-methyl lignan. Favela-Hernández et al. (2012) purified and characterized three lignans from a chloroform extract from leaves of *L. tridentata*, which were named dihydroguaiaretic acid, 4-epi-larreatricin and 3′-demethoxy-6-O-demethylisoguaiacin. Similarly, Schmidt et al. (2012), through an extract of dichloromethane from *L. tridentata,* purified nine lignans known as meso-dihydroguaiaretic acid, meso-dihydroguaiaretic acid, 3-O-methyldihydroguaiaretic acid, 4,4'-dihyd roxy-8,8'-dehydro-7,7'-epoxylignan, 3,4,4'-trihydroxy-7,7'-epoxylignan, 3,3',4'-trihydroxy-4-meth oxy-7,7'-epoxylignan, 3,3',4,4'-tetrahydroxy-7,7'-epoxylignan, 3,3'-di-O-demethylisoguaiacin, and 3-O-demethylisoguaiacin. A study carried out by Bashyal et al. (2017) demonstrated that a methanolic extract of *L. tridentata* presented a high quantity of lignans, such as NDGA and 3'-O-methylnordihydroguairetic acid. Additionally, the following compounds were isolated and characterized: 3'-demethoxy-6-O-demethylisoguaiacin, nor-isoguaicin, 3'-demethoxyisoguaiacin, 6,3'-di-O-demethoxy-isoguaiacin, 3'-hydroxy-4-epi-larreatricin y (7R, 7'R)-7, 7'-bis(4', 3, 4-trihydroxyphenyl)-(8R, 8'S)-8, and 8'-dimethyltetrahydrofuran. In other study, justicidin B lignan was identified in nine different fractions obtained from a water-alcohol extract from *L. tridentata's* leaves (Skouta et al., 2018).

11.3.2 Triterpenes Identified in *L. tridentata*

Another group of compounds widely distributed in these plants are triterpenes (Figure 11.2), which have a pentacyclic structure of 30 carbons comprised by isoprene units (Agra et al., 2015). These compounds are synthetized in the plants via the mevalonate pathway, and currently more than 2,000 different structures in natural products have been identified (Thimmappa et al., 2014). Generally, these compounds can be classified in six groups based on their structure. They are known as euphanes, taraxanes, oleananes, lupaneps, ursanes, and baccharanes, which can have glycoside units in their structure. Likewise, these compounds have presented a wide array of diversity in their biological activities (Xia et al., 2014; Yin, 2012).

Hui-Zheng et al. (1988) identified two triterpenes from *L. tridentata* stem methanolic extract: 3β-(3,4-dihydroxycinnamoyl)-erythrodiol and 3β-(4-hydroxycinnamoyl)-erythrodiol. Similarly, Jitsuno and Mimaki, (2010) isolated 25 triterpenes from aerial parts of *L. tridentata* and were identified as: 3-[(O-(4-O-sulfo-b-D-glucopyranosyl)-(1→3)-a-L-arabinopyranosyl)oxy]olean-1 2-en-28-oic acid b-D-glucopyranosyl ester sodium salt, 3-[(O-b-D-glucopyranosyl-(1→3)-2-O-sulfo-a-L-arabinopyranosyl)oxy]olean-12-en-28-oic acid b-D-glucopyranosyl ester sodium salt, 3-[(O-b-D-glucopyranosyl-(1→3)-a-L-arabinopyranosyl)oxy]olean-12-en-28-oic acid, 3-[(O-b-D-glucopyranosyl-(1→3)-a-L-arabinopyranosyl)oxy]olean-12-en-28-oic acid b-D-glucopyranosyl ester, 3-[(O-b-D-xylopyranosyl-(1→3)-b-D-glucuronopyranosyl)oxy]olean-12-en-28-oic acid b-D-glucopyranosyl ester, 3-[(O-b-D-glucopyranosyl-(1→3)-O-[a-L-rhamnopyranosyl-(1→2)] -a-L-arabinopyranosyl)oxy]olean-12-en-28-oic acid, 3-[(O-b-D-glucopyranosyl-(1→3)-O-[a-L-rh amnopyranosyl-(1→2)]-a-L-arabinopyranosyl)oxy]olean-12-en-28-oic acid b-D-glucopyranosyl ester, 3-[(O-b-D-glucopyranosyl-(1→3)-O-[a-L-rhamnopyranosyl-(1→2)]- a-L-arabinopyranosyl) oxy]olean-12-en-28-oic acid O-b-D-glucopyranosyl-(1→6)-b-D-glucopyranosyl ester, 3-[(b-D-Xy lopyranosyl)oxy]olean-12-ene-28,29-dioic acid 28-b-D-glucopyranosyl ester, 3-[(O-b-D-Gluco pyranosyl-(1→3)-b-D-xylopyranosyl)oxy]olean12-ene-28,29-dioic acid, 3-[(b-D-xylopyranosyl) oxy]olean-12-ene28,29-dioic acid, 3-[(O-b-D-Glucopyranosyl-(1→3)-a-L-arabinopyranosyl)oxy]- 30-noroleana-12,20(29)-dien-28-oic acid, 3-[(a-L-Arabinopyranosyl)oxy]-30-noroleana-12,20(2 9)-dien28-oic acid b-D-glucopyranosyl ester, 3-[(O-(3-O-Acetyl-b-D-glucopyranosyl)-(1→3)-a -L-arabinopyranosyl)oxy]-30-noroleana-12,20(29)-dien-28-oic acid b-D-glucopyranosyl ester, 3-[(O-(6-O-Acetyl-b-D-glucopyranosyl)-(1→3)-a-L-arabinopyranosyl)oxy]-30-noroleana-12,20(2 9)-dien-28-oic acid b-D-glucopyranosyl ester, 3-[(O-(4-O-Sulfo-b-D-glucopyranosyl)-(1→3)-a-L-arabinopyranosyl)oxy]-30-noroleana-12,20(29)-dien-28-oic acid b-D-glucopyranosyl ester sodium salt, 3-[(O-(4-O-Sulfo-b-D-glucopyranosyl)-(1→3)-O-[a-L-rhamnopyranosyl-(1→2)]-a-L-arabinopyranosyl)oxy]-30-noroleana-12,20 (29)-dien-28-dioic acid b-D-glucopyranosyl ester sodium salt, 3-[(O-b-D-Glucopyranosyl-(1→3)-O-[a-L-rhamnopyranosyl- (1→2)]-a-L-arabinopy ranosyl)oxy]-30-noroleana-12,20(29)-dien28-oic acid O-b-D-glucopyranosyl-(1→6)-b-D-gluc opyranosyl ester, 3-[(O-b-D-Glucopyranosyl-(1→3)-O-[a-L-rhamnopyranosyl- (1→2)]-a-L-arab inopyranosyl)oxy]-20b-hydroxy-30-norolean-12- en-28-oic acid b-D-glucopyranosyl ester, 3-[(O -b-D-Glucopyranosyl-(1→3)-O-[a-L-rhamnopyranosyl- (1→2)]-a-L-arabinopyranosyl)oxy]-20 a-hydroxy-29-norolean-12- en-28-oic acid b-D-glucopyranosyl ester, 3b-hydroxy-30-noroleana12 ,20(29)-dien-28-oic acid b-D-glucopyranosyl ester, 3-[(O-b-D-glucopyranosyl-(1→3)-a-L-arabi nopyranosyl)oxy]- 30-noroleana-12,20(29)-dien-28-oic acid b-D-glucopyranosyl ester, 3-[(O-a-L-rhamnopyranosyl- (1→2)-a-L-arabinopyranosyl)oxy]-30-noroleana-12,20(29)-dien28-oic acid b-D-glucopyranosyl ester, 3-[(O-b-D-glucopyranosyl-(1→3)-O-[a-L-rhamnopyranosyl-(1→2)] -a-L-arabinopyranosyl)oxy]-30-noroleana-12,20(29)- dien-28-oic acid y 3-[(O-bD-glucopyrano syl-(1→3)-O-[a-L-rhamnopyranosyl-(1→2)]-a-Larabinopyranosyl)oxy]-30-noroleana-12,20(29)-di en-28-oic acid b-D-glucopyranosyl ester.

11.3.3 Other Bioactive Compounds Identified in *L. tridentata*

Besides phenolic compounds and triterpenes, other chemical compounds have been identified in *L. tridentata*. Aarland et al. (2015) analyze the phytochemical profile of hexane, ethyl acetate, and methanol extracts from *L. tridentata*. The results demonstrated the presence of anthraquinones, alkaloids, and saponines in the analyzed extracts (Figure 11.2). In a more specific manner, Bañuelos-Valenzuela et al. (2018) managed to identify two terpenes (Figure 11.2) in an ethanolic extract of *L. tridentata.* Mentioned bioactive compounds were thymol and carvacrol, whose concentrations varied between 0.30 to 4.30 mg/mL and 4.43 to 7.80 mg/mL, respectively. In the same context, Jardine et al. (2010) observed the presence of different terpenes in *L. tridentata;* among those found were pinene, camphene, limonene, ocineme, camphor, isoborneol, and cymene.

11.4 BIOLOGICAL ACTIVITIES

11.4.1 Antimicrobial Activity

One of the most relevant biological activities of *L. tridentata* is its antimicrobial effect (Figure 11.3), which has been evidenced against Gram-positive and Gram-negative bacteria (Table 11.1).

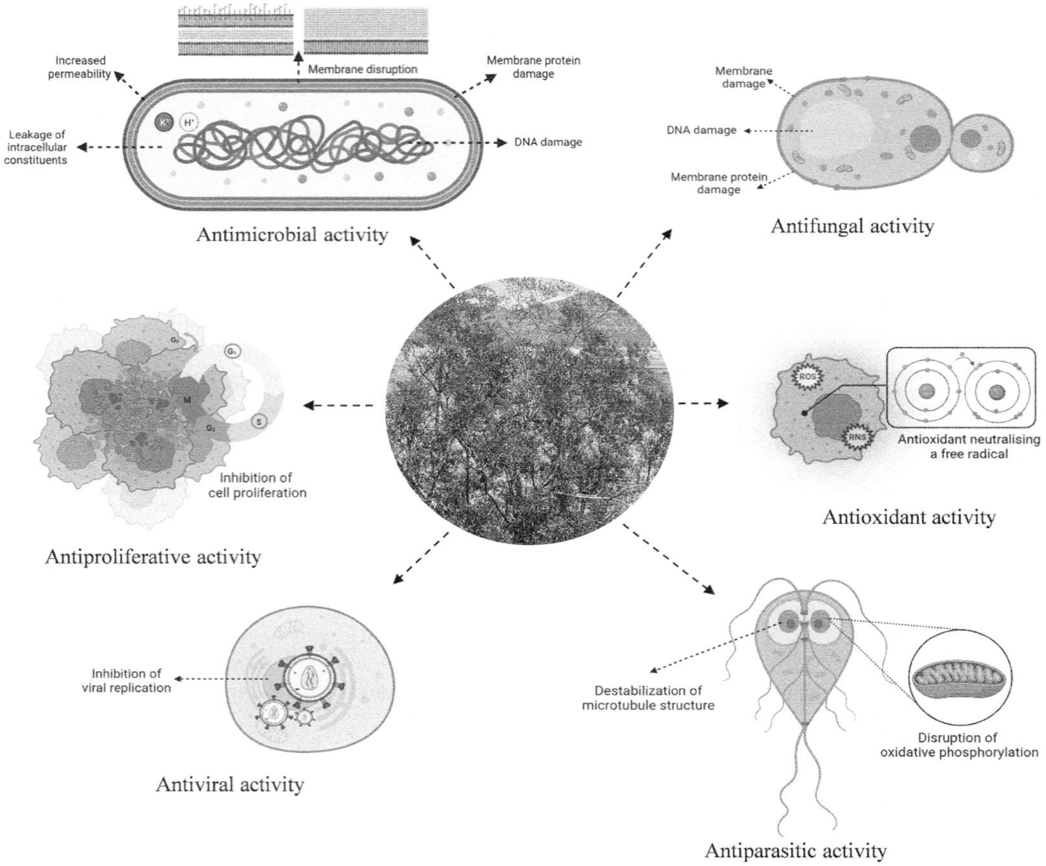

Figure 11.3 Biological activities of *Larrea tridentata.*

TABLE 11.1 ***LARREA TRIDENTATA*** **AND ITS BIOLOGICAL ACTIVITIES**

Activity	Used Plant Parts	Extract/Tested Compound	Effect	Reference
Antimicrobial	Branches	Ethanol	Inhibitory effect (growth inhibition zone) *S. aureus* (15–20 mm), *B. subtillis* (15–20 mm) and *E. faecalis* (15–20 mm)	Dimayuga and García (1991)
	Leaves	Ethanol	Inhibitory effect (growth inhibition zone) Salmonella enterica (1.57 mm)	Itza et al. (2019)
	Leaves	Ethanol	Inhibitory effect (MIC) *S. dysenteriaea* (14 mg/mL), *Y. enterocolitic* (19 mg/mL), *L. monocytogenes* (10 mg/mL), *P. vulgaris* (10 mg/mL), *C. perfringens* (12 mg/mL), *N. asteroids* (16.3 mg/mL), and *N. brasiliensis* (16.6 mg/mL)	Verástegui et al. (1996)
	Whole plant	Methanol	Inhibitory effect (MIC) *S. aureus* (1.25 mg/mL), *E. coli* (10 mg/mL), and *P. aeruginosa* (40 mg/mL)	Navarro et al. (1996)
	Leaf and flower	95% ethanol/distilled water/glycerol	Inhibitory effect (MIC) *S. aureus* (1.7 µL/mL)	Snowden et al. (2014)
	Whole plant	Ethanol/distilled water/glycerol	Inhibitory effect (MIC) *S. aureus* (day 0 = 60 µg/mL; day 15= 183 µg/mL)	Ruiz et al. (2017)
		Ethanol	Inhibitory effect (MIC) *E. coli* (100 µg/mL), *S. aureus* (100 µg/mL), *S. typhimurium* (100 µg/mL), *S. licuefaciens* (100 µg/mL), *P. aeruginosa* (100 µg/mL), *K. oxytoca* (100 µg/mL), *P. mirabilis* (100 µg/mL), and *P. vulgaris* (100 µg/mL)	Joublanc et al. (2008)
	Whole plant	Chloroform and methanol	Inhibitory effect (MIC) *S. pneumoniae* (8031.25 µg/mL), *E. faecalis* (MIC_{50}: 250 µg/mL), and *S. aureus* (80125 µg/mL)	Bocanegra-García et al. (2009)

(*Continued*)

TABLE 11.1 (CONTINUED) *LARREA TRIDENTATA* AND ITS BIOLOGICAL ACTIVITIES

	Leaves	Dichloromethane and ethyl acetate	Inhibitory effect (MIC) *S. aureus* (31.3 μg/mL), *Staphylococcus* coagulase-negative (62.5-250 μg/mL) and *E. faecalis* (125-375 μg/mL)	Martins et al. (2013b)
	Aerial parts	Dichloromethane	100% inhibition of *B. subtilis* at 1,000 μg/mL	Dentali and Hoffmann (1992)
	Leaves	Chloroform	Inhibitory effect (MIC) Lignan 3/ sensitive and resistant *S. aureus* (25 μg/mL), *E. faecalis* (12.5 μg/mL), *E. cloacae* (12.5 μg/mL), *E. coli* (50 μg/mL), and *M. tuberculosis* MDR (12.5 μg/mL) Lignans 1 and 2/ *S. aureus* MR (50 μg/mL), *E. cloacae* (12.5 μg/mL), sensitive *M. tuberculosis* (50 μg/mL), and *M. tuberculosis* MDR (12.5-50 μg/mL) Flavonoids 4 and 5/ M. *tuberculosis* MDR (25–50 μg/mL)	Favela-Hernández et al. (2012)
	Whole plant	Ethanol	Inhibitory effect (growth inhibition zone) *E. coli* (17 mm), *A. baumannii* (19 mm), *Pseudomonas* sp. (15 mm) and *S. aureus* (12 mm)	Delgadillo Ruiz et al. (2017)
Antifungal	Leaves	Acetone 70%	Fungicidal effect 100% inhibition of *Phytium* sp., *C. coccodes*, *C. truncatum*, *A. alternata*, *F. solani*, *R. solani*. *F. oxysporum* (8 strains). Fungistatic effect 70% inhibition of *F. verticilloides*	Osorio et al. (2010)
	Young leaves and stems	Ethanol	97% inhibitory effect on mycelial growth of *F. oxysporum* f. sp. *radicis-lycopersici*	Peñuelas et al. (2017)

(Continued)

TABLE 11.1 (CONTINUED) *LARREA TRIDENTATA* AND ITS BIOLOGICAL ACTIVITIES

	Leaves and Branches	Fermented and non-fermented methanolic extract	100% inhibition of *P. capsici*	Díaz-Díaz et al. (2013)
	Leaves	Aqueous	Inhibitory effect on mycelium growth at different times. 48, 72, and 96 h: *P. capsici* (100%), *A. flavus* (100%) 48, 72, and 96 h: *Rhizopus* sp. 60.30%, 59.33%, and 66.5%, respectively.	Galván et al. (2014)
	Whole plant	Alcoholic	Inhibition of *P. capsici* at 2, 5, and 10 mg/mL	Mojica-Marin et al. (2011)
	Leaves and branches	Methanol	Inhibitory effect on mycelial growth of *Pythium* sp. at 500 µl/L	Lira-Saldívar et al. (2003)
	Leaves	Hydroalcohol, acetone, methanol, ethanol	Inhibitory effect (growth inhibition zone) *A. flavus* (ethanolic extract: 12–21 mm; methanolic extract: 7–20 mm) and *Penicillium* sp. (ethanolic extract: 8-18 mm) MIC of hydroalcoholic extract *A. flavus* (7 mg/mL), *Penicillium* sp. (5 mg/mL)	Moreno et al. (2011)
		Ethanol	Inhibition of mycelial growth of *P. cinnamomi* (MIC_{50} 6.96 ppm, MIC_{90} 11.19 ppm)	Castillo-Reyes et al. (2015)
	Leaves	Ethyl acetate 300-500 g/mL NDGA 300 g/mL methyl-NDGA	Inhibition of mycelium growth NDGA: *A. flavus* (81.92%), *A. parasiticus* (100%) methyl-NDGA: *A. flavus* (100%), *A. parasiticus* (100%)	Vargas-Arispuro et al. (2005)
Antiparasitic	Leaves	Dichloromethane:methanol	Inhibition of *E. histolytica*, *G. lamblia*, and *T. vaginalis* (IC_{50}: 100 to 118 µg/mL)	Camacho-Corona et al. (2015)

(*Continued*)

TABLE 11.1 (CONTINUED) *LARREA TRIDENTATA* AND ITS BIOLOGICAL ACTIVITIES

	Aerial parts	Methanol	Inhibitory effect (IC_{50}) *E. histolytica* (90.38 µg/mL, *G. lamblia* (98.47 µg/mL) and *T. vaginalis* (130.57 µg/mL) Percentage of inhibition *E. histolytica* (95.21%), *G. lamblia* (95.52%), and *T. vaginalis* (91.26%)	Garza, 2014
	Stem and leaves	Dichloromethane	Inhibitory effect (IC_{50}) *T. brucei* rhodesiens (2.8 mg/mL), *T. cruzi* (14.6 mg/mL), and *L. donovani* (5.2 mg/mL)	Schmidt et al. (2012)
		Methanol-water	NDGA activity: *E. histolytica, G. lamblia,* and *N. fowleri* (EC_{50}: 36 to 103 µM) 3'-O-methyl-NDGA activity: *N. fowleri* (EC_{50}= 38 µM), *E. histolytica* (EC_{50}= 171 µM) Nor-3'-demethoxyisoguaiacin activity: *G. lamblia* (EC_{50} = 49 µM) Nor-isoguaicin activity: *E. histolytica, G. lamblia,* and *N. fowleri* (EC_{50}: 74 µM to 83 µM) 3-methoxy-6, 7, 4'-trihydroxyflavonol activity: *G. lamblia* (EC_{50} = 153 µM), *N. fowleri* (EC_{50} = 235 µM)	Bashyal et al. (2017)
	Leaves	Hydromethanolic	Anthelmintic activity on sheathed (32.1%) and examined (68.4%) larvae of *Haemonchus contortus* at 200 mg/mL. EC_{50} 36 mg/mL	Garcia et al. (2018)
Antiviral	Leaves	Chloroform:methanol	Inhibition of tat transactivation of HIV-1	Gnabre et al. (1995a)
	Leaves	Ethyl acetate	Inhibition of cytopathic effects of HIV-1 in EMF-SS cells of human lymphoblasts.	Gnabre et al. (1996)

(Continued)

TABLE 11.1 (CONTINUED) *LARREA TRIDENTATA* AND ITS BIOLOGICAL ACTIVITIES

		3'-O-methyl NDGA 3'-O-methyl-NDGA and M4N	Inhibition of HIV tat-regulated transactivation in human epithelial cells (IC_{50} 25 μM) and inhibition of tat-regulated SEAP production	Hwu et al. (1998)
		M4N	Inhibition of the expression of α -ICP4 (IC_{50} 43.5 μM), essential for HSV replication, in Vero cells. Inhibition of the binding of Sp1 to the promoter α-ICP4.	Chen et al. (1998)
		Mal-4, M4N, tetra-acetyl NDGA	Inhibition of the gene expression of promoter P97 of HPV16 (IC_{50}: Mal-4, M_4N, and tetra-acetyl NDGA 37, 28, and 11 μM, respectively.	Craigo et al. (2000)
	Leaves	Methanol	Antiviral effect of purified compounds against DENV at 120 μg/mL	Maldonado (2016)
		NDGA	92% reduction of secreted NS1 in Huh-7 hepatoma cell line	Soto-Acosta et al. (2014)
Antioxidant	Whole plant	Methanol	Inhibition of 82.87% of the DPPH radical	Ramírez et al. (2016)
	Leaves	Methanol and obtained by fermentation in solid state	Reduction of phosphomolybdate ion (150–1300 nmol alpha-tocopherol/g dry sample)	Martins et al. (2013a)
	Leaves	Ethanol:water	Cytoprotective activity against *in vitro* oxidative stress in human SH-SY5Y cells.	Moran-Santibañez (2019)
	Leaves	Ethanol:water	DPPH: IC_{50} = 111.7 μg/mL ABTS: IC_{50} = 8.49 μg/mL Superoxide: IC_{50} = 0.43 g/mL Nitric oxide: IC_{50} = 230.4 mg/mL FRAP: 1.01 11 μg/mL	Skouta et al.(2018)

(*Continued*)

TABLE 11.1 (CONTINUED) *LARREA TRIDENTATA* AND ITS BIOLOGICAL ACTIVITIES

	Leaves	Methanol	Antioxidant effect against intracellular ROS in HL-60 cell line Epoxylignan 1: IC_{50} = 1.3 µg/mL 2: IC_{50} = 7.5 µg/mL Flavonoid 5: IC_{50} = 1.6 µg/mL Flavonoid 12: IC_{50} = 1.3 µg/mL 3'-demethoxy-6-O-demethylisoguaicin: IC_{50} = 1.6 µg/mL NDGA: IC_{50} = 0.7 µg/mL	Abou-Gazar et al. (2004)
	Leaves	Methanol, ethanol, and acetone	93-95% inhibition of DPPH radical 70% methanolic extract/ FRAP: 2.55 mM FE(II)/g, 90% methanolic extract / FRAP: 2.73 mM FE (II)/g	Martins et al. (2012)
	Leaves	Water	97% inhibition of ABTS radical (4.11 µET/g) 92% inhibition of DPPH radical 57% LOI inhibition FRAP: 0.73 mM Fe(II)/g	Aguirre-Joya et al. (2018)
Antiproliferative	Leaves	Methanolic	Antiproliferative effect in murine cells (RAW 264.7: IC_{50} = 19.85 µg/mL) and human cell (A-549, LS-180, HeLa and 22Rv-1, BxPc-3: IC_{50} = 64.28 to 163.73 µg/mL)	Salido et al. (2016)
		NDGA	Inhibition on cell viability of human leukemic cell lines HL-60 (IC_{50} = 10.3 µM) and U-937 (IC_{50} = 12.5 µM)	Leon et al. (2016)
		NDGA	Inhibition of IGF-1R autophosphorylation induced by IGF-1 and inhibition of DHT-induced proliferation in LAPC-4 prostate cancer cells	Ryan et al. (2008)

(*Continued*)

TABLE 11.1 (CONTINUED) *LARREA TRIDENTATA* AND ITS BIOLOGICAL ACTIVITIES

		NDGA	*in vitro* growth inhibition of small cell lung cancer (SHP77, H69, and H345) and non-small cell lung cancer (A549, H460, and H157) (IC_{50} = 10 to 65 µM) Most sensitive cells: H69 (IC_{50} = 15 µM) and H345 (IC_{50} = 10 µM)	Soriano et al. (1999)
		NDGA	*in vitro* growth inhibition of MCF-7/neo (23-55%) and MCF-7/HER2-18 (18.9-38%) cells at 10 and 15 µM	Zavodovskaya et al. (2008)

MIC: minimal inhibitory concentration; MBC: minimal bactericidal concentration; NDGA: Nordihydroguaiaretic acid; IC_{50}: Half-maximal inhibitory concentration; EC_{50}: Half-maximal inhibitory concentration.

For instance, Dimayuga and García (1991) demonstrated that the pathogenic microorganisms *Staphylococcus aureus*, *Bacillus subtilis*, and *Streptococcus faecalis* have a high sensitivity to the ethanolic extract of *L. tridentata*, producing halo inhibitors of between 15 and 20 mm for all microorganisms at a concentration of 20 mg/mL of the extract. Itza et al. (2019) demonstrated that *L. tridentata* ethanolic extract was able to inhibit the growth of *Streptococcus pyogenes, S. aureus, Enterococcus faecalis, B. subtilis, Klebsiella oxytoca, Escherichia coli, Salmonella enterica*, and *S. gallinarum*. The concentration of 35% of the extract was the most effective for bacteria inhibition, being *S. aureus* (inhibition halo: 2.7 mm) and *E. faecalis* (inhibition halo: 2.33 mm) the most susceptible. Moreover, different studies have shown that alcoholic and hydroalcoholic extracts have an inhibitory effect against several microorganisms such as *S. aureus, E. coli, Pseudomonas aeruginosa, Yersinia enterocolitica, Listeria monocytogenes, S. dysenteriae, Proteus vulgaris, Clostridium perfringens, Nocardia asteroids*, and *Nocardia brasiliensis*. In general, it was observed that the extracts of *L. tridentata* presented a strong inhibitory effect against *S. aureus* because they showed the lowest rates of minimum inhibitory concentration (MIC), which ranged between 60 to 1250 μg/mL. Likewise, it was observed that *L. tridentata* strongly inhibits the bacterial replication of *S. aureus* at concentrations of 6–20 μL/mL, as it promotes 10^6 reductions (CFU/mL) in comparison to the control treatment (Verástegui et al., 1996; Navarro et al., 1996; Snowden et al., 2014; Ruiz et al., 2017). In a similar context, Joublanc et al. (2008) demonstrated that an ethanolic extract of *L. tridentata* exhibited inhibitory potential against the following bacteria: *E. coli, S. aureus, S. typhimurium, Serratia licuefaciens, P. aeruginosa, K. oxytoca, Proteus mirabilis*, and *P. vulgaris* where the MIC values were 100 μg/mL. Along with this, this author reported that the extract of *L. tridentata* presented a lethal effect against the tested microorganisms, whose minimum bactericidal concentration (MBC) values were of 200 μg/mL.

In order to understand features of the nature of bioactive compounds that provide antimicrobial potential in *L. tridentata*, Bocanegra-García et al. (2009) investigated the antimicrobial effect of *L. tridentata* associated with the type of solvent used in the extraction (hexane, chloroform, methanol, and water) against pathogenic microorganisms related to lower respiratory tract infections (clinical isolates). It was observed that the chloroform and methanol extracts presented a greater antimicrobial effect against the clinical isolates, with *S. pneumoniae* (MIC_{80}: 32.25 μg/mL, for both extracts), *E. faecalis* (MIC_{50}: 250 μg/mL, for both extracts), and *S. aureus* (MIC_{80}: 125 μg/mL, for both extracts) being the most sensitive to these treatments. To better understand this biological activity, Martins et al. (2013b) tested the antimicrobial potential of the methanolic extract of *L. tridentata* and its fractions (hexane, dichloromethane, ethyl acetate, and ethanol). The testing was carried out against different Gram-positive and Gram-negative microorganisms resistant to antibiotics and against clinical and reference isolates. Overall, it was observed that the extract and factions presented a higher antimicrobial potential against Gram-positive microorganisms, and it was likewise evidenced that dichloromethane and the ethyl acetate fractions were the most active. The microorganisms with the highest susceptibility to the extract and active fractions (dichloromethane and the ethyl acetate) were *S. aureus* (MIC: 31.3-125 μg/mL), coagulase negative *Staphylococcus* (MIC: 62.5-250 μg/mL), and *E. faecalis* (MIC: 125-375 μg/mL). In the same context, it was demonstrated that extract of *L. tridentata* was an excellent source of chemical compounds with antimicrobial potential. Lignans (three compounds) and flavonoids (four compounds) purified from the chloroform fraction were responsible for the antimicrobial potential in the extract, as they showed strong antimicrobial activities against clinical isolates of *S. aureus, E. faecalis, Enterobacter cloacae, E. coli*, and *Mycobacterium tuberculosis* isolates, whose MIC ranged from 12.5 to <50 μg/mL. The lignan 3'-demethoxy-6-O-demethylisoguaiazine was the most active compound isolated from chloroform fraction, while dihydroguayretic acid, 4-epi-larreatricin, and 5,4′-dihydroxy-7-methoxyflavone, were less active

(Favela-Hernández et al., 2012). In another study, Delgadillo-Ruiz et al. (2017) demonstrated that the ethanolic extract of *L. tridentata* is a source of lower polarity compounds (thymol and carvacrol), which suggests that terpenes may also influence the antimicrobial potential of the plant, as the extract possessed an inhibitory effect against *E. coli, Acinetobacter baumannii, Pseudomonas* sp., and *S. aureus*, presenting halo inhibition of 17, 19, 15, and 12 mm, respectively.

The results described above demonstrate that *L. tridentata* could represent a good option for the development of pharmacological treatments against bacterial infections of clinical relevance, as their effect has been proven against clinical relevance bacteria which are resistant to antibiotics and clinical isolates. The above may be feasible because different authors have established that natural products with inhibitory effects in concentrations lower than 1000 µg/mL may be considered as potential antimicrobial agents for the development of phytopharmaceuticals or the purification or active compounds. Additionally, the antimicrobial action mechanism in extracts of *L. tridentata* has not been established. Nevertheless, considering the bioactive compounds identified in active extracts, it can be suggested that they can inhibit the bacterial growth through different modes of action. The compounds identified are mainly phenolic compounds. Part of these compounds are lignans, which have been demonstrated to cause damage to the cytoplasmic membrane, affecting the proton-motive force and the proteins in the transportation system of the ATP-binding cassette. In addition, lignans induce damage to DNA, altering the cellular division in Gram-positive and Gram-negative microorganisms that lead to bacterial death (Bürli et al., 2004; Maruyama et al., 2007; Hwang and Lim, 2015; Favela-Hernández et al., 2015). Another type of phenolic compound present in extracts of *L. tridentata* are flavonoids, which have been shown to have the capacity to cause membrane disruption and inhibit the synthesis of nucleic acid, the electron transport chain synthesis, and the ATP synthesis (Simoes et al., 2009; Górniak et al., 2019). The antimicrobial effect of phenolic compounds can be explained based on their chemical structure, which presents amphipathic characteristics. Likewise, these structures present different functional groups, hydroxyl and methoxyl, which allow them to interact with polar and non-polar components of bacterial structures, causing their deaths (Pandey and Kumar, 2013; Górniak et al., 2019). Additionally, another type of compound identified in the active extracts of *L. tridentata* are terpenes. In general, these types of compounds have demonstrated different effects against bacterial cells, such as degradation of the cell wall, alterations to the cytoplasmic membrane, cytoplasm clotting, damage to the membrane proteins, an increase in permeability with subsequent leakage of intracellular components, reduction of the proton-motive force and the intracellular ATP, and interaction with the bacterial DNA (Nazzaro et al., 2013; López-Romero et al., 2015; Khameneh et al., 2019). This is attributed to the structural characteristics of the compounds, which are of a non-polar nature. Likewise, these structures may possess functional groups, such as hydroxyl groups, which increase their reactivity. The above allows the interaction with lipids in the bacterial membrane, causing its destabilization. Additionally, these characteristics, along with their low molecular weight, allow them to cross the bacterial membrane by passive diffusion, causing damage to the intracellular components (Swamy et al., 2016; Guimarães et al., 2019). Based on the above, it could be suggested that the chemical compounds in *L. tridentata* perform different modes of antimicrobial action, which leads to bacterial death.

11.4.2 Antifungal Activity

Other biological activity widely reported for *L. tridentata* is antifungal potential (Figure 11.3; Table 11.1). In this sense, Osorio et al. (2010) carried out *in vitro* tests on eight pathogenic strains

(*Pythium sp., Colletotrichum truncatum, C. coccodes, Alternaria alternata, Fusarium verticillioides, F. solani, F. sambucinum, and Rhizoctonia solani*) and ten isolates of *F. oxysporum* using the polyphenolic extract of *L. tridentata* leaves, with 70% acetone as the extraction solvent. The extract at 0.70 mg/L, showed fungicidal effect (100%) on *Pythium sp., C. coccodes, C. truncatum, A. alternata, F. solani*, and *R. solani*. On one hand, the extract showed fungistatic activity (75% inhibition) against *F. verticillioides*. Also, the extract inhibited 100% the growth of eight clinical isolates of *F. oxysporum*. Similarly, ethanolic extract of *L. tridentata* presented an inhibitory effect of 97% on the mycelial growth of *F. oxysporum* f. sp. *radicis-lycopersici* at 500 ppm (Peñuelas et al., 2017). Díaz-Díaz et al. (2013) evaluated the antifungal capacity of *L. tridentata* when using fermented and non-fermented methanolic extract. Results showed that the one without fermenting at 100 ppm was the most efficient, as it inhibited 100% of *Phytophthora capsici* growth. On the other hand, the fermented methanolic extract was able to completely inhibit the growth of the fungus at 1,000 ppm, which is consistent with the fact that fermentation of these extracts is not necessary to maintain control of the *in vitro* mycelial growth of *P. capsici*. In addition, they suggested that the Soxhlet method is the best for extracting bioactive substances from *L. tridentata*. Galván et al. (2014) evaluated the effect of aqueous extracts of *L. tridentata* leaves on the mycelial growth of phytopathogenic fungi (*P. capsici, Aspergillus flavus, Rhizopus* sp., *Alternaria solani*, and *Botrytis* sp.). The *in vitro* test was performed at two different concentrations (10% and 20%), and mycelial growth was recorded at 24, 48, 72, and 96 hours (h). The most relevant results were obtained on the first three fungi mentioned. At 48, 72 and 96 h, both concentrations inhibited 100% of the mycelial growth of *P. capsici* and *A. flavus*. In contrast, with respect to the *Rhizopus* sp. strain, at 48 h, the 20% extract inhibited 60.30% of the mycelium growth; at 72 h, the 10% extract inhibited 59.33%; and at 94 h, the 20% extract inhibited 66.5%.

Several studies have demonstrated that antifungal potential of *L. tridentata* may vary depending on the plant location. For example, bioassays conducted by Mojica et al. (2011) showed that collected samples from Nuevo León, San Luis Potosí, and Coahuila at 2, 5, and 10 mg/mL completely inhibited the mycelial growth of *P. capsici*, while one collected in Durango was only effective against fungus at 5 and 10 mg/mL. In the same regard, they highlighted that the lowest mycelial growth occurred with the specimen from Coahuila at 0.5 mg/mL (11.16 mm) after six days of incubation. Similar research compared the activity of *L. tridentata* ecotypes from the Chihuahuan Desert and the Sonoran Desert. For that, methanolic and ethanolic extracts from resins, leaves, and branches of *L. tridentata* were obtained. The greater inhibitory effect against *Pythium* sp., was observed when evaluating the samples collected in the Sonoran Desert (85.07%) compared to those of the Chihuahua Desert (80.92%). Thus, fungicidal effect of the *L. tridentata* extracts was efficient regardless of the solvent used; however, the methanolic extracts from both deserts showed a notable effect, as they completely inhibited the *in vitro* mycelial growth of involved phytopathogen at low doses (500 μl/L) (Lira-Saldívar et al., 2003).

A possible explanation for the considerable variations between specimens from the same region is longevity of leaves and/or their thickness (Gisvold, 1974; Mabry, 1978). Firn and Jones (2003) mentioned that even though plants store and produce various secondary metabolites, the properties of each compound can be modified depending on their environment. Although the content of bioactive compounds of *L. tridentata*, as well as its antifungal potential, are affected by the location, which favors the conditions for its growth as well as the timing of collection, the chemical nature of the solvent used for the extraction could play an important role in the type and quantity of the extracted compounds (Castillo-Reyes et al., 2015).

Moreno et al. (2011) evaluated the antifungal activity of *L. tridentata* leaves, using four solvents (hydroalcohol, acetone, methanol, and ethanol). Antifungal effect was evaluated by agar diffusion and agar extract dilution technique. In the first method, ethanolic extract inhibited the growth of *A. flavus*, as it presented inhibition halos that varied between 12 and 21 mm of diameter, while the methanolic extract showed inhibition halos between 7 and 20 mm. Additionally, *Penicillium* sp. was affected by ethanolic extract, with inhibition halos between 8 and 18 mm in diameter. The MIC of the hydroalcoholic extract were 7 mg/mL and 5 mg/mL for *A. flavus* and *Penicillum* sp., respectively. In the same context, Castillo-Reyes et al. (2015) demonstrated the ability of *L. tridentata* to inhibit *P. cinnamomi in vitro* using extracts with water, ethanol, lanolin, and cocoa butter. In this study, ethanolic extract exhibited greater inhibition potential on mycelial growth, showing MIC_{50} and MIC_{90} values for 6.96 and 11.19 ppm, respectively. In another investigation by Vargas-Arispuro et al. (2005), crude extracts of leaves and branches of *L. tridentata* on *A. flavus* and *A. parasiticus* were evaluated using fractions of ethanol, ethyl ether, ethyl acetate, and n-butanol. Overall, crude extract of the leaves (500 µg/mL) presented the highest toxicity against both fungi (70%–90% inhibition, respectively), and the ethyl acetate fraction showed better antifungal activity. Subsequently, this fraction was subjected to column chromatography, and the activities of the new obtained fractions were evaluated. Fraction 4, obtained with chloroform: acetone (90:10, v/v), inhibited 86% and 65% of the colonial growth of *A. flavus* and *A. parasiticus*, respectively. The identified compounds responsible for the activity of the fraction were NDGA and 3-methylnordihydroguaiaretic acid (methyl-NDGA). NDGA at 300 g/mL, inhibited 81.92% of the mycelial growth of *A. flavus*, while at 500 g/mL, 100% inhibition of *A. parasiticus* was achieved. On the other hand, methyl-NDGA at 300 g/mL promoted a complete growth inhibition of both fungi. The information described above shows that *L. tridentata* could be a viable source to obtain phytopathogenic inhibitors that can be used in agricultural crops.

Based on previous research, compounds from *L. tridentata* with the most influence on antifungal activity of have been lignans; however, the antifungal mechanism of action of *L. tridentata* is not clearly defined. However, having access to the secondary metabolites present in the plant can suggest a different mode of action. Terpenes have the property of damaging biomembranes, which are attributed to their lipophilic characteristics. Other compounds, such as alkaloids, interact with DNA and cause mutations in microorganism. On the other hand, flavonoids and tannins form nucleophilic complexes with proteins, causing their inactivation. Similarly, phenolic compounds can inactivate fungal proteins, through the interaction with amino or sulfhydryl groups of amino acids (Rodríguez-Guadarrama et al., 2018). It should be noted that the bioactivity of the layer probably interferes with the permeability of the fungal cell membrane and allows interaction with membrane proteins, which causes structural and functional deformation. This event can lead to dysfunction and subsequent membrane disruption. In turn, research has reported that phenolic compounds could fade the pH gradient and the electrical potential components of the proton motive force, interfering with the cell's energy generation and conservation of the ATP system, inhibiting the binding of enzymes to the membrane, and avoiding the use of the substrate for energy production (Hapon et al., 2018). Based on the above, it is suggested that the compounds present in *L. tridentata* provide biocidal properties capable of inhibiting fungal growth throughout several mechanisms of action.

11.4.2 Antiparasitic Activity

The antiparasitic effect of *L. tridentata* has been reported by several authors (Figure 11.3; Table 11.1). *In vitro* tests carried out by Camacho-Corona et al. (2015) reported the antiparasitic capacity of *L. tridentata* against strains of *Entamoeba histolytica*, *Giardia lamblia*, and *Trichomonas vaginalis* using extract from the leaves with dichloromethane:methanol (1:1) as a solvent, which presented a moderate inhibitory activity against the protozoa with IC_{50} values that ranged between 100 and 118 μg/mL. Similarly, the methanolic extract of *L. tridentata* (300 μg/mL) inhibited *E. histolytica*, *G. lamblia*, and *T. vaginalis* by 95.21, 95.52, and 91.26%, respectively with IC_{50} values of 90.38 μg/mL, 98.47 μg/mL, and 130.57 μg/mL (Garza, 2014). On the other hand, Schmidt et al. (2012) used the stem and leaves of *L. tridentata*, with dichloromethane as a solvent to evaluate its antiparasitic activity against *Trypanosoma brucei rhodesiense*, *T. cruzi*, *Leishmania donovani*, and *Plasmodium falciparum*, for which an IC_{50} of 2.8, 14.6, 5.2, and 2.9 mg/mL, respectively, were obtained. In this research, the active components – nine lignans (three dibenzylbutanes, four epoxylignes, two aryltetralines), six flavonoids, and one ferulic acid ester (3'-Oxohexylferulated) – were identified. The most effective was NDGA, with IC_{50} values between 4.5 and 33.1 μM, while the epoxylignanes obtained IC_{50} values of 12.0 to 183 μM, the aryltetralines IC_{50} = 11.4 to 185 μM, and the flavonoids IC_{50} = 8.0 to 199 μM. In another study, the activity of *L. tridentata* against the pathogens *E. histolytica*, *G. lamblia*, and *Naegleria fowleri* in their trophozoite phase was evaluated. To this matter, Bashyal et al., (2017), fractioned an aqueous methanolic extract with hexane, ethyl acetate, and *n*-butanol. The ethyl acetate fraction exhibited antiparasitic activity at 50 μg/mL, being selected for further study. The fractionation of the extracts allows the isolation of nine pure compounds (eight lignans and one flavonol). NDGA was reported to be active against all three pathogens, with EC_{50} values ranging from 36 to 103 μM. 3'-O-methylNDGA exhibited high activity against *N. fowleri* (EC_{50}= 38 μM) and moderate activity against *E. histolytica* (EC_{50}= 171 μM). Nor-3'-demethoxyisoguaiacine showed greater effectiveness against *G. lamblia* (EC_{50} = 49 M) compared to the two remaining protozoa. Additionally, norisoguaicin showed a similar activity against the three pathogens, presenting EC_{50} values of 74 μM to 83 μM, while 3-methoxy-6,7,4'-trihydroxyflavonol showed moderate activity against *G. lamblia* (EC_{50} = 153 μM) and *N. fowleri* (EC_{50} = 235 μM). In turn, the authors investigated NDGA and 3'-O-methylnordihydroguaiaretic acid because of their potent antiparasitic activity against *N. fowleri*, carrying out a cytotoxicity study on human umbilical vein endothelial cells, derived from the endothelium of veins from the umbilical cord. These lignans were able to inhibit cell viability with an EC_{50} of 86 and 59 μM, respectively, concluding that they are 2.4 and 1.6 times less toxic to cells compared to *N. fowleri* (Bashyal et al., 2017).

The effectiveness of *L. tridentata* as an anthelmintic has also been demonstrated. Garcia et al. (2018) carried out an *in vitro* study with hydromethanolic extracts from leaves at different concentrations (12.5, 25, 50, 100, and 200 mg/mL) on sheathed and unsheathed larvae of *Haemonchus contortus*. The plant was reported to possess dose-dependent anthelmintic activity, with 200 mg/mL of the extract being able to reduce 32.1% and 68.4% of sheathed and unsheathed larvae, respectively. In turn, the EC_{50} was determined to be 36 mg/mL. In addition to the above, it is important to mention that, although *H. contortus* infection occurs mainly in ruminants, there is a possibility that the parasite could be transmitted zoonotically to humans, as a clinical case of infection caused by this helminth in a 60-year-old woman from Iran was reported (Ghadirian and Arfaa, 1973).

Based on the exposed investigations, the capacity of *L. tridentata* in the inhibition of certain pathogenic parasites is evidenced; however, the specific compound or compounds that provide this activity in all cases is unknown, as well as the mechanism of action. However, based on the bioactive compounds identified in the active extracts, various authors have suggested the possibility of exhibiting different mechanisms of action. Phenolic compounds interfere with the mechanism of energy generation by uncoupling oxidative phosphorylation, causing alterations in the glycoproteins found on the parasite cell surface, causing its death (Bauri et al., 2015). Within the phenolic compounds are flavonoids, which have been reported to have antiparasitic activity. Sharma et al. (2007) reported activity against *P. falciparum* by catechin enzymes involved in fatty acid biosynthesis. A different study carried out by Bolaños et al. (2014) reported that (-)-epicatechin promoted protein changes on the cytoskeleton of *E. histolytica*, as well as deregulation of enzymes that participate in energy metabolism (glyceraldehyde-phosphate and fructose-1,6-bisphosphate aldolase), while affectation in *T. cruzi* has been involved in the activity of arginine kinase (Paveto et al., 2004) and NADH-oxidase (Scotti et al., 2010). On the other hand, licochalcone A has been shown to act in the inhibition of *Leishmania* species, causing structural changes in the mitochondria. The probable mechanism of action of this compound is by inhibiting mitochondrial dehydrogenase (Kayser et al., 2003; Ramírez et al., 2010).

Similarly, it has been demonstrated that for *L. donovani*, kaempferol promoted the inhibition of the enzymes pyruvate kinase, dihydroorotase, and cytidine deaminase, which affected the pyrimidine biosynthesis pathway and, consequently, contributed to the parasite's death (Scotti et al., 2015; Tiwari et al., 2016). Likewise, it has been reported that flavonoids with condensed tannins showed anthelmintic activity (Klongsiriwet et al., 2015). For their part, tannins themselves have the ability to interfere in parasites' ability to generate energy through the uncoupling of oxidative phosphorylation, which causes their death (Bauri et al., 2015).

Based on the compounds found in *L. tridentata*, the majority of identified compounds are lignans, however, the mechanism of action of their antiparasitic effect has not been determined. However, Gutiérrez-Gutiérrez et al. (2017) reported that four lignans –5′ -desmethoxy- β-pel tatin-A-methylether, acetylpodophyllotoxin, burseranin, and podophyllotoxin – isolated from *Bursera fagaroides* var. *fagaroides* act on *G. lamblia* affecting the expression and stability of tubulin polymerization, destabilizing the structure of microtubules, thus affecting the growth, morphology, and adhesion of *G. lamblia*. This may suggest that the extracts made from *L. tridentata* could probably present a similar behavior depending on the structure of the chemical compounds. Under this premise, *L. tridentata* could represent a viable alternative for the development of new antiparasitic agents, as different modes of action are suggested in which its bioactive compounds could inhibit the protozoa and helminths.

11.4.3 Antiviral Activity

L. tridentata could represent an alternative for the treatment of viral diseases, as various investigations have shown its effectiveness against viruses of clinical relevance (Figure 11.3; Table 11.1). Gnabre et al. (1995a) reported that the leaf extract (chloroform-methanol 1:1) showed effect against human immunodeficiency virus type 1 (HIV-1) *in vitro* via inhibition of tat transactivation. Also, the ability of this plant to inhibit the cytopathic effects of HIV-1 in human lymphoblasts (CEM-SS cells) has been investigated. For this, an XTT (2,3-bis(2-methoxy-4-nitro-5-sulfophenyl)-5-carboxanilide-2H-tetrazolium) assay was carried out using crude fractions of *L. tridentata* leaves. From the ethyl acetate soluble fraction, two fractions were identified that contained a mixture of lignans, which showed the ability to reduce the cytopathic effect of HIV-1 at concentrations of 0.187 and 0.75 μg/mL, obtaining a cell viability of 7% and 58%, respectively (Gnabre et al., 1996).

Another study reported that one of the methylated derivatives of NDGA, 3′-O-methylnordihydroguaiaretic acid (3′-O-methyl-NDGA, Mal-4), has antiviral activity (EC_{50} of 25 μM) against HIV via tat-transactivation inhibition (Gnabre et al., 1995b). It has also been reported that by methylating NDGA with dimethyl sulfate, tetra-O-methylnordihydroguaiaretic acid (M_4N) is obtained, which has the potential to inhibit HIV tat-transactivation (Gnabre et al., 2015).

Hwu et al. (1998) demonstrated that NDGA methylations, using dimethyl sulfate and potassium carbonate in acetone, allow for various compounds to be obtained, among them

3'-O-methyl-NDGA and M_4N; the first compound was able to inhibit the HIV tat-transactivation with IC_{50} of 25 μM. Also, they described that each of the methylated NDGA (100 μM), except for 3',3''-dimethyl-NDGA, inhibited by over 85%, a reporter gene regulated by the long-terminal promoter repeat (LTR) of HIV, which is regulated by tat. Finally, it is reported that M4N (3 μM) was 11 times more effective in inhibiting HIV tat-transactivation compared to 4'-O-methyl-NDGA, and 3.4 times better than 3′-O-methyl-NDGA. Therefore, it is concluded that M_4N acts as a better inhibitor of tat-transactivation regulator than Mal-4.

In the same context, M_4N showed inhibitory effect in the replication of the herpes simplex virus (HSV). Chen et al. (1998) infected Vero cells with a plasmid containing the promoter of the α-ICP4 gene, which is essential for HSV replication and is one of the first genes expressed during lytic transcription of the virus. It was found that M_4N was able to inhibit the promoter activity of α-ICP4 with an IC_{50} of 43.5 μM. At the same time, an electrophoretic mobility shift assay was performed in the presence of M_4N (100 μM), in which the binding of the Sp1 protein to the binding sites most proximal to the α-ICP4 promoter was examined, and the results showed a significant inhibition of the fast- and slow-moving bands, with 88% and 45% inhibition, respectively. Likewise, it was observed that M_4N suppress the growth of HSV-1 and HSV-2 at a low concentration (IC_{50}= 4-11.7 μM), while in cells, the virus generates resistance against acyclovir.

Craigo et al. (2000) evaluated the effect of NDGA derivatives such as Mal-4, M_4N, and tetra-acetyl NDGA on the inhibition of gene expression of the P_{97} promoter of human papilloma virus type 16 (HPV16). To do this, they used the C-33 A cervical carcinoma cell line, which was infected with a recombinant plasmid containing the promoter of the P_{97} gene and the luciferase gene as an indicator of gene expression. The results showed that Mal-4 and M_4N inhibited the P_{97} promoter, exerting a dose-dependent effect, while tetra-acetyl NDGA turned out to be more potent as an inhibitor of HPV gene expression, as there was an initial fall on P_{97} activity at low concentrations. The IC_{50} values for Mal-4, M_4N, and tetra-acetyl NDGA were 37, 28, and 11 μM, respectively.

On the other hand, Maldonado (2016) demonstrated the antiviral activity of a series of purified compounds (three flavonoids and two lignans) from the leaves of *L. tridentata* using an indirect immunofluorescence assay in BHK-21 cells. The compounds Lt 1, Lt 2 (5,4'-dihydroxy-3,7,8,3'-tetramethoxyflavone), Lt 3 (5,8,4'-trihydroxy-3,7-dimethoxyflavone), Lt 5 (5 ,4'-dihydroxy-3,7,8-trimethoxyflavone) showed excellent activity against dengue virus (DENV) at concentrations of 120 μg/mL. Lt 1 was the most effective while Lt 4 (dihydroguaiaretic acid) was the least effective.

Also, the effect of NDGA on DENV infection in the hepatoma cell line Huh-7 has been investigated. For this, Huh-7 cells were infected with the DENV virus and treated with NDGA. Subsequently the levels of secreted NS1-viral protein and viral yield were quantified in the supernatant through ELISA. The levels of secreted NS1 showed a 92% of reduction after 24 h of treatment with NDGA (100 μM). Similarly, NDGA effectively reduced the viral yield. However, after 48 h of treatment, the effect on NS1 secretion and viral yield was decreased, which is probably associated with NDGA metabolization during prolonged incubation. Finally, they confirmed that the inhibitory effect of NDGA on DENV infection occurs during replication (Soto-Acosta et al., 2014).

The previous studies show that *L. tridentata* could represent a good alternative for the development of pharmacological agents to treat viral diseases, as its effect on different viruses of clinical relevance was evidenced, which have been reported to have generated resistance to antivirals currently available on the market. In addition, in the investigations carried out, the isolated compounds to which the antiviral activity of *L. tridentata* has been attributed are phenolic compounds, specifically lignans, which include NDGA and its methylated derivatives, for which the molecular mechanisms of antiviral activity have not been fully established; however, several authors have reported the inhibition of viral transcription of the gene dependent on Specificity Protein One (Sp1). Sp1 is a crucial transcription factor, controlling more than 1,000 genes and is found in one or more motifs in the promoters of various clinically important viruses, including HIV (contains three Sp1 motifs), HSV (eight Sp1 motifs), and HPV (one Sp1 motif).

Regarding HIV – one of the most studied diseases – it has been reported that the Sp1 motifs interact with the transactivator tat (Gnabre et al., 2015). Tat has the ability to be a vital regulator of HIV gene expression, as it binds to the transactivation response (TAR) element, where it activates the transcription of the DNA long terminal repeat (LRT) promoter HIV-1 proviral (Gnabre, 1999). The lignans of *L. tridentata* bind to DNA sequences rich in GC boxes (guanine/cytosine), and Sp1 has an affinity for these regions. Under this premise, Sp1 could be the main target for lignans. Blocking this region with the compounds prevents Sp1 from binding to its corresponding site in the HIV LTR promoter and, consequently, viral replication is inhibited. Therefore, these lignans can inhibit the basal transcription of the HIV gene, as well as the transactivation of the gene regulated by tat. It should be noted that, among the methylated lignans identified, the most active have been Mal-4 and M_4N. These suppress HIV basal transcription and tat-dependent and -independent transactivation. These molecules present different degrees of methylation, which is related to their lipophilicity; this allows them to more easily penetrate the host cell lipid bilayer and the viral layer to disrupt specific viral replication (Gnabre et al., 2015).

In addition, it has been reported that the non-structural glycoprotein NS1 secreted by the DEVN virus could play a role in its replication. In a DEVN-infected cell model, it has been shown that the lack of glycosylation of NS1 impairs its secretion and reduces the proliferation of infectious virus (Fan et al., 2014). A study conducted by Maldonado (2016) concluded that some flavonoids isolated from *L. tridentata* act as antivirals against DENV, especially against serotype 2. This could probably be associated with the ability of the compounds to interact with envelope proteins of the virus, preventing its entry into the cell.

To date, there are no studies that demonstrate the antiviral activity of the bioactive ingredients of *L. tridentata* against SARS-COV2; however, because of its outstanding antiviral activity against viruses of clinical relevance, it is an important candidate in the search for new agents against this pathogen. It may be used in *in silico* studies to find out the inhibitory capacity of *L. tridentata* compounds on molecular targets of SARS-COV-2 and subsequently carry out *in vitro* tests to find out their potential.

11.4.4 Antioxidant Activity

Studies on *L. tridentata* have highlighted its antioxidant potential, which could represent a good resource for the search for new options against oxidative stress (Figure 11.3; Table 11.1). By 1,1-diphenyl-2-picrylhydracil (DPPH) assay, Ramirez et al. (2016) studied the antioxidant activity of the methanolic extract of *L. tridentata*. It was reported that the extract (0.1 mg/mL) showed 82.87% DPPH inhibition, which highlights its high capability of capturing free radicals. Using the phosphomolybdate method, Martins et al. (2013a) evaluated the antioxidant effect of extracts obtained by solid state fermentation and the *L. tridentata* leaves methanolic extract. In general, it was observed that all the extracts presented antioxidant activity, whose values were between 150 and 1300 nmol alpha-tocopherol/g of dry sample.

Morán-Santibañez (2019) analyzed the capability of extracts from *L. tridentata* leaves (ethanol-water, 6:4 v/v) to reduce the oxidative stress induced by H_2O_2 in human SH-SY5Y human cells. The extract (15 µg/mL) showed a cytoprotective effect via a decrease of ROS levels, activation of caspase 3 and 7, and cleavage of poly (ADP-ribose) polymerase (PARP), which reflects its antiapoptotic capability. This suggests *L. tridentata* extract could be used as a good natural antioxidant. Another study analyzed the antioxidant activity of the ethanol-water leaf extract (60:40 v/v), ethanolic extract, and aqueous extract. The authors reported that the extracts inhibited the DPPH radical in the following way: ethanol-water extract IC_{50} = 111.17 μg/mL; ethanolic IC_{50} = 135.4 μg/mL; aqueous IC_{50} = 572.7 μg/mL. While in the ABTS assay, the IC_{50} values were as follows: ethanol-water = 8.49 μg/mL; aqueous 9.75 μg/mL; ethanolic = 35.84 μg/mL.

Another study analyzed the antioxidant potential of the ethanol-water extract (60:40 v/v), ethanolic extract, and aqueous extract of *L. tridentata* leaves using different methods. In the DPPH assay, the following IC_{50} values were obtained: ethanol-water = 111.17 μg/mL; ethanolic = 135.4 μg/mL; aqueous = 572.7 μg/mL. By ABTS (3-ethylbenzothiazoline-6-sulfonic acid), the following IC_{50} values were obtained: ethanol-water = 8.49 μg/mL, ethanolic = 9.75 μg/mL, aqueous = 35.84 μg/mL. In turn, the following IC_{50} values were obtained in the superoxide assay: ethanol-water = 0.43 g/mL, ethanolic = 2.1 mg/mL, aqueous = 10.1 mg/mL. In the nitric oxide assay, the IC_{50} values were: 230.4 mg/mL, 520.7 g/mL, and 551.3 μg/mL for the ethanol-water, aqueous, and ethanolic extract, respectively. Additionally, the capability of the extracts to absorb UV-visible radiation at 330 nm was analyzed, reporting AI values of 1.7, 1.1, and 0.6 for the ethanol-water, ethanolic, and aqueous extracts, respectively. On the other hand, in the ferric ion-reducing antioxidant power (FRAP) assay, using the extracts at 250 μg/mL, it was reported that the ethanol-water, ethanolic, and aqueous extracts presented values of 1.01, 0.95, and 0.44, respectively. Based on the results described, it was concluded that the extract with the best antioxidant properties was ethanol-water (Skouta et al., 2018).

In the same way, the antioxidant potential of various compounds isolated from the methanolic extract of *L. tridentata* leaves was investigated, finding three new lignans and 10 previously reported flavonoids. All compounds were analyzed by the DFCH (2,7-dichlorodihydrofluorescein) assay using the HL-60 cell line, with the aim of knowing the effect on intracellular ROS (Volk and Moreland, 2014). Epoxylignans 1 ((7S,8S,7'S,8'S)-3,3',4'-trihydroxy-4-methoxy-7,7'-epoxylignan) and 2 (Meso-(rel 7S,8S,7'R, 8'R)-3,4,3',4'-tetrahydroxy-7,7'-epoxylignan), and the flavonoids 5 (4-epi-larreatricin) and 12 (3"-hydroxy-4-epi-larreatricin) showed strong antioxidant activity with IC_{50} values of 1.3, 7.5, 1.6, and 1.3 μg/mL, respectively. The compounds 3'-demethoxy-6-O-demethylisoguaicin (IC_{50} = 1.6 μg/mL) and NDGA (IC_{50} = 0.7 μg/mL) were also potent antioxidants (Abou-Gazar et al., 2004).

Martins et. al (2012) investigated the antioxidant capability of a series of *L. tridentata* leaves obtained with methanol, ethanol, and acetone at different concentrations (90%, 70%, 50%, and 30% v/v). The obtained results showed that all the extracts exhibited antioxidant effect in the DPPH and FRAP assays. In the DPPH, the extracts presented similar effects with inhibitions between 93% to 95%. However, by FRAP, the extracts obtained with 70% and 90% of methanol presented high FRAP values, being 2.55 and 2.73 mM FE (II)/g dry weight, respectively. Three phytoestrogens were also isolated in this study: NDGA, kaempferol, and quercetin.

Aguirre-Joya et al. (2018) evaluated the antioxidant effect of the aqueous extract of *L. tridentata* leaves by ABTS, DPPH, lipid oxidation inhibition (LOI), and FRAP. The results showed that the extract presented excellent antioxidant and free radical scavenging activity, as it inhibited ABTS radical by 97% (4.11 μM trolox equivalent/g), DPPH by 92%, and LOI by 57%, while in the FRAP assay, a value of 0.73 mM Fe (II)/g of dry weight was obtained. Furthermore, the main antioxidants in the extracts were identified by HPLC-MS: the lignan NDGA and two flavonoids (kaempferol and quercetin) (Aguirre-Joya et al., 2018).

The foregoing demonstrates the antioxidant capability of *L. tridentata*, as the extracts from the leaves and stems showed relevant results. However, even though the mechanism of action by which the plant exerts its antioxidant effect is not fully defined, it has been observed that this property could be associated with the compounds present in the plant, such as phenolic ones. It has been shown that these compounds act as powerful antioxidants that have the ability to protect against oxidative damage because, due to their structure, they have the ability to donate hydrogen atoms that interact with ROS and reactive nitrogen species (RNS) species. In addition, it is known that the antioxidant activity of polyphenols is also associated with their ability to chelate metal ions involved in the production of free radicals and to inhibit enzymes associated with their production, such as cytochrome P450, lipoxygenases, cyclooxygenases, and xanthine oxidase (Pereira et al., 2009; Ramírez et al., 2016). In turn, flavonoids are within the phenolic compounds, which stabilize ROS through a reaction with the reactive compound, this occurs because of the high reactivity of the hydroxyl group of the flavonoids, causing the inactivity of the radicals. In the same way, they act through interaction with various enzyme systems, in which some effects may result from a set of radical scavenging and interaction with enzyme functions. These reduce α-tocopherol radicals, inhibit oxidases, and mitigate nitrosative stress (Nijveldt et al., 2001; Carocho and Ferreira, 2013).

Flavonoids are known to chelate iron, which prevents the development of free radicals. For example, quercetin has iron chelating and stabilizing properties, which causes the inhibition of lipid peroxidation. For their part, flavones and catechins seem to be the most potent flavonoids against ROS, as they can directly eliminate superoxides, while others eliminate peroxynitrites (Nijveldt et al., 2001). Similarly, phenolic acids (hydroxycinnamic and hydroxybenzoic) have been reported as good antioxidants, exerting their effect as chelators and free radical scavengers, especially on OH^- and ROO^- radicals, O_2^- and NO_3^- anions (Carocho and Ferreira, 2013). On the other hand, NDGA is one of the lignan-type compounds with the highest antioxidant potential in *L. tridentata*; however, its mode of action is still unknown. Still, the antioxidant mechanism of NDGA could be associated with the direct elimination of ROS and the induction of antioxidant enzymes through the Nrf2 pathway, the latter being a factor that directs antioxidant and cytoprotective responses to oxidative stress (Guzman-Beltran et al., 2013). Furthermore, NDGA has also been shown to inhibit lipoxygenase enzymatic pathways (Fujimoto et al., 2004). In addition to the above, the way in which various bioactive compounds of *L. tridentata* could act against oxidative stress has been shown. Under this premise, *L. tridentata* could be used to generate an antioxidant agent capable of inhibiting oxidative stress and mitigating its effects.

11.4.5 Antiproliferative Activity

Other biological activity reported for *L. tridentata* is antiproliferative potential (Figure 11.3, Table 11.1). Salido et al. (2016) investigated the *in vitro* antiproliferative effect of the methanolic extract of *L. tridentata*. By MTT assay, they demonstrated that the extract effectively inhibits the proliferation of murine (RAW 264.7 macrophages) and human (A-549, LS-180, HeLa and 22Rv-1, BxPc-3) cancer cells. The murine cell line was inhibited with an IC_{50} of 19.85 µg/mL, while in human cells, the IC_{50} values were between 64.28 and 163.73 µg/mL. On the other hand, the noncancerous cell lines L929 and ARPE-19 presented greater resistance to the extract. Additionally, the extract was more selective than cisplatin, which was used as a positive control.

On the other hand, the potential of NDGA to inhibit the proliferation of cancer cells has been extensively investigated. Leon et al. (2016) evaluated the antiproliferative effect of NDGA on human leukemia cell lines HL-60 and U-937 using a neutral red assay and a colorimetric assay (that allows monitoring the ability of cells to incorporate the dye in lysosomes). The results show that NDGA (30 µM) almost completely decreased cell viability at 24, 48, and 72 h. At 24 h of treatment, IC_{50} values of 10.3 and 12.5 μM were obtained for HL-60 and U-937, respectively. Additionally, the effect of NDGA on peripheral blood mononuclear cells (PMBC) was evaluated, observing that the compound did not reduce the viability of these cells. This demonstrates that NDGA has greater selectivity on leukemic cells.

In another investigation, the antiproliferative effect of NDGA against LAPC-4 prostate cancer cells was evaluated. For this, the cells were deprived of androgens for three days, then 1 nM of dihydrotestosterone (DHT) was added together with other androgens for seven days. After the incubation time, the growth of LAPC-4 doubled. NDGA showed a rapid effect at inhibiting IGF-1 induced IGF-1R autophosphorylation. Whereas DHT increased IGF-1R protein expression as well as mRNA levels. Furthermore, NDGA at 10 μM was reported to inhibit cell proliferation caused by DHT (Ryan et al., 2008).

Using MTT assay, Soriano et al. (1999) evaluated the inhibitory effect of six agents –including NDGA – on the *in vitro* growth of small cell lung cancer (SHP77, H69, and H345) and non-small cell lung cancer (A549, H460, and H157). Results showed that NDGA was a potent inhibitor of all cell lines, with IC_{50} values between 10 to 65 μM. In turn, the cell lines H69 (IC_{50} = 15 μM) and H345 (IC50 = 10 μM) were particularly more sensitive. For their part, Zavodovskaya et al. (2008) analyzed the *in vitro* effect of NDGA on MCF-7/neo cells, a human breast cancer cell line sensitive to the antiestrogen tamoxifen, and on MCF-7/HER2-18 cells that overexpress the HER2 receptor and is resistant to tamoxifen. When performing a CyQuant cell proliferation assay, NDGA was effective in inhibiting the growth of both cell lines by 23% (MCF-7/neo) and 9% (MCF-7/HER2-18) at a concentration of 10 μM, while at 15 μM it inhibited 55% (MCF-7/neo) and 38% (MCF-7/HER2-18).

This evidences the antiproliferative potential of *L. tridentata*, mainly the lignan NDGA. Although there are several studies on NDGA, its mechanism of antiproliferative action has not been fully established. Various authors have reported that NDGA inhibits cell proliferation through induction of cell cycle arrest (G_1 and S) (Zavodovskaya et al., 2008; Gao et al., 2010), suppression of tumor growth by inhibition of metabolic enzymes (Manda et al., 2020) and blockage of phosphorylation of receptor tyrosine kinases (RTKs), such as IGF-1R, and HER2 (Ryan et al., 2008; Zavodovskaya et al., 2008). RTKs play a critical role in many cellular processes, including cell proliferation, differentiation, metabolism, and survival (Hsu and Hung, 2016). Therefore, it is suggested that, in the inhibition of HER2 autophosphorylation, NDGA acts directly on the enzymatic component of the receptor, unlike most RTK inhibitors, which block the ligand binding site (Regad, 2015; Youngren et al., 2005). On the other hand, NDGA could act as a chemopreventive by inhibiting the lipoxygenase activity (Nakadate et al., 1982).

Yoshida et al. (2007) have reported that NDGA increases the susceptibility of prostate and colorectal tumor cells to TRAIL-induced apoptosis by regulating death receptor expression. Likewise, Rowe et al. (2008) showed that NDGA can sensitize breast cancer cells to the therapeutic agent trastuzumab (monoclonal antibody against HER2). In a study with breast cancer cells, NDGA was reported to induce apoptosis via a caspase-independent pathway involved in the translocation of apoptosis-inducing factor (AIF) from mitochondria to the nucleus (Zavodovskaya et al., 2008).

On the other hand, polyphenols are one of the most studied groups of compounds in *L. tridentata*. It is known that these compounds inhibit the development and promotion of cancer, interfering in the four phases of carcinogenesis – initiation, promotion, progression, and invasion (Quintana, 2015). In the appearance of tumors, polyphenols can intervene in their proliferation routes, inhibiting the oxidizing agents that stimulate tumor development (Delgado, 2015); they have been shown to inhibit enzymes important in tumor progression, such as xanthine oxidase

(Chang et al., 1993) and cyclo-oxygenase (COX) (Mutoh et al., 2000). Similarly, flavonoids have been shown to have antiproliferative activity via cell cycle inhibition (Delgado, 2015). In addition, these compounds can induce cell death by apoptosis, which is associated with their ability to act as pro-oxidants, causing an increase in intracellular levels of ROS, which leads to the formation of H_2O_2 (Yang et al., 2000). Other authors have reported that this could be related to an alteration in the expression of heat shock proteins (Rong et al., 2000), as well as changes in molecular signaling pathways (Yin et al., 1999).

The findings suggest that bioactive compounds from *L. tridentata* could be good candidates for the development of new anticancer agents. However, more targeted studies are needed to determine the potential of these compounds at *in vivo* cancer models.

11.5 TOXICITY AND GENERATION OF PHARMACEUTICAL PRODUCTS

The use of therapeutic drugs for the control or treatment of diseases has been documented throughout history. However, any chemical entity in high enough doses can become toxic. Therefore, during the drug generation process, the study of the toxicological effects of new drug candidates becomes necessary (Dorato and Buckley, 2006).

The determination of the therapeutic index (TI) – which is defined as the relationship between the highest exposure to the drug that does not produce toxicity and the exposure that produces the desired efficacy – contributes to the characterization and optimization of the safety and efficacy of drug candidates, thus allowing the identification of those chemical entities that have an adequately balanced safety–efficacy profile for a given indication (Muller and Milton, 2012). Ideally, drug candidates should be chemical entities that are safe and effective in treating the target condition. Therefore, the toxic effects of drug candidates must be sufficiently defined to allow the initiation of clinical trials. Naturally, toxicological studies have three main objectives: 1) determination of the toxicological spectrum over a wide range of doses in laboratory animals, 2) extrapolation of responses to other species, with special emphasis on the potential for undesirable effects in humans, and 3) determination of safe levels of exposure. To meet these objectives, during the pre-clinical phase, different cells or animal models can be used, while during the clinical phases the safety and efficacy of the drug is obtained from experimentation on humans (Dorato and Buckley, 2006).

During the drug discovery and development phases – early discovery phase, non-clinical development, and in clinical trials – toxicology testing should be a must. During the discovery phase, short-term toxicological trials should be evaluated. Whereas, during the preclinical phases (non-clinical development), toxicokinetic parameters, genotoxicity, and toxicity $\leq$30 days must be determined. In the clinical phase in humans, aspects such as drug metabolism, toxicokinetics, and toxicity $\leq$6 months should be explored during Phase 1. In clinical Phase 3, oncogenicity, environmental assessment, and toxicity $\geq$6 months should be established. Finally, in the launch phase, the study of new formulations and new delivery methods can be evaluated (Dorato and Buckley, 2006). One of the most important factors contributing to the failure of pharmaceutical products is the adverse effects observed in the preclinical and clinical phases. Therefore, early identification of toxicity induced by new candidates avoids spending time and money on compounds that are likely to fail in their development stage (Bendels et al., 2019).

The toxicological effects of compounds, formulations, or extracts of *L. tridentata* have been reported; however, so far, the necessary data are still lacking to consider pharmaceutical products or formulations obtained from this plant as safe for administration in humans. Smith et al. (1994) reported that chaparral tea consumption was associated with the development of renal cystic cell carcinoma and acquired renal cystic disease in a 56-year-old woman. Although the patient claimed to have consumed three to four cups of chaparral tea daily for a period of approximately three months, the exact doses consumed were unknown, and it was also unknown if the patient was consuming any other substance or drug with renal toxic effects or with renal carcinogenic capacity. The only additional information reported was consumption of five to six cups of taheebo tea (*Tabebuia impetiginosa/avellanedae*) per day for a period of six months.

Sheik et al. (1997) reviewed 18 Food and Drugs Administration (FDA) case reports of adverse events associated with chaparral ingestion between 1992 and 1994. In 13 reports, evidence of hepatotoxicity with increased serum values of the liver chemical test were registered. These abnormalities were developed after three to 52 weeks of chaparral ingestion and resolved from 1 to 17 weeks after discontinuing consumption. Liver injury predominantly was presented as drug-induced or toxic cholestatic hepatitis. In four of the 13 individuals in which hepatotoxicity was recorded, there was progression to cirrhosis. In addition, in two individuals, acute fulminant hepatic failure was developed requiring liver transplantation. Although this study reports liver toxicity caused by *L. tridentata*, the reports do not clearly indicate the doses used by individuals or the frequency of consumption, data that are necessary to know the toxic or permissive doses more clearly of a drug or medicinal preparation.

NDGA is one of the bioactive compounds found in greater quantity in the leaves and twigs of *L. tridentata* (up to 10% dry weight). In various studies, the ability of NDGA to induce cystic nephropathy in rats has been reported (Goodman et al., 1970; Evan and Gardner, 1979; Gardner et al., 1987). In addition, NDGA has been reported to induce hypersensitivity when administered through the skin route, which is why it was withdrawn from the market by the FDA (Olsen et al., 1991; Kulp-Shorten et al., 1993). Lambert et al. (2002) reported the toxicological potential of NDGA in Balb/c rats, demonstrating that NDGA may contribute to *L. tridentata*-induced hepatotoxicity. In this study it was shown that intraperitoneal administration of NDGA induces 50% lethality in mice (LD_{50}) at a dose of 75 mg/kg. This study also reports an increase in serum levels of the liver damage marker alanine aminotransferase, which suggests possible damage to this organ. Additionally, glucuronidation is reported to be a possible detoxification mechanism for NDGA, via the formation of mono- and diglucuronides of NDGA after intravenous administration. The results reported in this study can be used as a basis for determining effective non-toxic doses and the therapeutic index of NDGA, in addition to the study of pharmacokinetic parameters and the effect of long-term administration of NDGA.

On the other hand, inventor Sinnott R.A.'s patent (registration key US5945106A) makes reference to the method of making a non-toxic extract made from *L. tridentata*. This extract is made from the extraction of plant material using an organic solvent and is subsequently saturated with ascorbic acid to reduce the toxicity of the NDGA present in the plant material. Furthermore, additional amounts of ascorbic acid are added to prevent the natural oxidation of NDGA. The extract is useful for the treatment of viral diseases caused by viruses of the Herpesviridae family or viruses that require the Sp1 class of proteins to initiate viral replication. Similarly, it can be used to reduce inflammation in diseases where inflammatory reactions are mediated by the effects of leukotrienes (Sinnott, 1996). The results reported in this study can also be considered for the preparation of less toxic formulations from NDGA or products made from extracts or active fractions of *L. tridentata*.

Portilla-de Buen et al. (2008) reported that the aqueous infusion prepared from 2.5 g of the plant in 230 mL of drinking water of *L. tridentata* did not induce toxicity in Wistar rats that received 1 mL of the infusion orally, three times a day for 95 days. No change in weight was observed in the mice that received the aqueous infusion; in addition, the results of the blood counts and blood chemistries remained normal. There were no differences between groups for weight gain or loss or water consumption at any time point. In addition, CBC and blood chemistry results remained within normal ranges throughout the study. This shows that, at low doses, the extracts, preparations, or compounds of *L. tridentata* probably do not induce toxicity.

Heron and Yarnell (2001) reported that low-dose herbal formulation containing less than 10% of *L. tridentata* tincture or topical application of *Ricinus* oil extract for topical use is safe for use. In the study, patients were prescribed either formulation for use over a 22-month period. Thirteen patients were prescribed the tincture, while another 23 patients (20 female and 3 male) were prescribed the oil for topical use. None of the patients showed signs of organ damage during the follow-up period. Therefore, it was concluded that the ingestion of low doses of tincture or the topical application of extracts in *Ricinus* oil is safe.

Together, the results shown here demonstrate that the toxicological effect of compounds, formulations, or extracts made from *L. tridentata* are little known and contradictory. For this reason, it is necessary to explore the toxicological effects of drug candidates obtained from this plant at each of the research levels –from the discovery phase, non-clinical phase, and clinical trials – to guarantee the efficacy and safety of the product under study.

CONCLUSIONS

L. tridentata is a medicinal plant that has demonstrated different biological activities that are associated with the presence of bioactive compounds. In this sense, this plant can be used to obtain fractions or purified active compounds that allow obtaining future pharmacological therapies. However, an extensive evaluation of toxicological studies is required to guarantee its safe and effective application in humans.

REFERENCES

Aarland, R. C., Peralta-Gomez, S., Sanchez, C. M., Parra-Bustamante, F., Villa-Hernandez, J. M., Leon-Sanchez, F. D. D., Perez-Flores, L. J., Rivera-Cabrera, F., & Mendoza-Espinoza, J. A. (2015). A pharmacological and phytochemical study of medicinal plants used in Mexican folk medicine. *Indian Journal of Traditional Knowledge, 14*(4), 550–557.

Abou-Gazar, H., Bedir, E., Takamatsu, S., Ferreira, D., & Khan, I. A. (2004). Antioxidant lignans from *Larrea tridentata*. *Phytochemistry, 65*(17), 2499–2505.

Agra, L. C., Ferro, J. N., Barbosa, F. T., & Barreto, E. (2015). Triterpenes with healing activity: A systematic review. *Journal of Dermatological Treatment, 26*(5), 465–470.

Aguilar, C. N., Aguilera-Carbo, A., Robledo, A., Ventura, J., Belmares, R., Martinez, D., ... & Contreras, J. (2008). Production of antioxidant nutraceuticals by solid-state cultures of pomegranate (*Punica granatum*) peel and creosote bush (*Larrea tridentata*) leaves. *Food Technology and Biotechnology, 46*(2), 218–222.

Aguilera, Y., Martin-Cabrejas, M. A., & de Mejia, E. G. (2016). Phenolic compounds in fruits and beverages consumed as part of the mediterranean diet: Their role in prevention of chronic diseases. *Phytochemistry Reviews, 15*, 405–423.

Aguirre-Joya, J. A., Pastrana-Castro, L., Nieto-Oropeza, D., Ventura-Sobrevilla, J., Rojas Molina, R., & Aguilar, C. N. (2018). The physicochemical, antifungal and antioxidant properties of a mixed polyphenol based bioactive film. *Heliyon, 4*(12), e00942.

Arteaga, S., Andrade-Cetto, A., & Cárdenas, R. (2005). *Larrea tridentata* (Creosote bush), an abundant plant of Mexican and US-American deserts and its metabolite nordihydroguaiaretic acid. *Journal of Ethnopharmacology, 98*(3), 231–239.

Asim, M., Saba, N., Jawaid, M., Nasir, M., Pervaiz, M., & Alothman, O. Y. (2018). A review on phenolic resin and its composites. *Current Analytical Chemistry, 14*(3), 185–197.

Azmir, J., Zaidul, I. S. M., Rahman, M. M., Sharif, K. M., Mohamed, A., Sahena, F., ... & Omar, A. K. M. (2013). Techniques for extraction of bioactive compounds from plant materials: A review. *Journal of Food Engineering, 117*(4), 426–436.

Balderas, J. M. M., Garza, E. J. T., Rodríguez, E. A., Chávez-Costa, A. C., Camacho, E. A. R., Olivo, A. M., & Ávalos, J. G. M. (2018). Estructura y diversidad de un matorral desértico micrófilo de *larrea tridentata* (dc.) coville en el noreste de México. *Interciencia: Revista de ciencia y tecnología de América, 43*(6), 449–454.

Bañuelos-Valenzuela, R., Delgadillo-Ruiz, L., Echavarría-Cháirez, F., Delgadillo-Ruiz, O., & Meza-López, C. (2018). Composición química y FTIR de extractos etanólicos *de Larrea tridentata, Origanum vulgare, Artemisa ludoviciana* y *Ruta graveolens. Agrociencia, 52*(3), 309–321.

Bashyal, B., Li, L., Bains, T., Debnath, A., & LaBarbera, D. V. (2017). *Larrea tridentata*: A novel source for anti-parasitic agents active against Entamoeba histolytica, Giardia lamblia and Naegleria fowleri. *PLoS Neglected Tropical Diseases, 11*(8), e0005832.

Bauri, R., Tigga, M. N., & Kullu, S. S. (2015). A review on use of medicinal plants to control parasites. *Indian Journal of Natural Products and Resources (IJNPR) [Formerly Natural Product Radiance (NPR)], 6*(4), 268–277.

Bendels, S., Bissantz, C., Fasching, B., Gerebtzoff, G., Guba, W., Kansy, M., ... & Roberts, S. (2019). Safety screening in early drug discovery: An optimized assay panel. *Journal of Pharmacological and Toxicological Methods, 99*, 106609.

Bernhard, H. O., & Thiele, K. (1981). Additional flavonoids from the leaves of *Larrea tridentata. Planta Medica, 41*(1), 100–101.

Bernhoft, A. (2010). A brief review on bioactive compounds in plants. *Bioactive Compounds in Plants-benefits and Risks for Man and Animals, 50*, 11–17.

Bocanegra-García, V., del Rayo Camacho-Corona, M., Ramírez-Cabrera, M., Rivera, G., & Garza-González, E. (2009). The bioactivity of plant extracts against representative bacterial pathogens of the lower respiratory tract. *BMC Research Notes, 2*(1), 95.

Bolaños, V., Díaz-Martínez, A., Soto, J., Rodríguez, M. A., López-Camarillo, C., Marchat, L. A., & Ramírez-Moreno, E. (2014). The flavonoid (–)-epicatechin affects cytoskeleton proteins and functions in *Entamoeba histolytica. Journal of Proteomics, 111*, 74–85.

Brinker, F. (1993). *Larrea tridentata* (DC) Coville (chaparral or creosote bush). *British Journal of Phytotherapy, 3*(1), 10–30.

Bürli, R. W., McMinn, D., Kaizerman, J. A., Hu, W., Ge, Y., Pack, Q., ... & Moser, H. E. (2004). DNA binding ligands targeting drug-resistant Gram-positive bacteria. Part 1: Internal benzimidazole derivatives. *Bioorganic & Medicinal Chemistry Letters, 14*(5), 1253–1257.

Camacho-Corona, M. d. R., García, A., Mata-Cárdenas, B. D., Garza-González, E., Ibarra-Alvarado, C., Rojas-Molina, A., ... & Gutiérrez, S. P. (2015). Screening for antibacterial and antiprotozoal activities of crude extracts derived from Mexican medicinal plants. *African Journal of Traditional, Complementary and Alternative Medicines, 12*(3), 104–112.

Carocho, M., & Ferreira, I. C. (2013). A review on antioxidants, prooxidants and related controversy: Natural and synthetic compounds, screening and analysis methodologies and future perspectives. *Food and Chemical Toxicology, 51*, 15–25.

Castillo-Reyes, F., Hernandez-Castillo, F. D., Clemente-Constantino, J. A., Gallegos-Morales, G., Rodríguez-Herrera, R., & Aguilar, C. N. (2015). *In vitro* antifungal activity of polyphenols-rich plant extracts against Phytophthora cinnamomi Rands. *African Journal of Agricultural Research, 10*(50), 4554–4560.

Chang, W. S., Lee, Y. J., Lu, F. J., & Chiang, H. C. (1993). Inhibitory effects of flavonoids on xanthine oxidase. *Anticancer Research, 13*(6A), 2165–2170.

Chen, H., Teng, L., Li, J.-N., Park, R., Mold, D. E., Gnabre, J., ... & Huang, R. C. C. (1998). Antiviral activities of methylated nordihydroguaiaretic acids. 2. Targeting herpes simplex virus replication by the mutation insensitive transcription inhibitor tetra-O-methyl-NDGA. *Journal of Medicinal Chemistry, 41*(16), 3001–3007.

Chhikara, N., Devi, H. R., Jaglan, S., Sharma, P., Gupta, P., & Panghal, A. (2018). Bioactive compounds, food applications and health benefits of *Parkia speciosa* (stinky beans): A review. *Agriculture & Food Security, 7*(1), 1–9.

Chirikdjian, J. J. (1973). Flavonoide von *Larrea tridentata*/Flavonoids of *Larrea tridentata. Zeitschrift für Naturforschung C, 28*(1–2), 32–35.

Cianciosi, D., Forbes-Hernández, T. Y., Afrin, S., Gasparrini, M., Reboredo-Rodriguez, P., Manna, P. P., ... & Battino, M. (2018). Phenolic compounds in honey and their associated health benefits: *A review. Molecules, 23*(9), 2322.

Cör, D., Knez, Ž., & Knez Hrnčič, M. (2018). Antitumour, antimicrobial, antioxidant and antiacetylcholinesterase effect of Ganoderma lucidum terpenoids and polysaccharides: *A review. Molecules, 23*(3), 649.

Craigo, J., Callahan, M., Huang, R. C. C., & DeLucia, A. L. (2000). Inhibition of human papillomavirus type 16 gene expression by nordihydroguaiaretic acid plant lignan derivatives. *Antiviral Research, 47*(1), 19–28.

Debnath, B., Singh, W. S., Das, M., Goswami, S., Singh, M. K., Maiti, D., & Manna, K. (2018). Role of plant alkaloids on human health: A review of biological activities. *Materials Today Chemistry, 9*, 56–72.

Delgadillo Ruíz, L., Bañuelos Valenzuela, R., Delgadillo Ruíz, O., Silva Vega, M., & Gallegos Flores, P. (2017). Composición química y efecto antibacteriano *in vitro* de extractos de *Larrea tridentata, Origanum vulgare, Artemisa ludoviciana* y *Ruta graveolens. Nova Scientia, 9*(19), 273–290.

Delgado Ciruelo, L. (2015). Mecanismos de acción implicados en la bioactividad de flavonoides. *Caenorhabditis elegans* y líneas celulares como sistemas modelo. Doctoral Tesis. Universidad de Salamanca.

Dentali, S., & Hoffmann, J. (1992). Potential antiinfective agents from *Eriodictyon angustifolium* and *Salvia apiana. International Journal of Pharmacognosy, 30*(3), 223–231.

Díaz-Díaz, A., Hernández-Castillo, F. D., Belmares-Cerda, R. E., Gallegos-Morales, G., Rodríguez-Herrera, R., & Aguilar-González, C. N. (2013). Efecto de extractos de Larrea tridentata y Flourensia cernua en el desarrollo de plantas de tomate inoculadas con Phytophthora capsici. *Aptitud combinatoria de características agronómicas, fisiológicas y de calidad de melón Combinatorial Ability of Agronomic Characteristics, Physiological and Melon Quality, Agraria, 10*(2), 49–58.

Dimayuga, R. E., & Garcia, S. K. (1991). Antimicrobial screening of medicinal plants from Baja California Sur, Mexico. *Journal of Ethnopharmacology, 31*(2), 181–192.

Dorato, M. A., & Buckley, L. A. (2006). Toxicology in the drug discovery and development process. *Current Protocols in Pharmacology, 32*(1), 10–13.

Downum, K. R., Dole, J., & Rodriguez, E. (1988). Nordihydroguaiaretic acid: Inter-and intrapopulational variation in the Sonoran desert creosote bush (*Larrea tridentata*, zygophyllaceae. *Biochemical Systematics and Ecology, 16*(6), 551–555.

Elakovich, S. D., & Stevens, K. L. (1985). Phytotoxic properties of nordihydroguaiaretic acid, a lignan from *Larrea tridentata* (Creosote bush). *Journal of Chemical Ecology, 11*(1), 27–33.

Evan, A. P., & Gardner Jr, K. D. (1979). Nephron obstruction in nordihydroguaiaretic acid-induced renal cystic disease. *Kidney International, 15*(1), 7–19.

Fan, J., Liu, Y., Yuan, Z. (2014). Critical role of Dengue Virus NS1 protein in viral replication. *Virologica Sinica, 29*(3), 162–169.

Favela-Hernández, J. M. J., Clemente-Soto, A. F., Balderas-Rentería, I., Garza-González, E., & Camacho-Corona, M. D. R. (2015). Potential mechanism of action of 3′-Demethoxy-6-O-demethyl-isoguaiacin on Methicillin Resistant Staphylococcus aureus. *Molecules, 20*(7), 12450–12458.

Favela-Hernández, J. M. J., García, A., Garza-González, E., Rivas-Galindo, V. M., & Camacho-Corona, M. D. R. (2012). Antibacterial and antimycobacterial lignans and flavonoids from *Larrea tridentata. Phytotherapy Research, 26*(12), 1957–1960.

Firn, R. D., & Jones, C. G. (2003). Natural products–a simple model to explain chemical diversity. *Natural Product Reports, 20*(4), 382–391.

Fujimoto, N., Kohta, R., Kitamura, S., & Honda, H. (2004). Estrogenic activity of an antioxidant, nordihydroguaiaretic acid (NDGA). *Life Sciences, 74*(11), 1417–1425.

Galván, J. V., Díaz, C. A. G., & Fernández, R. G. (2014). Efecto de los extractos acuosos de hojas de plantas de gobernadora (*Larrea tridentata*), hojasen (*Flourensia cernua*) y encino (*Quercus pungens*), sobre el crecimiento micelial in vitro de hongos fitopatógenos. *Acta Universitaria*, *24*(5), 13–19.

Gao, P., Zhai, F., Guan, L., & Zheng, J. (2010). Nordihydroguaiaretic acid inhibits growth of cervical cancer SiHa cells by up-regulating p21. *Oncology Letters*, *2*(1), 123–128.

García, J. E., Gómez, L., Mendoza-de-Gives, P., Rivera-Corona, J. L., Millán-Orozco, J., Ascacio, J. A., ... & Mellado, M. (2018). Anthelmintic efficacy of hydro-methanolic extracts of *Larrea tridentata* against larvae of *Haemonchus contortus*. *Tropical Animal Health and Production*, *50*(5), 1099–1105.

Garcia-Castillo, B. (2012). *Larrea tridentata* (Sessé & Mo. ex DC.) Coville. Tesis de Licenciatura, Universidad Autónoma Agraria Antonio Narro.

Gardner Jr, K. D., Reed, W. P., Evan, A. P., Zedalis, J., Hylarides, M. D., & Leon, A. A. (1987). Endotoxin provocation of experimental renal cystic disease. *Kidney International*, *32*(3), 329–334.

Garza González, J. N. (2014). *Evaluación de la actividad antiprotozoaria de plantas con antecedentes etnobotánicos en México*. Doctoral dissertation, Universidad Autónoma de Nuevo León.

Ghadirian, E., & Arfaa, F. N. M. N. (1973). First report of human infection with *Haemonchus contortus, Ostertagia ostertagi, and Marshallagia marshalli* (family *Trichostrongylidae*) in Iran. *The Journal of Parasitology*, *59*(6), 1144–1145.

Gisvold, O., & Thaker, E. (1974). Lignans from *Larrea divaricata*. *Journal of Pharmaceutical Sciences*, *63*(12), 1905–1907.

Gnabre, J. N. (1999). U.S. Patent No. 5,989,555. Washington, DC: U.S. Patent and Trademark Office.

Gnabre, J. N., Brady, J. N., Clanton, D. J., Ito, Y., Dittmer, J., Bates, R. B., & Huang, R. (1995a). Inhibition of human immunodeficiency virus type 1 transcription and replication by DNA sequence-selective plant lignans. *Proceedings of the National Academy of Sciences*, *92*(24), 11239–11243.

Gnabre, J. N., Ito, Y., Ma, Y., & Huang, R. C. (1996). Isolation of anti-HIV-1 lignans from *Larrea tridentata* by counter-current chromatography. *Journal of Chromatography A*, *719*(2), 353–364.

Gnabre, J., Bates, R., & Huang, R. C. (2015). Creosote bush lignans for human disease treatment and prevention: Perspectives on combination therapy. *Journal of Traditional and Complementary Medicine*, *5*(3), 119–126.

Gnabre, J., Huang, R. C. C., Bates, R. B., Burns, J. J., Caldera, S., Malcomson, M. E., & McClure, K. J. (1995b). Characterization of anti—HIV lignans. *Tetrahedron*, *51*(45), 12203–12210.

Goodman, T., Grice, H. C., Becking, G. C., & Salem, F. A. (1970). A cystic nephropathy induced by nordihydroguaiaretic acid in the rat. Light and electron microscopic investigations. *Laboratory Investigation: A Journal of Technical Methods and Pathology*, *23*(1), 93–107.

Górniak, I., Bartoszewski, R., & Króliczewski, J. (2019). Comprehensive review of antimicrobial activities of plant flavonoids. *Phytochemistry Reviews*, *18*(1), 241–272.

Guimarães, A. C., Meireles, L. M., Lemos, M. F., Guimarães, M. C. C., Endringer, D. C., Fronza, M., & Scherer, R. (2019). Antibacterial activity of terpenes and terpenoids present in essential oils. *Molecules*, *24*(13), 2471.

Gutiérrez-Gutiérrez, F., Puebla-Pérez, A. M., González-Pozos, S., Hernández-Hernández, J. M., Pérez-Rangel, A., Alvarez, L. P., ... & Castillo-Romero, A. (2017). Antigiardial activity of podophyllotoxin-type lignans from *Bursera fagaroides* var. fagaroides. *Molecules*, *22*(5), 799.

Guzman-Beltran, S., Pedraza-Chaverri, J., Gonzalez-Reyes, S., Hernandez-Sanchez, F., Juarez-Figueroa, U. E., Gonzalez, Y., ... Torres, M. (2013). Nordihydroguaiaretic acid attenuates the oxidative stress-induced decrease of CD33 expression in human monocytes. *Oxidative Medicine and Cellular Longevity*, *2013*, 1–14.

Hapon, M. V., Boiteux, J. J., Fernández, M. A., Lucero, G., Silva, M. F., & Pizzuolo, P. H. (2018). Effect of phenolic compounds present in Argentinian plant extracts on mycelial growth of the plant pathogen Botrytis cinerea Pers. *International Journal of Experimental Botany*, *86*, 270–277.

Heron, S., & Yarnell, E. (2001). The safety of low-dose *Larrea tridentata* (DC) Coville (creosote bush or chaparral): A retrospective clinical study. *The Journal of Alternative & Complementary Medicine*, *7*(2), 175–185.

Hsu, J. L., & Hung, M. C. (2016). The role of HER2, EGFR, and other receptor tyrosine kinases in breast cancer. *Cancer and Metastasis Reviews*, *35*(4), 575–588.

Hui-Zheng, X., Zhi-Zhen, L., Chohachi, K., Soejarto, D. D., Cordell, G. A., Fong, H. H., & Hodgson, W. (1988). 3β-(3, 4-Dihydroxycinnamoyl)-erythrodiol and 3β-(4-hydroxycinnamoyl)-erythrodiol from Larrea tridentata. *Phytochemistry, 27*(1), 233–235.

Hwang, D., & Lim, Y. H. (2015). Resveratrol antibacterial activity against *Escherichia coli* is mediated by Z-ring formation inhibition via suppression of FtsZ expression. *Scientific Reports, 5,* 10029

Hwu, J. R., Tseng, W. N., Gnabre, J., Giza, P., & Huang, R. C. C. (1998). Antiviral activities of methylated nordihydroguaiaretic acids. 1. Synthesis, structure identification, and inhibition of tat-regulated HIV transactivation. *Journal of Medicinal Chemistry, 41*(16), 2994–3000.

Hyder, P. W., Fredrickson, E. L., Estell, R. E., Tellez, M., & Gibbens, R. P. (2002). Distribution and concentration of total phenolics, condensed tannins, and nordihydroguaiaretic acid (NDGA) in creosotebush (*Larrea tridentata*). *Biochemical Systematics and Ecology, 30*(10), 905–912.

Itza Ortiz, M. F. (2019). *Phytobiotic Activity of Larrea tridentata, Origanum vulgare and Plectranthus amboinicus in gram positive and gram negative Bacterias.* Instituto de Ciencias Biomédicas.

Jardine, K., Abrell, L., Kurc, S. A., Huxman, T., Ortega, J., & Guenther, A. (2010). Volatile organic compound emissions from *Larrea tridentata* (creosotebush). *Atmospheric Chemistry and Physics, 10*(24), 12191–12206.

Jitsuno, M., & Mimaki, Y. (2010). Triterpene glycosides from the aerial parts of *Larrea tridentata. Phytochemistry, 71*(17-18), 2157–2167.

Joublanc, E. L., Martínez-Hernández, J. L., & Saldívar, R. H. L. (2008). Microbial activity of *Larrea tridentata* against pathogen strain and potential effect whit chitosan and *Agave lecheguilla* stracts. *Planta Medica, 74*(09), PA300.

Kayser, O., Kiderlen, A., & Croft, S. (2003). Natural products as antiparasitic drugs. *Parasitology Research, 90*(2) Supplement 2, S55–S62.

Khameneh, B., Iranshahy, M., Soheili, V., & Bazzaz, B. S. F. (2019). Review on plant antimicrobials: A mechanistic viewpoint. *Antimicrobial Resistance & Infection Control, 8*(1), 118.

Klongsiriwet, C., Quijada, J., Williams, A. R., Mueller-Harvey, I., Williamson, E. M., & Hoste, H. (2015). Synergistic inhibition of *Haemonchus contortus exsheathment* by flavonoid monomers and condensed tannins. *International Journal for Parasitology: Drugs and Drug Resistance, 5*(3), 127–134.

Konno, C., Lu, Z. Z., Xue, H. Z., Erdelmeier, C. A., Meksuriyen, D., Che, C. T., ... & Fong, H. H. (1990). Furanoid lignans from *Larrea tridentata. Journal of Natural Products, 53*(2), 396–406.

Kulp-Shorten, C. L., Konnikov, N., & Callen, J. P. (1993). Comparative evaluation of the efficacy and safety of masoprocol and 5-fluorouracil cream for the treatment of multiple actinic keratoses of the head and neck. *The Journal of Geriatric Dermatology, 1,* 161–168.

Lambert, J. D., Zhao, D., Meyers, R. O., Kuester, R. K., Timmermann, B. N., & Dorr, R. T. (2002). Nordihydroguaiaretic acid: Hepatotoxicity and detoxification in the mouse. *Toxicon, 40*(12), 1701–1708.

Lambert, J. D., Sang, S., Dougherty, A., Caldwell, C. G., Meyers, R. O., Dorr, R. T., & Timmermann, B. N. (2005). Cytotoxic lignans from *Larrea tridentata. Phytochemistry, 66*(7), 811–815.

Leon, D., Parada, D., Vargas-Uribe, M., Perez, A. A., Ojeda, L., Zambrano, A., ... & Salas, M. (2016). Effect of nordihydroguaiaretic acid on cell viability and glucose transport in human leukemic cell lines. *FEBS Open Bio, 6*(10), 1000–1007.

Li, A. N., Li, S., Zhang, Y. J., Xu, X. R., Chen, Y. M., & Li, H. B. (2014). Resources and biological activities of natural polyphenols. *Nutrients, 6*(12), 6020–6047.

Lira-Saldívar, R. H. L., Balvantín-García, G. F., Hernández-Castillo, F. D., Gamboa-Alvarado, R., Jasso-de-Rodríguez, D., & Jiménez-Díaz, F. (2003). Evaluation of resin content and the antifungal effect of *Larrea tridentata* (Sesse and Moc. Ex DC) Coville extracts from two mexican deserts against *Pythium* sp. Pringsh. *Revista Mexicana de Fitopatología, 21*(2), 97–101.

Lopez-Romero, J. C., Gonzalez-Rios, H., Borges, A., & Simoes, M. (2015). Antibacterial effects and mode of action of selected essential oils components against *Escherichia coli* and *Staphylococcus aureus. Evidence-Based Complementary and Alternative Medicine, 2015,* 1–9.

Mabry, T. J., Hunziker, J. H., & Difeo Jr, D. (1978). *Creosote bush: Biology and chemistry of Larrea in New World deserts.* Dowden, Hutchinson & Ross, Inc.

Maldonado, I. C. (2016). Interacción del virus dengue serotipo 2 con compuestos químicos de *Larrea tridentata* y fracciones primarias de *Sambucus nigra* en un estudio biodirigido. Doctoral dissertation, Autonomous University of Coahuila.

Manda, G., Rojo, A. I., Martínez-Klimova, E., Pedraza-Chaverri, J., & Cuadrado, A. (2020). Nordihydroguaiaretic acid: From herbal medicine to clinical development for cancer and chronic diseases. *Frontiers in Pharmacology, 11*, 151.

Martins, S., Aguilar, C. N., Garza-Rodriguez, I. D. L., Mussatto, S. I., & Teixeira, J. A. (2010a). Kinetic study of nordihydroguaiaretic acid recovery from *Larrea tridentata* by microwave-assisted extraction. *Journal of Chemical Technology & Biotechnology, 85*(8), 1142–1147.

Martins, S., Mussatto, S. I., Aguilar, C. N., & Teixeira, J. A. (2010b). Antioxidant capacity and NDGA content of *Larrea tridentata* (a desert bush) leaves extracted with different solvents. *Journal of Biotechnology, 150*, 500.

Martins, S., Mussatto, S. I., Aguilar, C. N., & Teixeira, J. A. (2011). *Effect of extraction solvents on the content of bioactivecompounds from Larrea tridentata leaves.* Universidade do Minho.

Martins, S., Aguilar, C. N., Teixeira, J. A., & Mussatto, S. I. (2012). Bioactive compounds (phytoestrogens) recovery from *Larrea tridentata* leaves by solvents extraction. *Separation and Purification Technology, 88*, 163–167.

Martins, S., Teixeira, J. A., & Mussatto, S. I. (2013a). Solid-state fermentation as a strategy to improve the bioactive compounds recovery from *Larrea tridentata* leaves. *Applied Biochemistry and Biotechnology, 171*(5), 1227–1239.

Martins, S., Amorim, E. L., Sobrinho, T. J. P., Saraiva, A. M., Pisciottano, M. N., Aguilar, C. N., ... & Mussatto, S. I. (2013b). Antibacterial activity of crude methanolic extract and fractions obtained from *Larrea tridentata* leaves. *Industrial Crops and Products, 41*, 306–311.

Maruyama, M., Yamauchi, S., Akiyama, K., Sugahara, T., Kishida, T., & Koba, Y. (2007). Antibacterial activity of a virgatusin-related compound. *Bioscience, Biotechnology, and Biochemistry, 71*(3), 677–680.

Mojica-Marin, V., Luna-Olvera, H. A., Morales-Ramos, L. H., Gonzalez-Aguilar, N. A., Pereyra-Alferez, B., Ruiz-Baca, E., & Elias-Santos, M. (2011). In vitro antifungal activity of Gobernadora (*Larrea tridentata* (DC) Coville) against *Phytophthora capsici* Leo. *African Journal of Agricultural Research, 6*(5), 1058–1066.

Morán-Santibañez, K., Vasquez, A. H., Varela-Ramirez, A., Henderson, V., Sweeney, J., Odero-Marah, V., ... & Skouta, R. (2019). Larrea tridentata extract mitigates oxidative stress-induced cytotoxicity in human neuroblastoma SH-SY5Y cells. *Antioxidants, 8*(10), 427.

Moreno-Limón, S., González-Solís, L. N., Salcedo-Martínez, S. M., Cárdenas-Ávila, M. L., & Perales-Ramírez, A. (2011). Efecto antifúngico de extractos de gobernadora (*Larrea tridentata* L.) sobre la inhibición in vitro de *Aspergillus flavus* y *Penicillium* sp. *Polibotánica, 32*, 193–205.

Muller, P. Y., & Milton, M. N. (2012). The determination and interpretation of the therapeutic index in drug development. *Nature Reviews Drug Discovery, 11*(10), 751–761.

Mutoh, T., Joad, J. P., & Bonham, A. C. (2000). Chronic passive cigarette smoke exposure augments bronchopulmonary C-fibre inputs to nucleus tractus solitarii neurones and reflex output in young guinea-pigs. *The Journal of Physiology, 523*(1), 223–233.

Nakadate, T., Yamamoto, S., Iseki, H., Sonoda, S., Takemura, S., Ura, A., Hosoda, Y., & Kato, R. (1982). Inhibition of 12-O-tetradecanoyl-phorbol-13-acetate-induced tumor promotion by nordihydroguaiaretic acid, a lipoxygenase inhibitor, and p-bromophenacyl bromide, a phospholipase A2 inhibitor. *Gan, 73*(6), 841–843.

Navarro, V., Villarreal, M. L., Rojas, G., & Lozoya, X. (1996). Antimicrobial evaluation of some plants used in Mexican traditional medicine for the treatment of infectious diseases. *Journal of Ethnopharmacology, 53*(3), 143–147.

Nazzaro, F., Fratianni, F., De Martino, L., Coppola, R., & De Feo, V. (2013). Effect of essential oils on pathogenic bacteria. *Pharmaceuticals, 6*(12), 1451–1474.

Nijveldt, R. J., Van Nood, E. L. S., Van Hoorn, D. E., Boelens, P. G., Van Norren, K., & Van Leeuwen, P. A. (2001). Flavonoids: A review of probable mechanisms of action and potential applications. *The American Journal of Clinical Nutrition, 74*(4), 418–425.

Olsen, E. A., Abernethy, M. L., Kulp-Shorten, C., Callen, J. P., Glazer, S. D., Huntley, A., ... & Wolf Jr, J. E. (1991). A double-blind, vehicle-controlled study evaluating masoprocol cream in the treatment of actinic keratoses on the head and neck. *Journal of the American Academy of Dermatology, 24*(5), 738–743.

Osorio, E., Flores, M., Hernández, D., Ventura, J., Rodríguez, R., & Aguilar, C. N. (2010). Biological efficiency of polyphenolic extracts from pecan nuts shell (*Carya Illinoensis*), pomegranate husk (*Punica granatum*) and creosote bush leaves (*Larrea tridentata* Cov.) against plant pathogenic fungi. *Industrial Crops and Products, 31*(1), 153–157.

Pandey, A. K., & Kumar, S. (2013). Perspective on plant products as antimicrobial agents: A review. *Pharmacologia, 4*(7), 469–480.

Paveto, C., Güida, M. C., Esteva, M. I., Martino, V., Coussio, J., Flawiá, M. M., & Torres, H. N. (2004). Anti-*Trypanosoma cruzi* activity of green tea (*Camellia sinensis*) catechins. *Antimicrobial Agents and Chemotherapy, 48*(1), 69–74.

Peñuelas-Rubio, O., Arellano-Gil, M., Verdugo-Fuentes, A. A., Chaparro-Encinas, L. A., Hernández-Rodríguez, S. E., Martínez-Carrillo, J. L., & Vargas-Arispuro, I. D. C. (2017). Extractos de Larrea tridentata como una estrategia ecológica contra *Fusarium oxysporum radicis-lycopersici* en plantas de tomate bajo condiciones de invernadero. *Revista mexicana de fitopatología, 35*(3), 360–376.

Pereira, D. M., Valentão, P., Pereira, J. A., & Andrade, P. B. (2009). Phenolics: From chemistry to biology. *Molecules, 14*(6), 2202–2211.

Portilla-de Buen, E., Ramos, L., Aguilar, A., Ramos, A., García-Martínez, D., Cárdenas, A., ... & Leal, C. (2008). *Larrea tridentata* en urolitiasis. Efecto en un modelo no metabólico en ratas. *Revista Médica del Instituto Mexicano del Seguro Social, 46*(5), 519–522.

Quintana, M. M. S. (2015). Evaluación de la actividad antiproliferativa de flavonoides sintéticos en células leucémicas humanas. Doctoral dissertation, Universidad de Las Palmas de Gran Canaria.

Ramírez, M. E., Mendoza, J. A., Arreola, R. H., & Ordaz, C. (2010). Flavonoides con actividad antiprotozoaria. *Revista Mexicana de Ciencias Farmacéuticas, 41*(1), 6–21.

Ramírez, M., Alvarado, M., & Rodríguez, J. G. (2016). Correlación de polifenoles totales, actividad antioxidante y potencial reductor de plantas nativas del semidesierto de Coahuila. *Investigación y Desarrollo en Ciencia y Tecnología de Alimentos, 1*, 151–156.

Regad, T. (2015). Targeting RTK signaling pathways in cancer. *Cancers, 7*(3), 1758–1784.

Rodríguez-Guadarrama, A. H., Guevara-González, R. G., Romero-Gómez, S. d. J., & Feregrino-Pérez, A. A. (2018). Antifungal activity of Mexican endemic plants on agricultural phytopathogens: A review. Paper presented at the 2018 XIV International Engineering Congress (CONIIN).

Rong, Y. U. A. N., Yang, E. B., Zhang, K., & Mack, P. (2000). Quercetin-induced apoptosis in the monoblastoid cell line U937 in vitro and the regulation of heat shock proteins expression. *Anticancer Research, 20*(6B), 4339–4345.

Rowe, D. L., Ozbay, T., Bender, L. M., & Nahta, R. (2008). Nordihydroguaiaretic acid, a cytotoxic insulin-like growth factor-I receptor/HER2 inhibitor in trastuzumab-resistant breast cancer. *Molecular Cancer Therapeutics, 7*(7), 1900–1908.

Ruiz, G., Turner, T., Nelson, E., Sparks, L., & Langland, J. (2017). Bacterial development of resistance to botanical antimicrobials. *Journal of Evolution and Health, 2*(2), 3.

Ryan, C. J., Zavodovskaya, M., Youngren, J. F., Campbell, M., Diamond, M., Jones, J., ... & Goldfine, I. D. (2008). Inhibitory effects of nordihydroguaiaretic acid (NDGA) on the IGF-1 receptor and androgen dependent growth of LAPC-4 prostate cancer cells. *The Prostate, 68*(11), 1232–1240.

Sagaste, C. A., Montero, G., Coronado, M. A., Ayala, J. R., León, J. Á., García, C., Rojano, B. A., Rosales, S., Montes, D. G. (2019). Creosote Bush (*Larrea tridentata*) Extract Assessment as a Green Antioxidant for Biodiesel. *Molecules, 24*(9), 1786.

Sakakibara, M., Timmermann, B. N., Nakatani, N., Waldrum, H., & Mabry, T. J. (1975). New 8-hydroxyflavonols from *Larrea tridentata*. *Phytochemistry, 14*(3), 849–851.

Salido, A. A. G., Assanga, S. B. I., Luján, L. M. L., Ángulo, D. F., Espinoza, C. L. L., & ALA, S. A. (2016). Composition of secondary metabolites in Mexican plant extracts and their antiproliferative activity towards cancer cell lines. *International Journal of Science, 5*, 63–77.

Schmidt, T. J., Rzeppa, S., Kaiser, M., & Brun, R. (2012). *Larrea tridentata*—Absolute configuration of its epoxylignans and investigations on its antiprotozoal activity. *Phytochemistry Letters, 5*(3), 632–638.

Scotti, L., Ferreira, E. I., Silva, M. S. D., & Scotti, M. T. (2010). Chemometric studies on natural products as potential inhibitors of the NADH oxidase from *Trypanosoma cruzi* using the VolSurf approach. *Molecules*, *15*(10), 7363–7377.

Scotti, L., Ishiki, H., Mendonca, F. J. B., Da Silva, M. S., & Scotti, M. T. (2015). *In-silico* analyses of natural products on leishmania enzyme targets. *Mini Reviews in Medicinal Chemistry*, *15*(3), 253–269.

SEINet Arizona – New Chapter Mexico. (2020). *Larrea tridentata* (Sessé & Moc. ex DC.) Coville. Available at: http://swbiodiversity.org/seinet/taxa/index.php?taxon=Larrea+tridentata&formsubmit=Search+Terms

Serrano, M. E. D., López, M. L., & Espuñes, T. D. R. S. (2006). Componentes bioactivos de alimentos funcionales de origen vegetal. *Revista Mexicana de Ciencias Farmacéuticas*, *37*(4), 58–68.

Shahidi, F., & Yeo, J. (2018). Bioactivities of phenolics by focusing on suppression of chronic diseases: A review. *International Journal of Molecular Sciences*, *19*(6), 1573.

Sharma, S. K., Parasuraman, P., Kumar, G., Surolia, N., & Surolia, A. (2007). Green tea catechins potentiate triclosan binding to enoyl-ACP reductase from Plasmodium falciparum (PfENR). *Journal of Medicinal Chemistry*, *50*(4), 765–775.

Sheikh, N. M., Philen, R. M., & Love, L. A. (1997). Chaparral-associated hepatotoxicity. *Archives of Internal Medicine*, *157*(8), 913–919.

Simoes, M., Bennett, R. N., & Rosa, E. A. (2009). Understanding antimicrobial activities of phytochemicals against multidrug resistant bacteria and biofilms. *Natural Product Reports*, *26*(6), 746–757.

Sinnott, A. R. (1996). Montoxic extract of Larrea tridentata and method of making the same. (Patent United States. No. US5945106A). LARREACORP, LTD.

Skouta, R., Morán-Santibañez, K., Valenzuela, C. A., Vasquez, A. H., & Fenelon, K. (2018). Assessing the antioxidant properties of *Larrea tridentata* extract as a potential molecular therapy against oxidative stress. *Molecules*, *23*(7), 1826.

Smith, A. Y., Feddersen, R. M., Gardner Jr, K. D., & Davis Jr, C. J. (1994). Cystic renal cell carcinoma and acquired renal cystic disease associated with consumption of chaparral tea: A case report. *The Journal of Urology*, *152*(6), 2089–2091.

Snowden, R., Harrington, H., Morrill, K., Jeane, L., Garrity, J., Orian, M., ... & Langland, J. (2014). A comparison of the anti-*Staphylococcus aureus* activity of extracts from commonly used medicinal plants. *The Journal of Alternative and Complementary Medicine*, *20*(5), 375–382.

Soriano, A. F., Helfrich, B., Chan, D. C., Heasley, L. E., Bunn, P. A., & Chou, T.-C. (1999). Synergistic effects of new chemopreventive agents and conventional cytotoxic agents against human lung cancer cell lines. *Cancer Research*, *59*(24), 6178–6184.

Soto-Acosta, R., Bautista-Carbajal, P., Syed, G. H., Siddiqui, A., & Del Angel, R. M. (2014). Nordihydroguaiaretic acid (NDGA) inhibits replication and viral morphogenesis of dengue virus. *Antiviral Research*, *109*, 132–140.

Swamy, M. K., Akhtar, M. S., & Sinniah, U. R. (2016). Antimicrobial properties of plant essential oils against human pathogens and their mode of action: An updated review. *Evidence-Based Complementary and Alternative Medicine*, *2016*, 1–21.

Thimmappa, R., Geisler, K., Louveau, T., O'Maille, P., & Osbourn, A. (2014). Triterpene biosynthesis in plants. *Annual Review of Plant Biology*, *65*(1), 225–257.

Tiwari, K., Kumar, R., & Dubey, V. K. (2016). Biochemical characterization of dihydroorotase of *Leishmania donovani*: Understanding pyrimidine metabolism through its inhibition. *Biochimie*, *131*, 45–53.

Treviño-Cueto, B., Luis, M., Contreras-Esquivel, J. C., Rodríguez, R., Aguilera, A., & Aguilar, C. N. (2007). Gallic acid and tannase accumulation during fungal solid state culture of a tannin-rich desert plant (Larrea tridentata Cov.). *Bioresource Technology*, *98*(3), 721–724.

Tungmunnithum, D., Thongboonyou, A., Pholboon, A., & Yangsabai, A. (2018). Flavonoids and other phenolic compounds from medicinal plants for pharmaceutical and medical aspects: An overview. *Medicines*, *5*(3), 93.

Vargas-Arispuro, I., Reyes-Báez, R., Rivera-Castañeda, G., Martínez-Téllez, M., & Rivero-Espejel, I. (2005). Antifungal lignans from the creosotebush (*Larrea tridentata*). *Industrial Crops and Products*, *22*(2), 101–107.

Ventura, J., Belmares, R., Aguilera-Carbo, A., Gutiérrez-Sanchez, G., Rodríguez-Herrera, R., & Aguilar, C. N. (2008). Fungal biodegradation of tannins from creosote bush (*Larrea tridentata*) and tar bush (*Fluorensia cernua*) for gallic and ellagic acid production. *Food Technology and Biotechnology, 46*(2), 213–217.

Verástegui, M. A., Sánchez, C. A., Heredia, N. L., & García-Alvarado, J. S. (1996). Antimicrobial activity of extracts of three major plants from the Chihuahuan desert. *Journal of Ethnopharmacology, 52*(3), 175–177.

Volk, A. P. D., & Moreland, J. G. (2014). ROS-containing endosomal compartments: Implications for signaling. In *Methods in Enzymology* (Vol. *535*, pp. 201–224): Elsevier.

Xia, Q., Zhang, H., Sun, X., Zhao, H., Wu, L., Zhu, D., ... & She, G. (2014). A comprehensive review of the structure elucidation and biological activity of triterpenoids from *Ganoderma* spp. *Molecules, 19*(11), 17478–17535.

Yang, G.-Y., Liao, J., Li, C., Chung, J., Yurkow, E. J., Ho, C.-T., & Yang, C. S. (2000). Effect of black and green tea polyphenols on c-jun phosphorylation and H_2O_2 production in transformed and non-transformed human bronchial cell lines: Possible mechanisms of cell growth inhibition and apoptosis induction. *Carcinogenesis, 21*(11), 2035–2039.

Yin, F., Giuliano, A. E., & Van Herle, A. J. (1999). Signal pathways involved in apigenin inhibition of growth and induction of apoptosis of human anaplastic thyroid cancer cells (ARO). *Anticancer Research, 19*(5B), 4297–4303.

Yin, M. C. (2012). Anti-glycative potential of triterpenes: A mini-review. *BioMedicine, 2*(1), 2–9.

Yoshida, T., Shiraishi, T., Horinaka, M., Nakata, S., Yasuda, T., Goda, A. E., ... & Sakai, T. (2007). Lipoxygenase inhibitors induce death receptor 5/TRAIL-R2 expression and sensitize malignant tumor cells to TRAIL-induced apoptosis. *Cancer Science, 98*(9), 1417–1423.

Youngren, J. F., Gable, K., Penaranda, C., Maddux, B. A., Zavodovskaya, M., Lobo, M., ... & Goldfine, I. D. (2005). Nordihydroguaiaretic acid (NDGA) inhibits the IGF-1 and c-erbB2/HER2/neu receptors and suppresses growth in breast cancer cells. *Breast Cancer Research and Treatment, 94*(1), 37–46.

Zavodovskaya, M., Campbell, M. J., Maddux, B. A., Shiry, L., Allan, G., Hodges, L., ... & Goldfine, I. D. (2008). Nordihydroguaiaretic acid (NDGA), an inhibitor of the HER2 and IGF-1 receptor tyrosine kinases, blocks the growth of HER2-overexpressing human breast cancer cells. *Journal of Cellular Biochemistry, 103*(2), 624–635.

Chapter 12

Toxicological Aspects of Medicinal Plants that Grow in Drylands and Polluted Environments

Rebeca Pérez-Morales, Miguel Ángel Téllez-López, Edgar Héctor Olivas-Calderón, and Alberto González-Zamora

CONTENTS

12.1 SECONDARY METABOLITES OF PLANTS

Plants produce numerous organic molecules that fulfill various cellular functions essential for physiological processes and for response mechanisms to various stimuli. The type of metabolite and its concentration are determined by the species, the genotype, the physiological state, the stage of development and the environmental factors present during the growth of the plant. Metabolism is defined as the set of chemical reactions that cells perform to synthesize complex substances from simpler ones or to degrade complex ones. Plants allocate a significant amount of carbon and energy to the synthesis of a wide variety of organic molecules that do not have a direct function in basic processes, which are called secondary metabolites (Buchanan et al., 2015).

DOI: 10.1201/9781003251255-12

Secondary metabolites are synthesized in small amounts and include a wide variety of compounds, including alkaloids, carotenoids, phenolic compounds, volatile compounds, steroids, phytoalexins, phytohormones, flavonoids, cyanogenic glycosides, glycosylates, isoprenoids, saponins, tannins and terpenoids that are particularly synthesized in very restricted groups of plants: that is, not all secondary metabolites are found in all groups of plants, some are synthesized only at the species level, while others at the genus level and very few at the level of taxonomic families (Delgado, 2015).

Secondary metabolites intervene in different functions related with the ecological interactions between the plant and its environment, among other functions they are known to act in the protection against predators, mainly providing unpleasant flavors for animals (Mithöfer and Boland, 2012) as well as a defense against viruses, fungi and bacteria acting as natural pesticides (Gorlenko et al., 2020); as attractants of pollinating insects or other animals that consume the fruits and/or seeds either from the volatile compounds and pigments in the floral structures and fruits (Stevenson et al., 2016); as animal repellents because metabolites can trigger toxic reactions or an unpleasant taste for some animals (Madariaga-Mazón et al., 2019); and as part of the regulation of processes such as avoiding water loss in plants (González-Zamora et al., 2013).

12.1.1 Uses and Applications of Secondary Metabolites

Humans have used secondary metabolites to develop active ingredients of medicines, nutraceutical products, insecticides, herbicides, perfumes or dyes, among other products (Erb and Kliebenstein, 2020). So, their commercial importance is enormous, as a large number of metabolic products of commercial importance are used in the pharmaceutical, food and cosmetic industries and as sources of numerous substances of agrochemical interest (Chiocchio et al., 2021). It is estimated that more than 100,000 secondary metabolites are produced by plants (Teoh, 2016), so the technology for their identification, characterization and production are increasingly sophisticated, to such a level that, in the past two decades, more than 1,600 chemical structures with biological activity have been described every year (Hussain et al., 2012).

12.1.2 Secondary Metabolites in Medicine

For millennia, plant species have been the main source of medicines to treat different diseases and are currently the subject of study in different areas of scientific knowledge because of their applications and uses. From the nineteenth century, with the discovery of salicylic acid from the extracts of *Salix alba* L. (white willow), research into plant products used in traditional medicine took a great boost. In the twentieth century, many drugs were developed from natural sources in plants that have served to treat some diseases (Mesa, 2017); one of the most important is the discovery and application of alkaloids, such as atropine and morphine, derived from the extracts of *Atropa belladonna* L. that have stimulating and analgesic properties. Another strong field of research was to find secondary metabolites for the development of anticancer drugs because of the severity of the disease and the high mortality rates it presents so that about 60% of anticancer compounds are natural products or derivatives (Seca and Pinto, 2018); other examples of important secondary metabolites in the treatment of diseases are glycosides used as heart stimulants. The trend in the twenty-first century is for research on the secondary metabolites of plants to be accentuated due to the emergence of new diseases, as evidenced during the recent pandemic triggered by the emergence of the SARS-Cov-2 virus coupled with new trends to consume products of natural origin free of chemical processing (Bhuiyan et al., 2020).

12.1.3 Secondary Metabolites in Agronomy

In agriculture, phytopathogenic bacteria and fungi, as well as insects, cause numerous economic losses (Clemensen et al., 2020). The excessive application of antibiotics, insecticides, fungicides and pesticides for the control of pests and pathogens has had an impact on the health of ecosystems and on the people who participate in these types of activities (García et al., 2017; Santillán-Sidón et al., 2020). In this context, the secondary metabolites of plants have been studied as an environmentally friendly alternative, some of the success stories that have been most widespread among people engaged in agricultural activities include 1) the application of nicotine extracts (*Nicotiana tabacum* L.) to control the growth of various groups of insects; this secondary metabolite exerts an insecticidal effect by interacting with acetylcholine receptors causing poisoning symptoms similar to those observed when applying organophosphate insecticides (Schorderet et al., 2019); 2) the use of pyrethrins produced by *Tanacetum cineraiaefolium* (Trevir.) Sch. Bip. that blocks the sodium channels present in the neuronal axons of a large group of insects and that causes them seizures and hyperactivity (Bekele, 2018); 3) the acetogenins of *Annona muricata* L. that have biopesticidal activity by blocking the energy production of mitochondria in insect cells (Pérez, 2012); 4) azadirachtin, obtained from the seeds of *Azadirachta indicates* A. Juss., which interferes with the reproduction of insects, altering sexual development and gametogenesis processes both in the larva and in the adult stage and that also affects the hormone-producing organs and the brain neurosecretory system (Bekele, 2018); 5) essential oils such as eugenol, terpineol and cinnamic alcohol obtained from various species (i.e., *Salvia rosmarinus* (L.) Schleid., *Eucalyptus globulus* Labill., *Syzygium aromaticum* (L.) Merr. & L.M. Perry, *Thymus vulgaris* L., *Melissa officinalis* L.), which have neuroinsecticidal activity by blocking binding sites to octopamine receptors in species such as *Camponotus pennsylvanicus* (De Geer, 1773) (black carpenter ant) and *Blattella germanica* Linnaeus, 1767 (German cockroach); citral, eugenol, farnesol and geraniol have shown insecticidal activity against *Musca domestica* Linnaeus, 1758 (housefly) (Bekele, 2018).

12.1.4 Secondary Metabolites in Industry

One of the most important groups of secondary metabolites in the industry are plant pigments. These are molecules that absorb light and reflect a part of the electromagnetic spectrum giving rise to a wide variety of colors. Since ancient times, humans have used natural dyes that were extracted from animals and minerals but mainly from plants, such as *Indigofera tinctorea* L., known in India for more than 4,000 years, to dye clothes, skins, religious and recreational objects (Yusuf et al., 2017). The best known pigments are chlorophylls, responsible for the photosynthesis process and give the green color to the leaves and/or stems in some groups of plants. Another group of pigments are responsible for the colors of flowers and fruits that act as attraction signals for various animals that facilitate the pollination and dispersion of seeds, in addition to protecting plants from damage from sunlight, phytopathogens and heavy metals. In recent years, there has been an increase in the production and use of natural dyes, such as food additives, medicines and cosmetic products, in order to reduce the side effects of synthetic products (Brockington et al., 2011).

12.2 FACTORS AFFECTING THE QUANTITY AND QUALITY OF SECONDARY METABOLITES

Most of the environmental factors, such as dryland or pollution, may produce stress in plants, inducing the production of reactive oxygen species (ROS). In response, the plant triggers a protective effect that promotes the biosynthesis of secondary metabolites. The amount and type of secondary metabolites synthesized by plants depends on the genetic variability and plasticity of the phenotypes; however, this can vary drastically depending on the species, organ, tissue, stage of development and environmental conditions (Lia et al., 2020). Different secondary metabolites can be synthesized through a special regulatory pathway and a special transport pathway in certain organs, tissues and cells and exert long-term effects on plant growth and survival in stressful environments (Kurepin et al., 2017). The main storage organs of secondary metabolites are the roots and stems; however, this is affected by the period, season and years of growth. Some structures in the leaves, such as glandular trichomes, are also sites where secondary metabolites accumulate (Xu et al., 2018). Flowers and accessory structures, such as nectaries, give off terpenes and aromatic compounds that present variations related to the stage of development of floral organs (Gupta et al., 2011). In the case of fruits and seeds, the stages of development have a significant influence on the content and composition of secondary metabolites (González-Zamora et al., 2015).

12.2.1 Physical Factors

There are several environmental factors that affect the production of secondary metabolites, such as water stress, the amount of radiation, seasonal variation, CO_2 concentration and the availability of salts in the soil, among others (Cheng et al., 2018). In plants that grow in conditions of high temperature and low humidity, it has been determined that the amount of secondary metabolites increases due to the level of aridity of the environment (González-Zamora et al., 2013). One of the main effects of water stress in plants is the accumulation of O_2, which can be detected by the presence of malondialdehyde (MDA), which is the end product of membrane lipid peroxidation (Rahimi et al., 2018). To prevent further damage, plants activate their antioxidant system, which includes enzymes that can convert O_2 into H_2O, such as superoxide dismutase (SOD), peroxidase (POD) and catalase (CAT), as well as protective enzymes such as ascorbate peroxidase (APX) and glutathione reductase (GR), in conjunction with phenols and flavonoids (Dumanović et al., 2021).

In stress conditions, the concentration of ROS and the levels of MDA increase; in contrast, the activity of the antioxidant enzyme system is inhibited (Huang et al., 2019). A study performed by Selmar and Kleinwachter (2013) proposed a model to explain the effects of water stress on the production of secondary metabolites. The model proposes that, in plants that inhabit conditions of water scarcity, the stomata are closed to minimize the effect of stress on the leaves, which allows there to be a lower concentration of CO_2 that is fixed through the Calvin cycle, while the concentration of reduced equivalents, such as NADPH + H^+, being lower are consumed and reoxidized, allowing for the accumulation of higher amounts of NADPH + H^+ and, thus, generating an over-reduced environment. This promotes the biosynthesis of terpenoids and phenols in addition to other reactions to consume NADPH + H^+. It has been determined that there are several genes involved in the response of plants to water stress, for example, dehydrin (DHN) (George et al., 2017), that increases their expression under conditions of water stress and activates other molecular responses, including the activation of genes that encode enzymes involved in the biosynthesis of some terpenoids such as geranyl diphosphate synthase (GPPS), farnesyl diphosphate synthase (FPPS), geranylgeranyl diphosphate synthase (GGPPS) and copalyl diphosphate synthase (CPS) (Caser et al., 2019).

Temperature stress directly affects the metabolism of plants, as it allows the accumulation of ROS and affects the folding of proteins. This effect is prevented by plants through the synthesis of chaperones, which maintain the folding state of proteins. Heat shock proteins (HSP) are a class of chaperones that are expressed during heat stress (Jacob, 2017); multiple families of genes involved in the expression of chaperones have been recognized. Among the effects of chaperone expression are the synthesis of antioxidants – the induction of kinases that respond to the stress of Ca-dependent protein kinases (Carra et al., 2017).

On the other hand, salinity stress causes changes that destabilize membrane function, redox homeostasis and nutrient balance, affecting primary metabolism that provides precursors to secondary metabolites (Xu et al., 2016). In the case of organisms living in conditions of high salt concentration, the biosynthesis of secondary metabolites is high, induced by salinity stress. The main secondary metabolites synthesized under conditions of high salinity are terpenoids, steroids, phenols, flavonoids and alkaloids that correlate with the antioxidant capacity of plants (Sytar et al., 2018). In dry areas, the low availability of water can generate an increase in soil salinity, so there may be two stressors affecting the synthesis of secondary metabolites simultaneously (Isah, 2019).

12.2.2 Chemical Factors

One of the most important environmental problems in many regions of the world is the contamination of the atmosphere, water and agricultural soils with heavy metals and other contaminants, which has an impact on the health of ecosystems (González-Zamora et al., 2020) and human populations that consume these products (Gandarilla-Esparza et al., 2021). Some heavy metals are essential trace elements for plants such as zinc (Zn), copper (Cu), nickel (Ni), molybdenum (Mo), manganese (Mn), chromium (Cr) and iron (Fe); however, some metals are toxic to plants, although they are in very low concentrations, as they are not essential, such is the case of lead (Pb), cadmium (Cd), mercury (Hg) and arsenic (As) (Shahid et al., 2017). The absorption and incorporation of heavy metals in plants causes dysfunctions in metabolism, mainly the excessive production of ROS, which can be neutralized by the enzymatic and non-enzymatic antioxidant pathways already explained above, which cause a significant increase in secondary metabolites (Asgari et al., 2017).

One of the main polluting elements with an effect on the health of people who consume plants is As. The presence of As is due to an increase in anthropogenic activities that generate it, as well as natural processes (volcanic eruptions) that cause the contamination of aquifers and groundwater (Al-Makishah et al., 2020). The absorption of As by plants causes phytotoxicity due to its accumulation within plant cells that produce oxidative stress with physiological and metabolic effects mainly due to the generation of ROS (Kalita et al., 2018), which cause damage at the cellular level and interfere with the synthesis of secondary metabolites (Rasheed et al., 2018). The stress caused by the high concentration of As alters enzymatic activity and levels of secondary metabolites, and changes occur in the enzymatic and non-enzymatic antioxidant system in response to this contaminant. It has been reported that As can be translocated to different parts of the plant; however, the roots function as phytochelating agents, so the greatest amount of metal remains in this area (Suriyagoda et al., 2018).

12.2.3 Interactions with Other Organisms

The microbiome associated with the root influences the induction of genes in plants that are involved in the synthesis of secondary metabolites, keeping the roots healthy and allowing tolerance for environmental stress (Jha and Subramanian, 2018). In the plant tissues, there is an association with several species of endophytic bacteria that synthesize a wide range of bioactive secondary metabolites, such as flavonoids, alkaloids, phenolic acids, steroids, quinones, tetraphyllones and xanthones (Tidke et al., 2017), which in most cases, benefit plants by promoting their growth and increasing their resistance to abiotic and biotic stress because the metabolism of plants interact with the metabolism of their associates contributing to the pathways of secondary metabolism partially or totally (Ludwig-Müller, 2015). The endophytic bacteria present in different plant organs can produce a broad spectrum of secondary metabolites (Zam et al., 2019) that include compounds with antibacterial activity against pathogenic bacteria (Hagaggi and Mohamed, 2020).

The interactions of plants with other non-bacterial organisms stimulate the production of secondary metabolites, as is the case of the monoterpenic indole alkaloids pathway that has been observed in plants suffering attacks from herbivores (Dugé de Bernonville et al., 2020), decreasing methylation levels, which also translates into a greater resistance capacity of plants to pathogens (González and Vera, 2019). Many of the secondary metabolites produced by plants are stored in vacuoles, as some are precursors of active compounds that can be activated in response to damage to the plant. Once damage occurs, lectin-type defensive proteins, protease inhibitors or secondary metabolites that are toxic to other organisms begin to be produced; an example of this is the production of isoflavonoid and sesquiterpene phytoalexins by a large number of plants as a defense response to microbial invasion (Zaynab et al., 2018).

12.3 EPIGENETIC REGULATION OF THE SYNTHESIS OF SECONDARY METABOLITES

Elicitors are natural or mineral substances that, when applied preventively to plants, help reduce or avoid damage caused by diseases, pests or adverse abiotic factors. The application of elicitors induces changes in plant metabolism due to the reprogramming of genes related to metabolism (Mejía-Teniente et al., 2013), increasing the synthesis of secondary metabolites in food and medicinal plants (Baenas et al., 2014). Gene reprogramming occurs primarily through methylation and can be quantified by gene expression analysis (Eichten et al., 2014). It has been determined that hypomethylation is associated with a higher level of gene expression as a plant response to induced stress; in contrast, hypermethylated genes are silenced; therefore, they are not expressed in mRNA and are not translated into functional proteins (Yang et al., 2018). Additionally, the regulation of secondary metabolites within the same pathway occurs simultaneously by regulatory genes in trans or cis by loci of specific quantitative traits (Kooke et al., 2019), for which it has been suggested that it has an important role in the evolution of secondary metabolism (Vidalis et al., 2016). Epigenetic modifications have a great effect on morphological and genetic variation. This epigenetic variation is important for plants in stressful environments, and these variations can be reversed when the stressor is removed, which does not occur with variants in the DNA sequence (Rando and Verstrepen, 2007); these mechanisms endow plants with broad phenotypic plasticity and better adaptation to environmental conditions (Kooke et al., 2015).

12.4 ENVIRONMENTAL POLLUTANTS AND THEIR EFFECTS ON PLANTS

12.4.1 Soil and Water Pollutants

The severity of soil deterioration depends on the pollution sources and whether the origin is anthropogenic or natural, in addition to soil properties such as pH, clay mineral, organic matter, texture and other conditions such as ionic strength, anions and bacteria that affect the adsorption of pollutants in soils (Nguyen et al., 2021). The presence of some contaminants in the soil, such as pesticides, is conditioned by their persistence and their sorption/desorption characteristics; these parameters vary for the same pesticide from one geographic location to another and with the depth of the soil (Katagi, 2013).

The presence of pesticides persists in different parts of the world. A study of agricultural soils reported that more than 80% of the soils contained dicloro-difenil-tricloroetano (DDT); the concentration of the sum of all metabolites and isomers ranged between 0.005 and 0.383 mg/kg; however, most of the collected plants growing in the soils sampled did not contain detectable residues of DDT. According to these results, the authors suggest that the contamination of these sites originated from the past use of DDT and not from impurities of other recently formulated and applied substances (Malusa et al., 2020). On the other hand, the problem of contamination by metals still represents a latent problem. A report on mining areas in Yunnan, China, shows concentrations of Pb and Zn ~50 times higher compared to the world average (Li et al., 2019). While some areas in Mexico are the hottest and driest in the country, it is characterized by conserving the cultivation of corn, beans, chili, tomato, watermelon and melon as well as beef production and dairy cattle. Additionally, mining is one of the most important economic activities; therefore, in some regions, different types of soil and water contamination can exist. Additionally, it has been documented that the high levels of contamination are mainly due to the use of pesticides for decades, as organochlorine compounds, organophosphates and phosphate fertilizers that favor the availability of As in the soil. Also, hydroarsenicism is a consequence of overexploitation of aquifers, and other factors involved in environmental deterioration, including poor management of waste in industries, which represents a danger to biodiversity, the environment, food security and human health (Ruelas-Inzunza et al., 2013; Bhattacharyya et al., 2021).

One study reported the presence of dieldrin in different superficial layers of the soil (0–8 and 8–23 cm), where the content of organic matter, the percentage of clay and the pH had the greatest influence on the concentration of this compound in the soil (García et al., 2017). Other findings reported that by adding phosphorus as a fertilizer to agricultural soils, the available As in the surface layer (0–30 cm) increased up to 662%; however, the metalloid did not exceed the permissible levels established in the normative standard (Hernández et al., 2013). Other studies have reported the presence of Cr, Ni, Cu, Zn, As, Cd, Hg and Pb in medicinal plant species and soil samples where the presence of metals is possibly due to excessive fertilization applied and other sources of contamination (Meng et al., 2022).

12.4.2 Pollutants Assimilated by Plants

During the past few decades, anthropocentric activity has caused a considerable increase in pollution. Among the most characterized are pesticides and heavy metals in the air, soil or water. Contamination by persistent pollutants allows plants to be exposed to them for long periods of time and increases the probability that these compounds will be assimilated by plants and bioaccumulate in different plant structures, affecting the quality of secondary metabolites produced. Accumulation of heavy metals in medicinal plants and spices has been reported, especially if they are grown or harvested in polluted areas. The accumulation of heavy metals in plants has been shown in various studies that report a low quality of secondary metabolites due to the contaminants present; however, it is important to consider the quality of the soil, the water used in irrigation and the levels of metals accumulated in the plants (Glavac et al., 2017).

On the other hand, pesticides are toxic substances that contaminate soils, bodies of water and crops; excessive use of pesticides can cause the destruction of biodiversity. In plants, pesticides cause oxidative stress, inhibit metabolic pathways, induce toxicity, prevent photosynthesis and negatively affect crop yield, including secondary metabolites production. Pesticides increased the production of ROS, such as superoxide radicals (O^{-2}), hydrogen peroxide (H_2O_2), singlet oxygen (O_2), hydroxyl radical (OH^-) and the hydroperoxyl radical (HO_2^-) that damage proteins, lipids, carbohydrates and the DNA of plants (Kim et al., 2017). Plants have response mechanisms, such as plant growth regulators (PGR), that promote growth and development in optimal and stress conditions because they act as chemical messengers that allow various physiological and biochemical responses for plants to survive in stress conditions. This response reduces pesticide-induced toxicity. Exogenous applications of PGR, such as brassinosteroid, cytokinins, salicylic acid, jasmonic acid, etc., mitigate the toxicity of pesticides by stimulating the antioxidant defense system and increasing tolerance to stress conditions. They provide resistance against pesticides by controlling the production of ROS, nutrient homeostasis, increase the production of secondary metabolites and activate antioxidant mechanisms that activate mechanisms of tolerance to pesticides (Jan et al., 2020).

The contamination of soil, water and air with toxic heavy metals due to various human activities is a crucial environmental problem in both developed and developing countries. Heavy metals could be introduced into medicinal plant products through contaminated soil, water, air resources, even by poor production practices. The cultivation of medicinal plants in environments contaminated with heavy metals can eventually affect the biosynthesis of secondary metabolites, causing significant changes in the quantity and quality of these compounds. Some aromatic and medicinal plants can absorb and accumulate metallic contaminants in harvestable foliage and are, therefore, considered a feasible alternative for the remediation of contaminated sites. But plants use different strategies and enzymatic and non-enzymatic antioxidant defense systems to deal with the overproduction of ROS caused by heavy metals that enter their cells through the foliar and/or root systems; the accumulation of pollutants can be carried out in the vacuoles of cells or in vesicles known as endosomes (Asgari et al., 2017).

12.4.3 Interactions between Secondary Metabolites and Polluting Compounds

Environmental pollutants can affect the function of secondary metabolites by acting as agonists or antagonists: in this sense an agonist is a substance that binds to a receptor and generates a specific response, while the antagonist limits the action of another compound (for example, ion transporters). It is also possible that contaminants have an additive effect – for example, in the case of cytotoxic secondary metabolites and some environmental contaminants that also have this effect – or a synergistic effect in which the presence of a substance significantly increases the effects of a secondary metabolite. Although all these interactions are possible, it is very difficult to characterize these effects in biological models or the human population; however, one approach is the quantification of contaminating compounds and secondary metabolites, which determines their safety and quality.

A controlled study determined the effects of contamination with 25 or 50 mM Ni and analyzed the synthesis and accumulation of bioactive molecules in *Hypericum perforatum* L., a medicinal plant used in the treatment of neurological disorders and recently identified as a possible treatment for cancerous tumors. The pharmacological activity has been attributed to hypericin, hyperforin and pseudohypericin. In this assay, seedlings grown in a Ni-supplemented environment lost the ability to produce or accumulate hyperforin and showed a 15- to 20-fold decrease in the concentration of pseudohypericin and hypericin. These results demonstrated that metal contamination can change the chemical composition of the plant and affect the quality, safety and efficacy of natural plant products (Murch et al., 2003).

Another study analyzed the physiology, oxidative stress and metabolite production in *H. perforatum* exposed to moderate concentrations of Cd and/or La (10 μM), in which an increase in ROS and MDA was observed as well as a decrease of growth, tissue water content, glutathione, ascorbic acid and affected nutrient content. Additionally, a decrease in hypericin and the expression of its putative gene (*hyp-1*) was observed as well as an accumulation of hyperforin, which shows that exposure to metals affects the quantity and quality of secondary metabolites (Babula et al., 2015).

In a controlled study, *Mentha piperita* L. plants were grown in soil contaminated with 23.7 mg/kg of As, 5 mg/kg of Cd, 136 mg/kg of Ni and 95 mg/kg of Pb, and the transfer of metals to the plants were evaluated, and their accumulation in roots, stems and leaves were compared to a control group without the addition of metals. The accumulation of Cd, Ni and Pb was observed in roots > stems > leaves, but no significant impact on growth, development and chlorophyll content was observed, compared to the control in the first month of exposure; however, after three months of exposure, phytotoxic effects were produced. The authors mention that the values of the transfer coefficients and the translocation factors were less than 1, which indicates that *M. piperita* immobilized the metals in the root, so they conclude that, during a short period of time, the plant has the ability to stabilize metals in the root (Dinu et al., 2021).

On other hand, a study determined the heavy metal content and nutrient status of some medicinal plants. The concentrations of Al, B, Ca, Cd, Cr, Cu, Fe, K, Mg, Mn, Na, Ni, Pb and Zn were quantified in commonly used parts of medicinal plants, such as the root, rhizome, seed, resin, gall and fruit, collected at nearby industrial regions, mining and farming sites. Overall heavy metal concentrations were found to be in slightly higher levels. This shows us that it is of crucial importance that the areas where medicinal plants are collected are clean, especially by means of heavy metals, as these plants can cause more harm than the benefits they may provide if they are contaminated (Karahan et al., 2020).

12.5 CONCLUSION

The production of secondary metabolites can be affected by various environmental stimuli, including water scarcity and exposure to environmental pollutants. Plants respond to both stimuli by various rescue mechanisms such as root chelation, assimilation and sequestration of the pollutant in vacuoles as well as the expression of the enzymatic and non-enzymatic antioxidant system; however, the amount of the pollutant and the period of exposure determine the effect on the production of secondary metabolites. The little availability of water, as well as the contamination of the soil and water can be common in many agricultural areas destined to the production of medicinal plants; however, the quantity, safety and quality of the secondary metabolites, with biological activity of interest, can be modulated through good management of possible contaminating compounds to avoid an adverse effect on the population that consumes these products in different formulations.

REFERENCE LIST

Al-Makishah, N. H., M. A. Taleb and M. A. Barakat. 2020. Arsenic bioaccumulation in arsenic-contaminated soil: a review. *Chemical Papers* 74: 2743–2757. https://doi.org/10.1007/s11696-020-01122-4

Asgari, B., M. Ghorbanpourb and S. Nikabadic. 2017. Heavy metals in contaminated environment: destiny of secondary metabolite biosynthesis, oxidative status and phytoextraction in medicinal plants. *Ecotoxicology and Environmental Safety* 145: 377–390. https://doi.org/10.1016/j.ecoenv.2017.07.035

Babula, P., B. Klejdus, J. Kovacik, J. Hedbavny and M. Hlavna. 2015. Lanthanum rather than cadmium induces oxidative stress and metabolite changes in *Hypericum perforatum*. *Journal of Hazardous Materials* 286: 334–342. https://doi.org/10.1016/j.jhazmat.2014.12.060

Baenas, N., C. García-Viguera and D. A. Moreno. 2014. Elicitation: a tool for enriching the bioactive composition of foods. *Molecules* 19 (9): 13541–13563. https://doi.org/10.3390/molecules190913541

Bekele, D. 2018. Review on insecticidal and repellent activity of plant products for malaria mosquito control. *Biomedical Research and Reviews* 2 (2): 1–7. https://doi.org/10.15761/BRR.1000114

Bhattacharyya, K., S. Sengupta, A. Pari, S. Halder, P. Bhattacharya, B. J. Pandian and A. R. Chinchmalatpure. 2021. Characterization and risk assessment of arsenic contamination in soil–plant (vegetable) system and its mitigation through water harvesting and organic amendment. *Environmental Geochemistry and Health* 43 (8): 2819–2834. https://doi.org/10.1007/s10653-020-00796-9

Bhuiyan, F. R., S. Howlader, T. Raihan and M. Hassan. 2020. Plants metabolites: possibility of natural therapeutics against the COVID-19 pandemic. *Frontiers in Medicine* 7: 444. https://doi.org/10.3389/fmed.2020.00444

Brockington, S. F., R. H. Walker, B. J. Glover, P. S. Soltis and D. E. Soltis. 2011. Complex pigment evolution in the Caryophyllales. *New Phytologist* 190 (4): 854–864. https://doi.org/10.1111/j.1469-8137.2011.03687.x

Buchanan, B., W. Gruissem and R. Jones. 2015. *Biochemistry and molecular biology of plants*. John Wiley & Sons, Hoboken, 1280 p.

Carra, S., S. Alberti, P. A. Arrigo, J. L. Benesch, I. J. Benjamin, W. Boelens, B. Bartelt-Kirbach, B. J. J. M. Brundel, J. Buchner, B. Bukau, J. A. Carver, H. Ecroyd, C. Emanuelsson, S. Finet, N. Golenhofen, P. Goloubinoff, N. Gusev, M. Haslbeck, L. E. Hightower, H. H. Kampinga, R. E. Klevit, K. Liberek, H. S. Mchaourab, K. A. McMenimen, A. Poletti, R. Quinlan, S. V. Strelkov, M. E. Toth, E. Vierling and R. M. Tanguay. 2017. The growing world of small heat shock proteins: from structure to functions. *Cell Stress Chaperones* 22: 601–611. https://doi.org/10.1007/s12192-017-0787-8

Caser, M., W. Chitarra, F. D'Angiolillo, I. Perrone, S. Demasi, C. Lovisolo, L. Pistelli, L. Pistelli and V. Scariot. 2019. Drought stress adaptation modulates plant secondary metabolite production in *Salvia dolomitica* Codd. *Industrial Crops & Products* 129: 85–96. https://doi.org/10.1016/j.indcrop.2018.11.068

Cheng, L., M. Han, L.-m. Yang, L. Yang, Z. Sun and T. Zhang. 2018. Changes in the physiological characteristics and baicalin biosynthesis metabolism of *Scutellaria baicalensis* Georgi under drought stress. *Industrial Crops & Products* 122: 473–482. https://doi.org/10.1016/j.indcrop.2018.06.030

Chiocchio, I., M. Mandrone, P. Tomasi, L. Marincich and F. Poli. 2021. Plant secondary metabolites: an opportunity for circular economy. *Molecules* 26 (2): 495. https://doi.org/10.3390/molecules26020495

Clemensen, A. K., F. D. Provenza, J. R. Hendrickson and M. A. Grusak. 2020. Ecological implications of plant secondary metabolites - phytochemical diversity can enhance agricultural sustainability. *Frontiers in Sustainable Foods Systems* 4: 547826. https://doi.org/10.3389/fsufs.2020.547826

Delgado, G. 2015. *Aspectos generales introductorios a la química de productos naturales*. *In*: Delgado, G. and A. Romo (eds.). *Temas selectos de química de productos naturales*. Instituto de Química, Universidad Nacional Autónoma de México, México. Pp: 1–29.

Dinu C, S. Gheorghe, A. G. Tenea, C. Stoica, G.-G. Vasile, R. L. Popescu, E. A. Serban and L. F. Pascu. 2021. Toxic metals (As, Cd, Ni, Pb) impact in the most common medicinal plant (*Mentha piperita*). *International Journal of Environmental Research Public Health* 18 (8): 3904. https://doi.org/10.3390/ijerph18083904

Dugé de Bernonville, T., S. Maury, A. Delaunay, C. Daviaud, C. Chaparro, J. Tost, S. E. O'Connor and V. Courdavault. 2020. Developmental methylome of the medicinal plant *Catharanthus roseus* unravels the tissue-specific control of the monoterpene indole alkaloid pathway by DNA methylation. *International Journal of Molecular Sciences* 21: 6028. https://doi.org/10.3390/ijms21176028

Dumanović, J., E. Nepovimova, M. Natić, K. Kuča and V. Jaćević. 2021. The significance of reactive oxygen species and antioxidant defense system in plants: a concise overview. *Frontiers in Plant Science* 11: 552969. https://doi.org/10.3389/fpls.2020.552969

Eichten, S. R., R. J. Schmitz and N. M. Springer. 2014. Epigenetics: beyond chromatin modifications and complex genetic regulation. *Plant Physiology* 165 (3): 933–947. https://doi.org/10.1104/pp.113.234211

Erb, M. and D. J. Kliebenstein. 2020. Plant secondary metabolites as defenses, regulators, and primary metabolites: the blurred functional trichotomy. *Plant Physiology* 184 (1): 39–52. https://doi.org/10.1104/pp.20.00433

Gandarilla-Esparza, D. D., E. Y. Calleros-Rincón, H. Moreno, M. F. González-Delgado, G. García, J. Duarte, A. González-Zamora, E. Ríos-Sánchez and R. Pérez-Morales. 2021. *FOXE1* polymorphisms and chronic exposure to nitrates in drinking water cause metabolic dysfunction, thyroid abnormalities, and genotoxic damage in women. *Genetics and Molecular Biology* 44 (3): e20210020. https://doi.org/10.1590/1678-4685-GMB-2021-0020

García, M., J. G. Luna, A. González, A. González, M. A. Gallegos, C. Vázquez, M. G. Cervantes and U. González. 2017. Relación de díeldrin y propiedades del suelo en la Comarca Lagunera, México. *Revista Mexicana de Ciencias Agrícolas* 8 (8): 1691–1703. https://doi.org/10.29312/remexca.v8i8.695

George, K. J., N. Malik, I. P. V. Kumar and K. S. Krishnamurthy. 2017. Gene expression analysis in drought tolerant and susceptible black pepper (*Piper nigrum* L.) in response to water deficit stress. *Acta Physiologiae Plantarum* 39: 104. https://doi.org/10.1007/s11738-017-2398-5

Glavač, N. K., S. Djogo, S. Ražić, S. Kreft and M. Veber. 2017. Accumulation of heavy metals from soil in medicinal plants. *Archives of Industrial Hygiene and Toxicology* 68 (3): 236–244. https://doi.org/10.1515/aiht-2017-68-2990

González, B. and P. Vera. 2019. Folate metabolism interferes with plant immunity through 1C Methionine Synthase-directed genome-wide DNA methylation enhancement. *Molecular Plant* 12 (9): 1227–1242. https://doi.org/10.1016/j.molp.2019.04.013

González–Zamora, A., E. Ríos-Sánchez and R. Pérez-Morales. 2020. Conservation of vascular plant diversity in an agricultural and industrial region in the Chihuahuan Desert, Mexico. *Global Ecology and Conservation* 22: e01002. https://doi.org/10.1016/j.gecco.2020.e01002

González-Zamora, A., E. Sierra-Campos, J. G. Luna-Ortega, R. Pérez-Morales, J. C. Rodríguez and J. L. García-Hernández. 2013. Characterization of different *Capsicum* varieties by evaluation of their capsaicinoids content by High Performance Liquid Chromatography, determination of pungency and effect of high temperature. *Molecules* 18 (11): 13471–13486. https://doi.org/10.3390/molecules181113471

González-Zamora, A., E. Sierra-Campos, R. Pérez-Morales, C. Vázquez-Vázquez, M. Á. Gallegos-Robles, J. D. López-Martínez and J. L. García-Hernández. 2015. Measurement of capsaicinoids in Chiltepin hot pepper: a comparison study between Spectrophotometric method and High Performance Liquid Chromatography analysis. *Journal of Chemistry* 2015: Article ID 709150. https://doi.org/10.1155/2015/709150

Gorlenko, C. L., H. Y. Kiselev, E. V. Budanova, A. A. Zamyatnin, and L. N. Ikryannikova. 2020. Plant secondary metabolites in the battle of drugs and drug-resistant bacteria: new heroes or worse clones of antibiotics? *Antibiotics* 9 (4): 170. https://doi.org/10.3390/antibiotics9040170

Gupta, N., S. K. Sharma, J. C. Rana and R. S. Chauhan. 2011. Expression of flavonoid biosynthesis genes vis-à-vis rutin content variation in different growth stages of *Fagopyrum* species. *Journal of Plant Physiology* 168 (17): 2117–2123. https://doi.org/10.1016/j.jplph.2011.06.018

Hagaggi, N. S. A. and A. A. A. Mohamed. 2020. Plant–bacterial endophyte secondary metabolite matching: a case study. *Archives of Microbiology* 202: 2679–2687. https://doi.org/10.1007/s00203-020-01989-7

Hernández, G., M. A. Segura, L. C. Álvarez, R. A. Aldaco, M. Fortis, and G. González. 2013. Comportamiento del arsénico en suelos de la región lagunera de Coahuila, México. *Terra Latinoamericana* 31(4): 295–303.

Huang, H., F. Ullah, D.-X. Zhou, M. Yi and Y. Zhao. 2019. Mechanisms of ROS regulation of plant development and stress responses. *Frontiers in Plant Science* 10: 800. https://doi.org/10.3389/fpls.2019.00800

Hussain, S., S. Fareed, S. Ansari, A. Rahman, I. Z. Ahmad and M. Saeed. 2012. Current approaches toward production of secondary plant metabolites. *Journal of Farmacy and BioAllied Sciences* 4 (1): 10–20. https://doi.org/10.4103%2F0975-7406.92725

Isah, T. 2019. Stress and defense responses in plant secondary metabolites production. *Biological Research* 52 (39): 1–25. https://doi.org/10.1186/s40659-019-0246-3

Jacob, P., H. Hirt and A. Bendahmane. 2017. The heat-shock protein/chaperone network and multiple stress resistance. *Plant Biotechnology Journal* 15 (4): 405–414. https://doi.org/10.1111/pbi.12659

Jan, S., S. Rattandeep, R. Bhardwaj, P. Ahmad and D. Kapoor. 2020. Plant growth regulators: a sustainable approach to combat pesticide toxicity. *3 Biotech* 10 (11): 466. https://doi.org/10.1007/s13205-020-02454-4

Jha, Y. and R. B. Subramanian. 2018. *Effect of root-associated Bacteria on soluble sugar metabolism in plant under environmental stress. In*: Ahmad, P., M. Abass, V. Pratap, D. Kumar, P. Alam and M. Nasser (eds.). *Plant metabolites and regulation under environmental stress.* Elsevier Academic Press, London. Pp: 407–414. https://doi.org/10.1016/B978-0-12-812689-9.00012-1

Kalita, J., A. K. Pradhan, Z. M. Shandilya and B. Tanti. 2018. Arsenic stress responses and tolerance in rice: physiological, cellular and molecular approaches. *Rice Science* 25 (5): 235–249. https://doi.org/10.1016/j.rsci.2018.06.007

Karahan, F., I. I. Ozyigit, I. A. Saracoglu, I. E. Yalcin, A. H. Ozyigit and A. Ilcim. 2020. Heavy metal levels and mineral nutrient status in different parts of various medicinal plants collected from Eastern Mediterranean Region of Turkey. *Biological Trace Element Research* 197 (1): 316–329. https://doi.org/10.1007/s12011-019-01974-2

Katagi, T. 2013. *Soil column leaching of pesticides. In*: Whitacre, D. (ed.). *Reviews of environmental contamination and toxicology.* Vol. 221. Springer, New York. Pp: 1–105. https://doi.org/10.1007/978-1-4614-4448-0_1

Kim, Y.-H., Al. L. Khan, M. Waqas and I.-J. Lee. 2017. Silicon regulates antioxidant activities of crop plants under abiotic-induced oxidative stress: a review. *Frontiers in Plant Science* 8: 510. https://doi.org/10.3389/fpls.2017.00510

Kooke, R., F. Johannes, R. Wardenaar, F. Becker, M. Etcheverry, V. Colot, D. Vreugdenhil and J. J. B. Keurentjes. 2015. Epigenetic basis of morphological variation and phenotypic plasticity in *Arabidopsis thaliana. Plant Cell* 27: 337–348. https://doi.org/10.1105/tpc.114.133025

Kooke, R., L. Morgado, F. Becker, H. van Eekelen, R. Hazarika, Q. Zheng, R. C. H. de Vos, F. Johannes and J. J. B. Keurentjes. 2019. Epigenetic mapping of the *Arabidopsis* metabolome reveals mediators of the epigenotype-phenotype map. *Genome Research* 29: 96–106. https://doi.org/10.1101/gr.232371.117

Kurepin, L. V., A. G. Ivanov, M. Zaman, R. P. Pharis, V. Hurry and N. P. A. Hüner. 2017. *Interaction of glycine betaine and plant hormones: protection of the photosynthetic apparatus during abiotic stress. In*: Hou, H. J. M., M. M. Najafpour, G. F. Moore and S. I. Allakhverdiev (eds.). *Photosynthesis: structures, mechanisms, and applications.* Springer International Publishing, Cham, Switzerland. Pp: 185–202. https://doi.org/10.1007/978-3-319-48873-8_9

Li, Z., J. Deblon, Y. Zu, G. Colinet, B. Li and Y. He. 2019. Geochemical baseline values determination and evaluation of heavy metal contamination in soils of Lanping Mining Valley (Yunnan Province, China). *International Journal of Environmental Research and Public Health* 16 (23): 4686. https://doi.org/10.3390/ijerph16234686

Lia, Y., D. Konga, Y. Fub, M. R. Sussmand and H. Wua. 2020. The effect of developmental and environmental factors on secondary metabolites in medicinal plants. *Plant Physiology and Biochemistry* 148: 80–89. https://doi.org/10.1016/j.plaphy.2020.01.006

Ludwig-Müller, J. 2015. Plants and endophytes: equal partners in secondary metabolite production? *Biotechnology Letters* 37: 1325–1334. https://doi.org/10.1007/s10529-015-1814-4

Madariaga-Mazón, A., R. B. Hernández-Alvarado, K. O. Noriega-Colima, A. Osnaya-Hernández and K. Martínez-Mayorga. 2019. Toxicity of secondary metabolites. *Physical Sciences Reviews* 4 (12): 20180116. https://doi.org/10.1515/psr-2018-01160116

Malusá, E., M. Tartanus, W. Danelski, A. Miszczak, E. Szustakowska, J. Kicińska and E. M. Furmanczyk. 2020. Monitoring of DDT in agricultural soils under organic farming in Poland and the risk of crop contamination. *Environmental Management* 66 (5): 916–929. https://doi.org/10.1007/s00267-020-01347-9

Mejía-Teniente, L., F. D. Durán-Flores, A. M. Chapa-Oliver, I. Torres-Pacheco, A. Cruz-Hernández, M. M. González-Chavira, R. V. Ocampo-Velázquez and R. G. Guevara-González. 2013. Oxidative and molecular responses in *Capsicum annuum* L. after hydrogen peroxide, salicylic acid and chitosan foliar applications. *International Journal of Molecular Sciences* 14 (3): 10178–10196. https://doi.org/10.3390/ijms140510178

Meng, C., P. Wang, Z. Hao, Z. Gao, Q. Li, H. Gao, Y. Liu, Q. Li, Q. Wang and F. Feng. 2022. Ecological and health risk assessment of heavy metals in soil and Chinese herbal medicines. *Environmental Geochemistry and Health* 44 (3): 817–828. https://doi.org/10.1007/s10653-021-00978-z

Mesa, A. M. 2017. Una visión histórica en el desarrollo de fármacos a partir de productos naturales. *Revista Mexicana de Ciencias Farmacéuticas* 48 (3): 16–27.

Mithöfer, A. and W. Boland. 2012. Plant defense against herbivores: chemical aspects. *Annual Review of Plant Biology* 63: 431–450. https://doi.org/10.1146/annurev-arplant-042110-103854

Murch, S. J., K. Haq, H. P. Vasantha and P. K. Saxena. 2003. Nickel contamination affects growth and secondary metabolite composition of St. John's wort (*Hypericum perforatum* L.). *Environmental and Experimental Botany* 49 (3): 251–257. https://doi.org/10.1016/S0098-8472(02)00090-4

Nguyen, K. T., M. B. Ahmed, A. Mojiri, Y. Huang, J. L. Zhou and D. Li. 2021. Advances in As contamination and adsorption in soil for effective management. *Journal of Environmental Management* 296: 113274. https://doi.org/10.1016/j.jenvman.2021.113274

Pérez, E. 2012. Plaguicidas botánicos: una alternativa a tener en cuenta. *Fitosanidad* 16 (1): 51–59.

Rahimi, Y., A. Taleeia and M. Ranjbar. 2018. Long-term water deficit modulates antioxidant capacity of peppermint (*Mentha piperita* L.). *Scientia Horticulturae* 237: 36–43. https://doi.org/10.1016/j.scienta.2018.04.004

Rando, O. J. and K. J. Verstrepen. 2007. Timescales of genetic and epigenetic inheritance. *Cell* 128: 655–668. https://doi.org/10.1016/j.cell.2007.01.023

Rasheed, H., P. Kay, R. Slack and Y. Yun. 2018. Arsenic species in wheat, raw and cooked rice: exposure and associated health implications. *Science of the Total Environment* 634 (1): 366–373. https://doi.org/10.1016/j.scitotenv.2018.03.339

Ruelas-Inzunza, J., C. Delgado-Alvarez, M. Frías-Espericueta and F. Páez-Osuna. 2013. *Mercury in the atmospheric and coastal environments of Mexico. In*: Whitacre, D. (ed.). *Reviews of environmental contamination and toxicology.* Vol. 226. Springer, New York. Pp: 65–99. https://doi.org/10.1007/978-1-4614-6898-1_3

Santillán-Sidón, P., R. Pérez-Morales, G. Anguiano, E. Ruiz-Baca, J. Rendón-Von Osten, E. Olivas-Calderón and C. Vazquez-Boucard. 2020. Glutathione S-transferase activity and genetic polymorphisms associated with exposure to organochloride pesticides in Todos Santos, BCS, Mexico: a preliminary study. *Environmental Science and Pollution Research* 27 (34): 43223–43232. https://doi.org/10.1007/s11356-020-10206-3

Schorderet, S., K. P. Kaminski, J.-L. Perret, P. Leroy, A. Mazurov, M. C. Peitsch, N. V. Ivanov and J. Hoeng. 2019. Antiparasitic properties of leaf extracts derived from selected *Nicotiana* species and *Nicotiana tabacum* varieties. *Food and Chemical Toxicology* 132: 110660. https://doi.org/10.1016/j.fct.2019.110660

Seca, A. M. L. and D. C. G. A. Pinto. 2018. Plant secondary metabolites as anticancer agents: successes in clinical trials and therapeutic application. *International Journal of Molecular Science* 19 (1): 263. https://doi.org/10.3390/ijms19010263

Selmar, D. and M. Kleinwachter, 2013. Influencing the product quality by deliberately applying drought stress during the cultivation of medicinal plants. *Industrial Crops and Products* 42: 558–666. https://doi.org/10.1016/j.indcrop.2012.06.020

Shahid, M., C. Dumat, S. Khalid, E. Schreck, T. Xiong and N. K. Niazi. 2017. Foliar heavy metal uptake, toxicity and detoxification in plants: a comparison of foliar and root metal uptake. *Journal of Hazardous Materials* 325: 36–58. https://doi.org/10.1016/j.jhazmat.2016.11.063

Stevenson, P. C., S. W. Nicolson and G. A. Wright. 2016. Plant secondary metabolites in nectar: impacts on pollinators and ecological functions. *Functional Ecology* 31 (1): 65–75. https://doi.org/10.1111/1365-2435.12761

Suriyagoda, L. D. B., K. Dittert and H. Lambers. 2018. Mechanism of arsenic uptake, translocation and plant resistance to accumulate arsenic in rice grains. *Agriculture, Ecosystems & Environment* 253 (1): 23–37. https://doi.org/10.1016/j.agee.2017.10.017

Sytar, O., S. Barki, M. Zivcak and M. Brestic. 2018. *The involvement of different secondary metabolites in salinity tolerance of crops. In*: Kumat, V. (ed.). *Salinity responses and tolerance in plants.* Vol. 2. Springer International Publishing AG, Berlin. Pp: 21–48. https://doi.org/10.1007/978-3-319-90318-7_2

Teoh, E. S. 2016. *Secondary metabolites of plants. In*: Teoh, E.S. *Medicinal orchids of Asia.* Springer, Berlin/Heidelberg, Germany. Pp: 59–73. https://doi.org/10.1007%2F978-3-319-24274-3_5

Tidke, S. A., K. L. Rakesh, D. Ramakrishna, S. Kiran, G. Kosturkova and R. A. Gokare. 2017. Current understanding of endophytes: their relevance, importance, and industrial potentials. *Journal of Biotechnology and Biochemistry* 3 (3): 43–59. https://doi.org/10.9790/264X-03034359

Vidalis, A., D. Živković, R. Wardenaar, D. Roquis, T. Aurélien and F. Johannes. 2016. Methylome evolution in plants. *Genome Biology* 17: 264. https://doi.org/10.1186/s13059-016-1127-5

Xu, C., X. Tang, H. Shao and H. Wang. 2016. Salinity tolerance mechanism of economic halophytes from physiological to molecular hierarchy for improving food quality. *Current Genomic* 17 (3): 207–214. https://doi.org/10.2174%2F1389202917666160202215548

Xu, J. Y., Y. L. Yu, R. Y. Shi, G. Y. Xie, Y. Zhu, G. Wu and M. J. Qin. 2018. Organ-specific metabolic shifts of flavonoids in *Scutellaria baicalensis* at different growth and development stages. *Molecules* 23 (2): 428. https://doi.org/10.3390/molecules23020428

Yang, D., Z. Huang, W. Jin, P. Xia, Q. Jia, Z. Yang, Z. Hou, H. Zhang, W. Ji and R. Han. 2018. DNA methylation: a new regulator of phenolic acids biosynthesis in *Salvia miltiorrhiza*. *Industrial Crops and Products* 124: 402–411. https://doi.org/10.1016/j.indcrop.2018.07.046

Yusuf, M., M. Shabbir and F. Mohammad. 2017. Natural colorants: historical, processing and sustainable prospects. *Natural Products and Bioprospecting* 7: 123–145. https://doi.org/10.1007/s13659-017-0119-9

Zam, S. I, A. Agustien, A. Syamsuardi and I. Mustafa. 2019. The diversity of endophytic bacteria from the traditional medicinal plants leaves that have anti-phytopathogens activity. *Journal of Tropical Life Science* 9 (1): 53–63. https://doi.org/10.11594/jtls.09.01.8

Zaynab, M., M. Fatima, S. Abbas, Y. Sharif, M. Umair, M. H. Zafar and K. Bahadar. 2018. Role of secondary metabolites in plant defense against pathogens. *Microbial Pathogenesis* 124: 198–202. https://doi.org/10.1016/j.micpath.2018.08.034

Index

D

E

M

N

O